BK11.1008∆680

Earthquake Prediction Techniques

Earthquake Prediction Techniques

Their Application in Japan

Edited by Toshi Asada
Translated by Masako Ohnuki

UNIVERSITY OF TOKYO PRESS

Publication of this book was assisted by a grant from the Ministry of Education, Science and Culture, Japan.

©UNIVERSITY OF TOKYO PRESS, 1982
UTP 3044–66248–5149
ISBN 0–86008–290–3

Printed in Japan
All rights reserved. No part of this publication may be reproduced or transmitted in any form or by any means, electronic or mechanical, including photocopy, recording, or any information storage and retrieval system, without permission in writing from the publisher.

CONTENTS

Preface
List of Authors

PART I EARTHQUAKES REPEAT THEMSELVES
  Chapter 1 The Development of Seismology in Japan [Toshi Asada]——3
    1.1 Early Seismology——3
    1.2 Seismology before World War II——4
    1.3 The Logic of Fault Models——5
    1.4 Sea-floor Spreading and Plate Tectonics——5
    1.5 Recent Progress in Observation——6
    1.6 Recent Progress in Seismology——7
    1.7 Earthquake Prediction——8
  Chapter 2 Great Earthquakes of the Past [Tatsuo Usami]——11
    2.1 Historical Records of Earthquakes——11
    2.2 Problems and Limitations of Historical Seismology——14
    2.3 Examples of Great Earthquakes——16
    2.4 Merits and Drawbacks of Imamura's Historical Seismology——21
    2.5 Examples of Ancient Earthquakes in Japan——24
  Chapter 3 Earthquake Scars [Tokihiko Matsuda]——31
    3.1 Earthquake Scars Left on Landforms——31
    3.2 Reading Crustal Movement—Methods and Their Accuracy——35
    3.3 From the Past to the Present—Prediction and Its Problems——39
    3.4 Active Faults and Disaster Prevention——50

PART II LONG-RANGE PRECURSORS
  Chapter 4 How Small Earthquakes Occur [Akio Takagi]——63
    4.1 Smaller Earthquakes Occur More Often——63
    4.2 Where Do Small Earthquakes Occur?——66
    4.3 How Smaller Earthquakes Occur——82
    4.4 Problems in Smaller Earthquake Research——87

  Chapter 5 Variations in Seismic Velocity [Toshikatsu Yoshii]——89
    5.1 Some Problems in Seismic Wave Variation——89

v

5.2 Causes of Seismic Wave Velocity Variations——90
5.3 Various Methods for Detecting Velocity Variations——92
5.4 Future Problems——99
5.5 Velocity Variation Research at the Starting Gate——101

Chapter 6 Survey Repetition [Hiroshi Sato]——103
6.1 Surveys and Earthquake Prediction——103
6.2 Surveys and Observation of Crustal Movement——104
6.3 Examples of Land Surface Movement Caused by Earthquakes——108
6.4 Steady Crustal Movement in Japan——111
6.5 Earth Movement before and after Great Earthquakes——115
6.6 Earthquake Prediction and Repeated Survey Operations——119
6.7 Critical Crustal Strain and Long-Range Earthquake Prediction——119
6.8 Crustal Movement Anomalies as Earthquake Precursors——123
6.9 Future Problems——126

PART III SHORT-TERM PRECURSORY PHENOMENA

Chapter 7 Continuous Observation of Crustal Movement [Shigeji Suyehiro]——133
7.1 The Meaning of the Continuous Observation of Crustal Movement——133
7.2 Continual Observation of Crustal Movement in the Past——135
7.3 The Embedded Volume Strainmeter System——142
7.4 The Development of an Observation Network in the Tokai and Southern Kanto Districts——149
7.5 Observation Results——151
7.6 Toward High-density Continuous Observation——161

Chapter 8 Changes in Groundwater Level and Chemical Composition [Hiroshi Wakita]——175
8.1 Earthquakes and Changes in Groundwater——175
8.2 New Precursory Phenomena——178
8.3 The Tashkent Earthquake——209
8.4 Earthquake Prediction in China——210
8.5 Present Geochemical Studies in Japan——214
8.6 Geochemical Studies for the Future——214

Chapter 9 Earthquakes and Electromagnetic Phenomena [Hitoshi Mizutani]——217
9.1 Historical Observations——218
9.2 Geomagnetic Changes Accompanying Earthquakes—Examples of Modern Observations——222
9.3 Earthquake-related Variation in the Earth's Current and the Earth's Electric Potential——228
9.4 Variations in Electric Resistance Accompanying Earthquakes——231

9.5 The Etiology of the Electromagnetic Phenomena Associated with Earthquakes——*236*
9.6 Problems in Earthquake Prediction Research Based on Electromagnetic Phenomena——*242*

PART IV THE ROAD TO ACTUAL EARTHQUAKE PREDICTION
Chapter 10 A Practical Strategy for Earthquake Prediction [Katsuhiko Ishibashi]——*249*
    10.1 Present Earthquake Prediction Strategy——*249*
    10.2 Earthquake Prediction in China——*255*
    10.3 The Intrinsic Nature of Earthquake Phenomena——*259*
    10.4 The Practical Procedure of Earthquake Prediction——*261*
    10.5 Prediction of a Tokai Earthquake——*267*
Chapter 11 The Evaluation of Short-Term Precursory Earthquake Phenomena [Akio Takagi]——*271*
    11.1 Foreshock Activity——*273*
    11.2 Continuous Observation of Crustal Movement——*276*
    11.3 Groundwater Level and Radon Concentration——*279*
    11.4 Changes in the Specific Resistivity of Rocks and in Crustal Electric Resistance——*280*
    11.5 Identification of Anomalous Changes——*280*
Chapter 12 The Development of the Earthquake Prediction Project and Its Problems [Tatsuo Usami]——*283*
    12.1 The Blueprint——*283*
    12.2 Outline of Earthquake Prediction Progress——*286*
    12.3 Earthquake Prediction Knowledge at the Time of the Blueprint and at Present——*289*
    12.4 The Rise of Amateurism——*294*
    12.5 Reflections and Suggestions——*299*

Index of Earthquakes——*305*
General Index——*311*

AUTHORS

ASADA, Toshi	Born 1919. Professor, Tokai University. Major field: Seismology.
ISHIBASHI, Katsuhiko	Born 1944. Senior Research Fellow, International Institute of Seismology and Earthquake Engineering, Ministry of Construction. Major field: Seismology.
MATSUDA, Tokihiko	Born 1931. Associate Professor, Earthquake Research Institute, University of Tokyo. Major field: Seismological Geology.
MIZUTANI, Hitoshi	Born 1942. Professor, Faculty of Science, Nagoya University. Major field: Earth and Planetary Physics.
SATO, Hiroshi	Born 1929. Professor, Faculty of Science, Hirosaki University. Major field: Geodesy.
SUYEHIRO, Shigeji	Born 1924. Director, Observation Dept., Japan Meteorological Agency. Major field: Observational Seismology.
TAKAGI, Akio	Born 1925. Professor, Faculty of Science, Tohoku University. Major Field: Seismology.
USAMI, Tatsuo	Born 1924. Professor, Earthquake Research Institute, University of Tokyo. Major field: Seismology.
WAKITA, Hiroshi	Born 1936. Associate Professor, Faculty of Science, University of Tokyo. Major field: Geochemistry.
YOSHII, Toshikatsu	Born 1942. Associate Professor, Earthquake Research Institute, University of Tokyo. Major field: Seismology.

PREFACE

The objective of this book is two-fold: to provide a reference book for researchers in the field, and to offer some insights into the natural science aspects of earthquake prediction. It is thus intended not just for those working in seismology, but for people in a variety of fields. While it may be a bit too technical for some, I have found that the layman's understanding of the natural sciences is surprisingly profound—sometimes more so than that of the scientist. Readers who are only interested in the broader issues in earthquake prediction will probably find it unnecessary to spend time on the more technical aspects of the book.

Although there are several good books on earthquake prediction, they all have one drawback. Due to the way the material is presented, the reader can easily come away thinking that earthquake prediction is a *fait accompli*, that there are only a few problems that remain unsolved.

The techniques of earthquake prediction have indeed become more sophisticated in recent years; as the authors of the papers in this volume demonstrate, our tools for measuring crustal movement and other phenomena which seem directly related to the occurrence of earthquakes have increased in number and in accuracy. However, there are still many gaps in our knowledge, and our techniques are still far from perfect reliability. At the same time, the whole phenomenon of earthquake prediction has social implications which we must not forget in our concentration on its scientific aspects.

The contributors to this book have all made their marks on earthquake prediction-related studies in Japan. Their specialties range from the combing of historical sources for information on earthquakes of the past to the cataloging and mapping of Japan's active faults, to the measurement of chemical changes in the earth's crust. A number of them are pioneers in their areas of specialization.

In writing and assembling this book we have tried to include reports from a broad range of research projects, emphasizing those in which progress in Japan has been especially rapid.

A few topics related to earthquake prediction have not been included in this book. Among them are earthquakes induced by the pressurized entry of water. Denver, Colorado, in the United States, for example, had never experienced an earthquake until the pressurized entry of water began, inducing minor earthquakes. These minor earthquakes gradually developed into earthquakes as large as M 6. Minor earthquakes can also occur when a dam is filled, and sometimes these earthquakes register as high as M 6.

Another subject not included is rock-crushing experiments. These have been performed since the 1960s, and terms like sealed pressure, pore pressure, effective pressure, and dilatancy have become essential to the vocabulary of seismology. The results of research in this area have enriched the academic foundations of earthquake prediction.

A third topic not covered in this book is the measurement of crustal stress under natural conditions. A great earthquake struck Iran in September 1978, for example, killing more than ten thousand people who lived in mud huts. Great earthquakes are rare in this particular area, occurring only once every several thousand years. And yet the imminence of another earthquake can be deduced by measuring crustal stress, although the method of measurement needs to be perfected.

Where earthquake prediction is concerned, however, there are some basics that are far more important than these items. One example is the fault model theory. Since this theory was developed, our understanding of earthquakes has become far more profound and realistic.

While emphasizing the development and application of earthquake prediction techniques, the authors have also tried to keep in mind the social implications of earthquake prediction, particularly as these are manifesting themselves in Japan—a highly earthquake-prone country where prediction has become a topic of mass media attention and of popular, as well as scientific, concern.

Earthquake prediction can be regarded as an applied science or technology, but it will always be an unfinished field. Prediction will approach perfection little by little, but it will never be reached. In other words, the possibility of failure will never be zero.

Although earthquake prediction may be regarded as a technique, it is not the kind that produces a product. Nor does it rely only on recently developed technologies involving software.

From the social point of view, earthquake prediction is obviously not just a natural science field. Seismologists cannot and should not be held solely responsible for the realization and practice of earthquake prediction, for earthquake prediction will only be successful if everyone involved takes responsibility and makes their utmost effort.

Earthquake prediction efforts in Japan began to get top priority in 1965.

Since that time the Earthquake Prediction Project has progressed smoothly. In 1976 the possibility of an earthquake in the Tokai region became the center of attention, and in 1978 the Special Large-Scale Earthquake Countermeasures Act was passed. At this point earthquake prediction was no longer solely the province of the natural sciences.

A closer look at this sequence of events illuminates the most fundamental problem facing earthquake prediction today. The public first became aware of the potential Tokai earthquake in 1969 when an announcement was made that the coastal areas of the Tokai district had been pushed several meters inland. Seven years then passed without incident. Interest in the area was rekindled in 1976 when evidence was found in an ancient document that there was considerable strain building up throughout the entire Suruga Bay area.

Scholars were in unanimous agreement on the likelihood of a Tokai earthquake during the next 20–30 years. To bolster their conviction, it was found that the subsidence of Cape Omaezaki in Shizuoka Prefecture had been accelerating by about one centimeter per year since 1973. Both scientists and concerned government agencies agreed to set up an observation network in the Tokai area which would send telemeter communications to the Japanese Meteorological Agency.

Given the existence of such a network, one is faced with the question of what to do if and when all the relevant recordings in the Meteorological Agency begin to register extreme anomalies. In this event, it is safe to assume that a large-scale earthquake will occur anywhere from several hours to several days later. Without adequate earthquake countermeasures such a situation would be disastrous. Information of such grave concern would surely be leaked to the public, resulting in utter confusion. Then, in the midst of the chaos, the earthquake itself would strike.

Now that earthquake countermeasures have been established, such a disaster can be avoided. But what about a situation in which a few suspicious precursors are noted rather than obvious anomalies? These earthquakes are more likely to be overlooked. This must be avoided—especially since the Tokai earthquake countermeasures are progressing so well. Now is the time for an all-out effort to improve the accuracy of earthquake prediction—specifically with reference to Tokai.

Achieving better accuracy is actually a simple matter. It involves setting up an observation network that uses the various kinds of instruments that have been developed to detect earthquake precursors and which are discussed in this book, and placing these instruments at very close intervals. The instruments would include buried strain gauges and other detectors of crustal movement, and apparatus for measuring radon and helium and the water level in deep wells. Ideally, instruments should be positioned on the ocean floor as well as on land. Observational data must be gathered frequently using various survey techniques.

Such observation networks should be developed as quickly as possible if better accuracy is to be obtained. Time here is, indeed, money. These networks are also necessary if much-needed data are to be available for purposes of comparison.

It is, of course, much easier to write about the need for such networks than it is to actually set them up. A consideration of the budget issues alone will illustrate the complexity of the problems. In Japan the request for funding earthquake prediction research must first compete with the other requests originating within the government agency. Then it must be approved by the Ministry of Finance. The results will vary according to government officials' degree of knowledge regarding earthquake prediction. And this is but one example of how success does not depend solely on the efforts of scholars.

There are more costly items in the budget—a bridge spanning Shikoku and Japan's main island, Honshu, for instance—and more spectacular ones—like the construction of a new railway for bullet trains, but as far as impact is concerned, money spent on earthquake prediction will have far greater effect for it concerns a much larger segment of the population.

Clearly the goal of accurate earthquake prediction cannot be achieved by natural scientists alone. To discuss earthquake prediction in terms of the political and social issues involved, however, is beyond my area of expertise. Thus it is critical that politicians, public administrators, journalists and others develop an accurate picture of the natural science side of the question. I will be happy if this book serves that purpose.

This informed and graceful translation by Masako Ohnuki has enabled us to make this work on earthquake prediction in Japan, originally written for Japanese readers, available to a worldwide audience. All the authors are grateful to Ms. Ohnuki, and to Ms. Lilla Weinberge who aided materially in the translation process. Our thanks are due also to the Ministry of Education, Science and Culture of Japan for a grant which made the translation possible. We are grateful to Ryuzo Yamada, Susan Schmidt, and Megumi Shimizu of the University of Tokyo Press, who encouraged us to publish this translation and aided us in adapting the contents for an international readership.

November 1, 1981

Toshi ASADA

Earthquake Prediction Techniques

Earthquake Prediction Techniques

PART I
EARTHQUAKES REPEAT THEMSELVES

Although the precise details of how earthquakes occur are not yet known, most seismologists are in overall agreement on why earthquakes occur.

The earth's crust is under continual stress and strain. When the strain reaches a certain point, the crust cannot resist the stress any longer and it ruptures. The stress is due to the spreading of the ocean floor, or oceanic lithosphere. As the ocean floor moves it creates stress in the continental crust pressing against it, and this stress causes strain within the continental crust.

When the crust ruptures, a sliding surface, or fault, is formed, and the stress that has accumulated is released. This cycle is repeated once a fault plane exists, and finally results in fault topography.

As the rupture point approaches, certain precursory phenomena occur. Earthquakes can be predicted by observing these phenomena. This sounds simple and clear-cut, but in practice it is not so easy.

What happens during an actual earthquake, then? Heterogeneity in the crust, particularly as it affects stress distribution, seems to be an important factor here, as does the possibility that the ocean floor spreads at an uneven rate. Decades of careful observation will be necessary before these elements are thoroughly understood.

Chapter 1   The Development of Seismology in Japan

Toshi Asada

1.1   Early Seismology

Around 1880, a group of foreign professors, James Ewing and John Milne of England among them, introduced the study of seismology to Japan. They were followed by Kiyokage Sekiya, Fusakichi Omori, and Akitsune Imamura. Sekiya was the first professor of seismology in Japan.*

The foreigners teaching during the early Meiji era already knew that earthquakes were a kind of elastic wave within which there were transverse and longitudinal waves, and they shared this knowledge with their colleagues in Japan. Despite this insight into the nature of seismic waves, however, upon the departure of the foreigners seismology in Japan fell into a state of confusion which lasted for 20 or 30 years. In spite of this confusion, by Imamura's time seismologists had succeeded in collecting extensive information on earthquakes. In the 1920s Hiroshi Nakano, of the Japan Meteorological Station, contributed greatly to the development of the elastic wave theory: his papers were widely referred to by theoretical seismologists throughout the world until well after World War II.

The most significant discovery made during the first quarter of the twentieth century was that earthquakes recur in the same places. On the Pacific coast great earthquakes can be expected once every 100 to 200 years, and in inland areas every 1,000 years or more, although Imamura did record several earthquakes that occurred in the same inland area several hundred years apart. Imamura repeatedly warned of the impending Kanto, Tonankai, and Nankai coastal earthquakes—all of which eventually took place during his lifetime. The fact that the Nankai earthquake followed the

---

*Sekiya became an assistant professor in 1881 and later a full professor, but died at an early age. He was succeeded by Omori in 1897. Imamura, although only two years younger than Omori, was not appointed a full professor until 1923.

Tonankai earthquake was exactly in accordance with his historically based prediction. (See Chapter 2 for a more detailed discussion of Imamura's work.)

1.2 Seismology before World War II

The years between the great Kanto earthquake (1923) and World War II form a distinct period in Japanese seismology. Construction of the Japan Meteorological Agency (JMA)'s strategically positioned earthquake observation network began in the 1920s and led to seismological studies on an advanced level. Kiyoo Wadati, for instance, based his research on deep focus earthquakes on data gathered by this network [1928].

Seismometry based on the elastic wave theory, including fault models, is at the forefront of modern seismology. In 1926 Japanese scientists had already begun to pursue this line of research. Takeo Matuzawa of the University of Tokyo was studying the crustal structure of Honshu based on data obtained by the JMA network [1928; 1929] and, at the same time, investigating converted waves on the surface of discontinuity [1928; 1929]. And this was only a small part of the research of the day. Similar areas of study were very popular in the western world in the 1960s, so it is interesting to find that this work was already being pursued in Japan more than five decades ago.

The so-called earthquake source mechanism, for instance, was formulated by Toshi Shida of Kyoto University and was based on the observation that the first-motion direction of seismic waves shows a distinct spatial pattern. His model was continually refined from both the theoretical and observational standpoints, and Hirokichi Honda established the authoritative version in the 1930s.

The Earthquake Research Institute was established after the Great Kanto earthquake (1923). Despite the advanced state of many aspects of Japanese seismology prior to World War II, the recording and analysis of seismic waves was not regarded as very important at the Earthquake Research Institute. Some fine research was done at the Institute, such as Chuji Tsuboi's study of critical strain [1933]—a study that made innumerable contributions to geophysics—but, overall, the contributions to the area of seismometry based on elastic wave theory were minimal.

Research on seismographs like the accelerometer was popular at the Earthquake Research Institute during this time, however—a tradition that was carried on after World War II by Takahiro Hagiwara. It seems particularly odd, therefore, that this institution never developed a seismological network for its own research projects. Imamura did establish such a network in the Kanto district between 1926 and 1930, but it had ceased to function by the time of World War II. The sensitivity and range of the seismographs of the day left much to be desired, of course, and they were unfit for the observation of microearthquakes. The decisive importance of such a network was recognized 30 years after the war when a telemeter observation network for microearthquakes was formed by all the universities participating in this research.

1.3 The Logic of Fault Models

The theory of the earthquake source mechanism established by Honda defines the source as two sets of couples. In this definition, the source is only a formal theoretical concept, not an actual fact.

During the 1960s the fault model hypothesis was introduced. In this model a fault plane is postulated and then the direction and amount of displacement or slip on the surface of this plane are given. By determining the position of the fault plane's surface relative to the earth's surface, the pattern of the seismic waves emanating from the fault plane can be calculated, as can the crustal movement caused by this "earthquake."

The fault plane of a magnitude 8 earthquake is enormous, extending more than 100 km. The active fault in this case could be more than 100 km long. Recent geological research on active faults is based on the success of this fault model hypothesis.

In a variation on this theory, ten years ago Keiichi Aki of M.I.T. explained in one of his popular articles that "an earthquake is a fault itself." In other words, when the earth's crust can no longer bear the stress, a fault begins to be formed, and this is the beginning of an earthquake. This is how modern seismologists conceive of an earthquake.

1.4 Sea-floor Spreading and Plate Tectonics

Old seismology textbooks were unique in that they usually did not discuss the causes of earthquakes. Seismologists postulated that an earthquake occurred when an irresistible force acted on the earth's crust, causing it to rupture, but they had no idea what created that force. It wasn't until the 1960s and 1970s, when the sea-floor spreading hypothesis and its implications for plate tectonics were finally accepted and understood, that the source of earthquakes became clear.

The notion that the sea floor, or lithosphere, is a moving plate tens of kilometers thick has now been proved many times over. The horizontal thrust of the sea floor against the continental crust therefore provides a logical source for the earthquake-causing force.

Seismic waves provide the information necessary to calculate the form, position, and amount of displacement of the fault that generates an earthquake. They also indicate the direction of the stress field causing the earthquake. On the basis of such calculations the direction of maximum compression in the axes of shallow earthquakes in Japan is known to be east–west.

Most recent findings in seismology support the theory of plate tectonics. The fact that earthquakes occur over and over again in the same locations, the fact that great earthquakes only occur along the Pacific coast, the existence of shallow- and deep-focus earthquakes—all of these, in addition to the direction of maximum compression mentioned above, add up to a convincing proof of the correctness of plate tectonics theory.

The idea that a change in the volume of the earth's crust causes earthquakes cannot be completely discounted. But, although the force that causes earthquakes from the point of view of plate tectonics is wide-ranging and operates from a considerable distance, it does seem to explain almost all shallow earthquakes.

1.5 Recent Progress in Observation

Since World War II seismology in general has made remarkable advances. Progress in Japanese seismology, particularly during the last decade, has been spectacular. This is mainly due to earthquake observation and prediction programs and the electronics and computer technology that supports them.

Repeated surveying is the most critical factor in earthquake prediction—a fact already recognized by Imamura decades ago. In addition to the detection of changes in ground level, it is now possible to measure the amount of horizontal strain in the crust thanks to the development of a geodimeter that can measure tens of kilometers to an accuracy of 10^{-6}.

In addition to repeated surveying seismographs and the observations of crustal movement, seismology depends on observations related to geochemistry and the electromagnetism of the earth. Such observations must not be limited to one location nor to one field; rather observation networks have to be established in each field. A recent example of such a network is the grand-scale bore-hole type volume strain gauge network built by the Meteorological Agency. This is the only network of its kind in the world.

The successful prediction of earthquakes in China led scientists in Japan, during the 1970s, to emphasize geochemistry and the observation of groundwater and its radon concentration. Deep wells are being dug for this purpose in strategic locations.

1.6 Recent Progress in Seismology

Seismology during the 1970s made great strides from the standpoint of earthquake prediction, both in Japan and elsewhere. Much of this progress resulted from the development of the fault model. The model itself is only an approximation, but it makes it possible to calculate the rough pattern of seismic waves and explains the crustal movement that accompanies earthquakes.

Our understanding of shallow earthquakes is a great deal clearer today than it was a decade ago, thanks to plate tectonic theory and its explanation of the forces that cause earthquakes.

Earthquake prediction is possible when the state of stress accumulation prior to the formation of a fault is known. Recent advances in seismology show that the accumulated strain results in deterioration and collapse of the crust. The extent and duration of such deterioration can be determined by the size of the ensuing earthquake.

A few hours before the fault forms—the displacement is thought to begin at a point on the fault plane and then to spread gradually with the approximate velocity of S waves—the crust moves very slowly in the direction of the fault formation. This movement causes the precursory phenomena prior to an earthquake.

This concept is based on recent research in rock demolition which has opened up new horizons in the physics of earthquakes. Given these developments, the elasticity theory of fault models becomes a fundamental factor in the consideration of earthquake phenomena.

Great progress in the area of seismicity is due to the use of telemeters in earthquake research networks, particularly those affiliated with universities. It is now known, for example, that deep-focus earthquake planes—which were already known to exist on the inclined plane that thrusts into the mantle—consist of two parallel planes 30 to 40 km apart. This finding is relevant to global tectonics, but is particularly important as it relates to the origin of earthquakes. Another intriguing finding is that earthquakes in Japan can be divided into two categories—shallow earthquakes that originate within the crust and deep-focus earthquakes that originate on the deep-focus earthquake plane and express the subduction of the oceanic crust. Shallow earthquakes are now known to occur only within the upper layer of the crust, called the 6 km/sec layer.

The only exception to this finding is in the area where the subducting lithosphere meets the continental crust. The reason for this exception is not yet known, but the relative thickness of the 6 km/sec layer does explain why damaging earthquakes in inland Japan are smaller in scale than those along the Pacific coast, and why, among inland earthquakes, those that take place in central areas are slightly larger than others.

Since 1950 explosion seismology has been used in Japan to determine the earth's crustal structure. Without this research the findings described above would not have been possible—a good example of how basic research can sometimes lead to unexpected results. These findings will soon be recognized in other parts of the world.

During the short time the telemeters of the bore-hole type volume strain gauge have been in operation, some interesting phenomena have been observed. Continuous observation using the conventional vault-type crustal movement observation technique has also yielded some significant information based on the accumulation of data over a long period of time. For instance, it is now known that some crustal movement propagates at very slow speed—40 to 50 km per year. Knowing the cause of this phenomenon would be of great benefit to seismology and earthquake prediction.

There has been a remarkable increase in knowledge concerning precursory phenomena both in Japan and abroad since the beginning of the 1970s. This does not mean that earthquake precursors have increased, of course. Rather, it is an indication that the accumulated record of such phenomena over the years is beginning to pay off. It is also due to an increase in reported incidents, spurred on no doubt by the recent popularity of earthquake prediction throughout the world. What makes this recent period unique is that the precursory phenomena are no longer seen as isolated incidents. The fundamental new developments in seismology that have been described have made it possible to understand them within a larger scientific context.

1.7 Earthquake Prediction

In this book, the present stage of earthquake prediction will be discussed in great detail. This section will outline the various factors involved.

The first step in earthquake prediction is speculation on where earthquakes might occur. At present such speculation is based on historical records, knowledge of active faults, the results of surveys, study of seismicity gaps where earthquakes have not occurred, and so forth. Pacific coast earthquakes, such as the Nankai or Tonankai earthquakes, can be predicted, as Imamura first proved, because they have recurred so often. Inland earthquakes, by comparison, only occur every several hundred or several thousand years and thus are more likely to take place in areas without any recorded history of earthquakes.

In the case of inland earthquake prediction, information about active faults is useful, but it is not enough. Sometimes surveys reveal strains in certain areas that might develop into earthquakes. If there is a rise in the land, leveling methods can detect it; if it is accompanied by precursory phenomena such as deterioration of the crust, the area can be marked as one in which an earthquake is likely to occur in the near future. Seismicity gaps where no small earthquakes occur are also considered earthquake-prone areas.

Finding an area where an earthquake with M 7 or 8 will occur in the future is not an easy task. Predicting a major earthquake in the Suruga Bay area in the near future is an unusually easy example of prediction, in fact. In areas of inland Japan, as opposed to those on the Pacific coast, it is much harder to say where earthquakes with a magnitude of 7 to 7.5 will occur within the next forty or fifty years.

Once it is known where earthquakes are likely to occur, the next question is when they will strike. If the location has been pinpointed accurately, various kinds of networks can be established to observe the precursory phenomena that make it possible to predict an earthquake four to five hours before it happens. The work of such networks includes the continuous observation of crustal movement (volume strain, tilt, expansion, etc.); the observation of microearthquakes and ground water level; and geochemical observations of radon concentration, electric currents, electric resistivity, etc. It becomes possible to detect precursory phenomena shortly before the earthquake occurs if extensive observation networks have been established in the area at close enough intervals. As earthquake prediction becomes more sophisticated it may become possible to accurately detect precursory phenomena with far less complicated observation networks. With today's knowledge, however, we cannot afford to be careless; not much is known yet about how much observation is actually necessary for successful prediction. Significant results can be anticipated from watching and waiting for precursory phenomena in areas where earthquakes are known to occur. Predicting how long it will be before an earthquake strikes is, of course, a more complicated task. At present, the only recourse is to watch for changes in the physical volume of the target area's crust over a long period of time.

Earthquake prediction is still in its infancy. Recently, the study of the various precursory phenomena has been exaggerated; people have gotten the idea that they are easy to detect and that earthquake prediction can be done routinely. Such a notion is, of course, false. It would be equally inaccurate, however, to dismiss the possibility of successful earthquake prediction.

Despite the spectacular recent progress in seismology and the great strides that have been made in understanding earthquake phenomena, our knowledge of how earthquakes occur is still inconclusive.

Where earthquake prediction is concerned, failure to maintain consistent research programs could result in irreversible failure. Earthquake prediction in Japan is based on the constant observations of the Meteorological Agency or the Geographic Research Institute: basic research must continue to be done, however, by seismologists and geophysicists—the overwhelming majority of whom are affiliated with universities and other research institutions—or it will not be possible to study earthquake prediction thoroughly.

References

Matuzawa, T., 1928: Observation of some of recent earthquakes and their time-distance curves (Part I). *Bull. Earthq. Res. Inst.*, Univ. Tokyo, **5**, 1–28.

——, 1929: Observation of some of recent earthquakes and their time-distance curves (Part II). *Bull. Earthq. Res. Inst.*, Univ. Tokyo, **6**, 177–229.

Tsuboi, C., 1933: Notes on the mechanical strength of the earth's crust. *Bull. Earthq. Res. Inst.*, Univ. Tokyo, **11**, 275–277.

Wadati, K., 1928: Shallow and deep earthquakes. *Geophys. Mag.*, **1**, 162–202.

Chapter 2   Great Earthquakes of the Past

Tatsuo Usami

2.1   Historical Records of Earthquakes

The term "ancient earthquakes" refers to all earthquakes that have already occurred. When it is used in connection with Japanese historical records, however, it usually refers only to earthquakes that took place before the end of the Edo Period (the 1860s) or, more precisely, to those before 1872. There are several reasons for this:
  (1) The Julian solar calendar was adopted in Japan in 1873;
  (2) Seismographs were first used to observe earthquakes in 1873;
  (3) Printed materials became available after 1873 and were easily accessible at the various stations and libraries concerned;
  (4) Records of earthquakes that occurred prior to 1872 are mostly uncollected; they were handwritten with a brush, making them difficult to read now.

The only historical records that are readily available are contained in the original and revised editions of *Materials for the History of Japanese Earthquakes* totaling four volumes and 4,000 pages (hereafter referred to as *The Historical Records*). These volumes, published between 1941 and 1949, include more than 6,000 earthquakes that took place from the beginning of history through the end of the Edo Period. Not all the historical earthquakes were included, but attempts to collect additional historical materials were almost nil. A few years ago I began collecting these materials myself, with the help of some historians; the task has been progressing smoothly. In the meantime the importance of ancient earthquakes has become increasingly apparent as our understanding of earthquake causes and processes has increased. Knowledge of ancient earthquakes can now be effectively applied to modern seismology.

The number of people collecting and studying ancient earthquake materials has been growing—a welcome trend indeed. It is regrettable, however, that these materials are not being widely shared. They should not be allowed to scatter after use by a few individual scholars. Thus, I have been requesting that libraries preserve copies or photocopies of such materials, bound with an accompanying annotation. About 8,000 printed pages of new historical materials have already been annotated and organized so far, and something new is being added almost daily. Efforts must be made to open communication with all individuals concerned throughout the country, and to arrange and print the materials collected so as to make them widely available to both scholars and communities.

Historical records include materials as various as official histories, clan histories, journals, various records, literature, petitions, memoranda, correspondence, death registers, genealogies, inscriptions, and legends. Some materials are exclusively about earthquakes; others, such as diaries and memoranda, only mention them. In terms of content there are two types—those that are based on actual observation and those that are based on hearsay. Some of the documents are originals, some are copies, and some are edited versions taken from other historical records. Both in form and in content, these materials are indeed varied.

The Historical Records consist mainly of official histories, government records, and the records of clans that were close to the center of power. They do not include information pertinent to the local history of towns and villages. The collection of just such information should be a central task in the future. In many cases, significant earthquake records can be found in seemingly unrelated materials—materials related to land taxes, industry, debts, transportation, and other areas that closely affect people's lives. But to sift through such materials looking for possible earthquake records is an enormous task that cannot possibly be completed successfully without the cooperation of local and national historians.

The basic problem in analyzing historical records is to determine their credibility. Equally important, as far as seismology is concerned, is whether or not all the relevant information about earthquakes has been recorded. When facts have been omitted by an author, the value of the records may be greatly affected. Recently I had an opportunity to examine *The Diary of the Nambu Clan,* which spans approximately 200 years. Some of its earthquake records are recorded precisely while others are not. An average of seven earthquakes per year were recorded between 1703 and 1709 in the *Diary*; during the same period Kakei Kita, the Clan Confucianist, recorded an annual average of ten earthquakes in his own diary, *The Diary of Kakei Kita.* Ignorance of such discrepancies could lead to an erroneous analysis of seismic activity during the period. Thus, in examining ancient earthquakes, it is necessary to use as much material on each earthquake as possible. It is also important, when poring over such a vast amount of material, to avoid choosing only those materials that tend to support the researcher's preconceptions.

As far as the credibility of the recorded facts is concerned, that can only be judged on a case-by-case basis. It is relatively easy to determine the times and dates of earthquakes, for instance, and thus to detect the occasional mixups of one great earthquake with another. If an earthquake is recorded as "great" at places far from the epicenter, it can be assumed that the report is based on hearsay rather than on direct experience. In cases such as these any seismologist with access to the archival records can judge their credibility.

There are times when the credibility of a record cannot be established, however. At the time of the Zenkoji earthquake on May 8, 1847, the Sanada Clan recorded 41,951 landslides. How was this figure determined? If there were records of the numbers of landslides by village, or of the scales of such landslides, the accuracy of the figure could be verified. In the absence of such records, however, there is no choice but to accept the figure given.

The credibility of historical records depends to a large extent on their source. There would seem to be little point in falsifying records of earthquakes, so it is assumed that this is not a factor. The most reliable sources are petitions, the appeals and damage reports found in local records, and records edited by clans and governors' offices. In the latter case there is always the possibility that the selection of materials has been influenced by the editor's bias. One must also pay attention to whether a report is based on firsthand experience or on hearsay. At the time of the Edo earthquake (Nov. 11, 1855), for instance, all the records of abnormal phenomena around Katsushika are based on hearsay. Each sentence begins with "It is said . . ." or "They say . . .". Letters are often dependable sources. Due to the alternate-year-of-residence system during the Edo Period, there was a great volume of correspondence between Edo and the feudal lords' localities; much of this correspondence is still in existence and can be quite helpful. Generally speaking, the more descriptive of the actual event the records are, the better. Historical records relating to the effects of earthquakes on land taxes and industry can be dependable if they refer to events that have just happened; those that refer to earthquakes that happened at some previous time are less likely to be reliable.

Recently a letter was found in Shizuoka Prefecture concerning the Hoei earthquake of October 28, 1707. It was a letter of apology written by someone who had falsely claimed that his house was crushed by the earthquake. If he had not been found out, he could have been eligible for loans and other financial compensation for his loss. Such a discovery cannot be shrugged off as just one more example of the immutability of human nature, because it calls into question the reliability of local damage reports which have always been thought to be quite accurate. Cases like this one are probably rare, but they do point out the need for critical thinking even when using materials as basic as local records. Even today there are discrepancies between the earthquake records of the police and those of the Department of Disaster Relief. In the latter cases, grants-in-aid may be involved. I usually consider the police reports, which are made a day or two after the earthquake, to be the more reliable of the two sources.

2.2 Problems and Limitations of Historical Seismology

Whenever there is a big earthquake, people always wonder what previous earthquakes were like and what measures were taken then. As a result, earthquakes have been studied and catalogued since ancient times (for example, *Examples of Great Earthquakes*). When seismology came of age in the Meiji Era (1868–1912), records of ancient earthquakes became important sources of information on the frequency and geographic distribution of earthquakes. The Earthquake Investigation Committee was founded in 1892, and research in this area then began in earnest. Minoru Tayama was assigned to the task of studying ancient earthquakes and, after a decade of effort, succeeded in collecting many of the necessary records. Tayama was a consultant to the Editorial Office of Historical Records, now the Historiographical Institute of the University of Tokyo. Based on the records he assembled, Fusakichi Omori published the *Catalogue of Great Earthquakes* in 1919.

During the Showa Era (from 1926 on), K. Musya took up where Tayama left off and, in 1949, published the four-volume *Historical Records* which contained much new material. For the first time the history of earthquakes in Japan was clearly outlined and most of the damaging earthquakes recorded in one place. Based on Musya's work, Akitsune Imamura studied the regional and chronological relationships of earthquakes and applied his findings to long-range earthquake predictions. His posthumous work was recently published as *Data on the Precursory Phenomena of Great Earthquakes*.

Although the collection of historical material was discontinued after Imamura's tenure, H. Kawasumi made good use of the existing collections, arriving at such earthquake focal elements as λ, ϕ, and the magnitude of major earthquakes, hypothesizing a 69-year cycle for earthquakes of intensity 5 or greater in southern Kanto, and drawing up what is known as the Kawasumi Map. Given these achievements, the significance of ancient earthquakes is receiving more recognition, and there is renewed interest in the collection and study of historical records; historical seismology is not yet a field of study in its own right, however.

Great earthquakes tend to recur in the same spots at intervals of between 100 and 1,000 years. Since modern seismology dates back only 100 years and reliable data have been collected for only 50 years, the importance of studying ancient earthquakes for the purposes of long-range prediction and disaster countermeasures goes without saying. As the editor of this book points out, collecting and organizing ancient documents is to earthquake prediction what experiments are to physics and chemistry. Speculations that are unsupported by historical evidence can lead to erroneous conclusions. Thus the extensive collection of historical materials should be a top priority. At the same time, a knowledge of modern seismology should inform the study of these materials. As seismology progresses it will create new demands for information which will necessitate the collection of additional historical materials. Modern seismology and historical seismology depend upon one another for their mutual growth.

At this time information is needed on the details of individual earthquakes so that earthquake source parameters as well as time series can be determined and future seismic activities predicted. Some precursory phenomena provide clues to short-range as well as long-range prediction. The distribution of tidal waves and other damaging abnormal phenomena, for example, can help in the prediction of future disasters. But it is not as easy to interpret such phenomena as one might think. The liquefaction of sand was first linked with earthquakes at the time of the Niigata earthquake (1964). Ancient documents include mention of liquefied sand but, until the Niigata earthquake, neither seismologists nor engineers had made the connection between such incidents in the past and their modern counterparts.

The credibility of historical materials has already been discussed, but the importance of recognizing their limitations must always be remembered. The more records there are to draw on, the more reliable the conclusions are likely to be. Therefore it should be realized that the collection and comprehensive study of new materials in itself constitutes research.

Ever since I began collecting historical materials I have emphasized the importance of primary sources such as local documents. The first step in studying an ancient earthquake is to estimate its magnitude. To do so, the condition of the buildings and structures of the time must be considered. The strength of big temples, shrines, and castles can be estimated from existing structures, but there is no way to estimate the condition of people's homes. Hopefully, in the future, engineers will do research in this area. To obtain the rate of damage, data on population and on the number of households exsiting at the time of the earthquake are needed. Unfortunately, there are few convenient sources for this information. The local records of prefectures, districts, cities, towns, and villages all have to be checked. The most convenient method I have found for estimating an earthquake's magnitude is to postulate that the strength of structures in the past is roughly equivalent to the strength of the present structures. I then subtract 0.5 to 1.0 from the seismic intensity obtained by applying the present scale of seismic intensity to an ancient earthquake.

The study of ancient earthquakes obviously has its limitations (± 1 in the case of seismic intensity, 30 to 40 km in the case of λ and ϕ). In part these limitations are due to the imprecision of the available records. More important, however, are the records that are lost or missing, for there is no way to recover the information they contained. The only hope is that additional historical material will be discovered.

It is particularly difficult to obtain information on earthquakes that took place prior to the Middle Ages (10th–16th centuries). There is evidence in the Kumano Channel of two earthquakes of magnitude greater than 7¾—one on Jan. 21, 1408 (136.9° E, 33.8° N, M 7.0), and one on April 4, 1520 (136.3° E, 33.6° N, M 7.0). Since there is insufficient evidence available to determine the actual magnitudes, they are represented as M 7. Unless new material turns up, there is no justification for postulating the scale of these two earthquakes as 8.

2.3 Examples of Great Earthquakes

Table 2.1 shows all of the great earthquakes that have occurred in the Kii Channel, which separates Shikoku from the Kii Peninsula of Honshū. The area around the channel is called the Nankai region.* This remarkable phenomenon is known throughout the world: an earthquake of the same scale occurs every 100 to 250 years at approximately the same place. The phenomena accompanying the earthquakes are always identical. Anyone can make a long-range prediction of the next earthquake given this information. Until several years ago the third earthquake listed in Table 2.1 was thought to have occurred in the Keiki district, around Kyoto, instead of the Nankai region. This made the interval between the second and the fourth earthquakes almost 500 years—which was, of course, too long relative to the other intervals. Such irregularity made the long-range prediction of the next earthquake difficult. Then, in 1968, an old document was found which made it clear that on March 6, 1100, an earthquake caused damage in western Japan. Shioe-sho (now Kochi City) in Tosa

---

* See the reference map inside the cover for this and other geographical places.

GREAT EARTHQUAKES OF THE PAST   17

Table 2.1 Earthquakes in the Nankai Region

Date	Epicenter Latitude	Epicenter Longitude	Magnitude (M)	Dimension of depression around Kochi City	Amount of uplift at Cape Muroto	Yunomine Hot Springs
29 Nov. 684	32.5°N	134.0°E	8.4	12 km²		
26 Aug. 887	33.0°N	135.3°E	8.6			
22 Feb. 1099	33.0°N	135.5°E	8.0	> 10 km²		
3 Aug. 1361	33.0°N	135.0°E	8.4			Outflow stopped
3 Feb. 1605	33.0°N	134.9°E	7.9			
28 Oct. 1707	33.2°N	135.9°E	8.4	20 km², 2 m >	1 ~ 2 m	Outflow stopped
24 Dec. 1854	33.0°N	135.0°E	8.4	1 ~ 1.5 m	1.2 m	Outflow stopped
21 Dec. 1946	33.0°N	135.6°E	8.1	15 km²	1.3 m	Outflow radically decreased

(now Kochi Prefecture) was inundated by the sea—a phenomenon that is characteristic of the great earthquakes in the Kii Channel. So it would seem that this earthquake had its epicenter in the Kii Channel. Either the Keiki earthquake recorded a year earlier was a different one altogether, or else the data are off by a year and the epicenter was actually in Nankai. In any case, there was a great earthquake in the Kii Channel in either 1099 or 1100. Based on this evidence Table 2.1 was drawn up; it clearly demonstrates the regularity of great earthquakes in this particular region.

Table 2.2 shows the great earthquakes that have occurred off the Pacific coast, west of the Kanto district. It is clear that the earthquake epicenters are generally moving from east to west and that the earthquakes off the Tokai coast south of Shizuoka and Mie Prefectures and those off the Nankai coast tend to occur in pairs. This pairing is shown in Table 2.3.

Table 2.2 Series of Great Earthquakes off the Coast of Kanto

Off Kanto	Off Tokai	Off Nankai
		29 Nov. 684
818		26 Aug. 887
	18 Dec. 1096→ca. 2 years→	22 Feb. 1099
		3 Aug. 1361
	20 Sept. 1498	
3 Feb. 1605	Same day	3 Feb. 1605
31 Dec. 1703→ca. 4years→	28 Oct. 1707 →Same day→	28 Oct. 1707
	23 Dec. 1854 →32 hours→	24 Dec. 1854
1 Sept. 1923→ca. 21years→	7 Dec. 1944 →ca. 2 years→	21 Dec. 1946

Table 2.3 Synchronous Occurrence of Nankai and Tokai Earthquakes

Nankai \ Tokai	Occurred	Did not occur	Total
Occurred	4 (3)	4	8 (7)
Did not occur	1 (2)		
Total	5		

It doesn't always occur, but it has been happening more often in recent years. The numbers in parentheses were estimated before the earthquake of 1099 was found to be a Nankai earthquake. Thus, ten years ago the chance of a Nankai earthquake occurring immediately after a Tokai earthquake was 3 in 5. With the discovery of the 1099 earthquake, however, the chance became 4 in 5. It then becomes more likely that there was an earthquake off the coast of Nankai around 1500. Such a likelihood spurs on the search for more ancient documents. As far as we know now, as a fifty-percent probability there was no Tokai earthquake recorded before a Nankai earthquake; new historical material may, of course, reveal that such earthquakes did in fact take place.

There has been a great deal of discussion of the earthquakes off the coast of Tokai. According to simple logic, as an 80 percent probability a Nankai earthquake will take place within two years after a Tokai earthquake. This reasoning, however, is not supported by the facts. Whether this negates the probability that a Tokai earthquake will occur within 20 or 30 years is not clear. Earthquake prediction based on modern interpretation of ancient earthquakes is inexact due to the scarcity of facts about ancient earthquakes—errors in epicenters and so forth—and to various types and emphases of modern interpretation. Where earthquakes are concerned, ten different people, whether they are amateurs or professionals, are likely to have ten different opinions. It is critical, therefore, to build up the research base necessary to minimize errors in seismic distribution and estimation of magnitude and epicenters, and to pinpoint the faults, geotectonic positions, and activities of ancient earthquakes.

Another Tokai earthquake is often predicted based on plate tectonics theory and on the interpretation of three facts:
(1) It has been 120 years since the last earthquake with an epicenter off the coast of Suruga Bay (Dec. 23, 1854).
(2) There has been a seismic gap from off of Cape Omaezaki to Suruga Bay during the past 50 years.
(3) Since 1884 a maximum shear strain of 3×10^{-5} has been accumulating in the Suruga Bay region.

With reference to (1), large-scale earthquakes have been occurring at intervals of 100 years and more in the Tokai region (from Izu to the Kumano Channel). If only Suruga Bay and the area off the Enshu Sea coast are taken into consideration, however, the interval is more than 300 years. In the entire Tokai region it has been only 36 years since the 1944 earthquake, and the 1854 earthquake took place only 126 years ago. It is hard to reconcile these facts with the prediction that another Tokai earthquake is due.

As for (2), although seismic gaps are known to be earthquake precursors, there are unanswered questions about where the seismic gap phenomenon is most likely to occur, what the relationship is between the scale of earthquakes that occur in seismic gaps and the size of the gaps, and how long such gaps persist. It is not known whether focal zones became seismic gaps immediately after the 1944 Tonankai earthquake and the 1946 Nankai earthquake, or whether this change only occurs 20 to 30 years before the next earthquake. A comprehensive study of the seismic gap must be carried out based on the observational records accumulated by the Japan Meteorological Agency over the past five decades. The fact that this has not been done already is an oversight on the part of seismologists.

In the case of (3), even though we know that the stress accumulating in Suruga Bay is considerable, it is not clear how this affects the specific area off the coast of Tokai.

It has been found that the Nankai Trough in Suruga Bay dislocated at the time of the 1854 Tokai earthquake. This fact is useful for predicting the potential damage that would occur along the coast of Suruga Bay in the case of another earthquake. It also raises the question of whether there was a similar dislocation at the time of the 1707 earthquake. If there was such a dislocation, then these two earthquakes would share a common characteristic. There was no such dislocation at the time of the 1944 Tonankai earthquake. This may point to another Tokai earthquake in the near future with its focal zone in Suruga Bay. If, however, there was no dislocation during the 1707 earthquake, then the anticipated Tokai earthquake would not necessarily be accompanied by dislocation and may in fact have already occurred, in 1944. The next one, then, would take place much later. Some new materials have surfaced in the search for additional records of the 1707 earthquake, but so far no solid evidence of dislocation in Suruga Bay has been found.

According to one record I discovered recently, 195 houses belonging to local officials and 532 houses belonging to lesser officials collapsed in the village of Iwamoto, near the mouth of the Fuji River, at the time of the 1707 earthquake. According to the scale of seismic intensity established by the Japan Meteorological Agency, this amount of damage would indicate an intensity of 7. Houses in the past were not as solidly built as the houses of today, however, so the intensity may well have been less than 6. The damage is not sufficient evidence, therefore, of dislocation in Suruga Bay. On the other hand, there was a strong aftershock on the following day. This incident is only recorded in documents that concern the eastern parts of Shizuoka, Yamanashi, and Kanto; documents from the western part of Shizuoka did not record the aftershock. In Yamanashi Prefecture, along the Fuji River, the aftershock was said to be much stronger than the main shock itself. Although this point remains to be clarified, the aftershock probably occurred in the extension of the Nankai Trough.

There was a similar case when one of the aftershocks of the 1923 great Kanto earthquake shook the Tanzawa Mountains on January 15, 1924. The great aftershock took place on an extension of the Sagami Trough. No evidence of movement along the Kozu–Matsuda Fault on this extension has been found. The epicenter of the Akita–Senpoku earthquake of March 15, 1914, was on the Senmaya Fault, which was formed at the time of the Rikuu earthquake of 1896. On the 28th of that month, a maximum aftershock occurred on the southern extension of that fault, totally demolishing some of the houses in the area. There are other examples but they are not as accurate. The earthquake in northern Gifu Province of April 9, 1858, which was caused, apparently, by movement of the Atotsugawa Fault, was followed two hours later by a major aftershock in Fukui. Judging from the damage distribution, this aftershock may well have occurred on an extension of the Atotsugawa Fault.

Generalizing from the above, it can be concluded that great earthquakes may either create an extensive movement of faults and troughs or occur within a part of a system of existing faults and troughs followed by major aftershocks on the extensions of these faults and troughs. Since there is not enough proof to support the postulate that Suruga Bay was included in the focal zone of the 1707 earthquake, it is probably safer at present to assume that the 1707 earthquake belongs in the latter category.

The 1858 earthquake mentioned above is considered to be related to the movement of the Atotsugawa Fault. More than 2,000 pages of material concerning this earthquake have recently been collected from various regions in Japan. An analysis of this material may help to clarify the relationship between the 1858 earthquake and the Atotsugawa Fault. Once this relationship is clear, given the fact that faults in inland Japan rupture once every few hundred to a thousand years, one can speculate that the Atotsugawa Fault will not be active again for some time. This would be a prediction, in other words, of an absence of earthquakes—an example that demonstrates the significance of thorough investigation supplemented by newly found historical documentation.

Generally speaking, ancient earthquakes can be grouped according to (1) chronology, (2) location, and (3) magnitude for purposes of long-range earthquake prediction. There is no special problem in determining the chronology of ancient earthquakes, but grouping them in terms of space and magnitude is not so easy. Oceanic trenches, troughs, faults, and tectonic lines can be used as guidelines for spatial groupings, but scholars sometimes disagree on the positions and continuity of faults. Furthermore, there are problems in determining the focal positions and magnitudes of ancient earthquakes. Support from the fields of geology and plate tectonics can help to resolve these issues. Grouping earthquakes in terms of magnitude presents problems that will only be clarified by the further collection and analysis of historical materials.

2.4 Merits and Drawbacks of Imamura's Historical Seismology

Among his many achievements during the early part of this century, Akitsune Imamura made a contribution to historical seismology by collecting and analyzing historical materials. Whenever he found a new piece of evidence, he published his interpretation of it along with the results of any related research in the first series of the journal *Zisin* [*Earthquake*]. He never published a complete collection of his findings, however. An unfinished manuscript, "On Past Seismic Activities in Japan," appeared in Vol. 8 (1936) of *Zisin. On Precursory Phenomena of Great Earthquakes—Posthumous Manuscripts of Dr. A. Imamura* was published in 1977 by Kokon Shoin Publishers. This book is particularly useful and accessible and will be used here to study Imamura's theory of ancient earthquakes.

Although Imamura was probably familiar with most of the historical materials available at that time, Musya's *Materials for the of History of Japanese Earthquakes* (3 vols.) was published late in Imamura's life (from 1941 to 1943), and its fourth volume, was published in 1949—after Imamura's death on Jan. 1, 1948. Thus the only material for Imamura to refer to that was well organized and easily accessible was M. Tayama's (1904). In examining Imamura's work, it is also important to remember the state of seismology at that time—the concept of magnitude was unknown, for instance—and the state of earthquake prediction in particular.

Let us first consider his prediction of the Nankai earthquake of Dec. 21, 1946. What, exactly, did he predict? The crux of the problem was the date. *Precursory Phenomena*, a paper presented to the Imperial Academy of Japan on Oct. 12, 1946, concludes: " . . . it is increasingly felt that the great earthquake just discussed will strike in the near future. This eventuality deserves the utmost attention from an academic standpoint as well as from the standpoint of disaster prevention." Given the information presented in Tables 2.1, 2.2, and 2.3, plus the fact that the Tonankai earthquake had already occurred in 1944, Imamura, who had himself postulated the epicenter migration theory, must have made both long-range and medium-range earthquake predictions. His private observation stations at seven locations on the Kii and Muroto Peninsulas were built to detect precursory phenomena for the purpose of short-range prediction. Unfortunately they were closed down due to lack of resources both during and and after World War II, making short-range prediction no longer possible. Although Imamura was confident that a great earthquake was coming, as is obvious from his paper, he would not make a definite prediction as to the date or extent of the earthquake other than to say it was "approaching in the near future." This is still a successful example, however, of what is now called intermediate-range prediction.

In his interpretation of ancient earthquakes Imamura made full use of his knowledge of geographical names, history, and the Ainu language. Whenever new historical material was found, he did not hesitate to revise his theories accordingly. Some of his ideas may have been rather arbitrary and unorthodox, however. For example, he interpreted a record of the Tsukushi great earthquake of 679 that noted: "The gaping fissure in the earth was about 20 feet wide and 30 thousand feet long . . ." as a record of volcanic fissure. He writes, "The place was Shimabara Peninsula in Kyushu, and the frightening natural phenomenon that gripped the earth in 1792, I concluded, was simply a recurrence of the natural disaster of 679." Recently, however, a record was found that describes the earthquake of 679 as being in the Tsukushi-Bungo area rather than in Shimabara.

Imamura's beliefs were so daring that they probably were not readily accepted by the academic societies of the day. His basic principle, however, "to find similar earthquakes in history and examine their relationship to one another," is still valid today. Given only the information at Imamura's disposal, we might well make the same mistakes today.

One of Imamura's better-known papers describes the shoreline of the Boso Peninsula before and after the Genroku earthquake in 1703. The shoreline was deduced from the heights at which certain boring shells were found. His paper does not show the plotted shoreline based on actual data, however. Rather, there are four smooth, beautifully drawn curves which are not convincing to the eyes of modern scientists. Imamura can be criticized for seeking to reach conclusions at the expense of accuracy and precise, orderly reasoning. Where accuracy is concerned, Imamura's findings are not scientifically acceptable today. This is probably why, when citing his papers, most authors are careful to begin with, "According to Imamura," or "We are told that . . .".

Imamura looked at earthquakes in terms of their time (chronology) and space (location) (Table 2.4). He divided the island arc of Japan along the Pacific coast into various areas, taking into consideration the distribution of both seismic intensity and tsunamis, recognizing that there were clear borderlines between such divisions. Naturally, there was no reliable standard for selecting the earthquakes in Table 2.4, since the concept of magnitude did not yet exist. Yet the earthquakes selected do influence his conclusion greatly. In any case, the fact that great earthquakes in Japan migrate from south to north and north to south, according to the historical period, is clearly demonstrated in Table 2.4. Based on this fact, Imamura predicted the Nankai earthquake of 1946. The prediction was made on the basis of a statistical table, but Imamura relied on his experience and intuition rather than on a statistical analysis to reach his conclusion. This is still the state of the art of long-range earthquake prediction based on historical earthquakes. The chronic tilting of the Kii and Muroto Penin-

Table 2.4 Large-scale Earthquakes on the Ocean Floor by Location

Off Saikaido coast	Off Nankaido coast (Beppu Bay)	Off Tokaido coast (Wakasa Bay)	Off Kanto coast	Off Sanriku coast	Off Hokkaido coast	Intervals (except off Saikaido coast) (unit: years)
	684					17
		(701)				117
			818			51
				869		18
	887					209
		1096				265
	1361					137
		1498				98
	(1596)					9
	1605		1605			6
				1611		66
1662						26
				1677		4
			1703			
	1707	1707				136
1769					1843	11
						0.0037
	1854	1854				0.0047
	(1854)					
					1894	40
				1896		2 (21)
				(1915)		
			1923			27 (8)
		1944				21

( ) indicates earthquakes with lesser validity.

sulas and the Tonankai earthquake of 1944 gave Imamura more confidence, and he began his search for the precursory phenomena that he knew would precede the Nankai earthquake.

Imamura classified inland earthquakes based on geological structures, folds, and faults. This standard has been further refined and is still in use. He also considered volcanic eruptions to be the same kind of energy release as earthquakes. Today seismology and volcanology are completely independent fields. A more comprehensive consideration of these phenomena may prove fruitful in the future.

Imamura's achievements have been examined in the light of what is now known. His research on ancient earthquakes and the application of that research to long-range earthquake prediction is a highly valuable contribution. His method of progressing from long-range prediction based on ancient earthquakes, to observations that solidify the prediction, to the careful minitoring of short-range precursors is still followed, in a varied form, by modern seismologists. Although his inaccuracies and flights of fancy cannot be ignored, they do point out problem areas. They may even serve to increase interest in ancient earthquakes. His findings are still questioned and criticized by modern seismologists. The public still regards Imamura's theories as the absolute truth: they went unchallenged at the time since he was the only scholar who was studying ancient earthquakes. There are people in rural areas who still remember and believe what he said when he visited their locality. Imamura's influence has been great and extends even beyond the realm of seismology *per se*.

2.5 Examples of Ancient Earthquakes in Japan

2.5.1 Hoei Earthquake

(Oct. 28, 1707, at approximately 14h00m. $\lambda = 135.7°$E, $\phi = 33.2°$N, M 8.4.)

This was one of the greatest earthquakes in Japanese history. The areas affected by violent shocks and by tsunamis were similar to those hit at the time of the Ansei Tokai–Nankai earthquakes of Dec. 23 and 24, 1854. For this reason, the Hoei earthquake is sometimes thought to have been two simultaneous earthquakes that took place off the coasts of Tokai and Nankai. The area that recorded an intensity of 6 extended from Suruga Bay to the western part of Shikoku Island; tsunamis hit the coast from Izu to Kyushu. The waves were 1.5 m high on Cape Muroto, 1.2 m in Kushimoto, and 1 to 2 m near Cape Omaezaki. Its rupture zone extended into Suruga Bay. Although a great many new historical materials have been discovered, it hasn't been possible to reach a firm conclusion. An examination of the great aftershock that reputedly hit Yamanashi Prefecture on the following day (Oct. 29) is expected to answer this question. About 50 days later, on Dec. 16, Mt. Fuji erupted violently and formed the Hoei Crater.

2.5.2 Ansei Tokai-Nankai Earthquakes

(Dec. 23, 1854, at 09h00m. $\lambda = 137.8°$E, $\phi = 34.1°$N, M 8.4. Dec. 24, 1854, at 16h00m. $\lambda = 135.6°$E, $\phi = 33.2°$N, M 8.4).

These two great earthquakes occurred 32 hours apart. Some historical material found recently confirmed that the tsunami source area for the Dec. 23 earthquake reached into Suruga Bay—a fact that has caused much apprehension since it was announced in the autumn of 1976. The earthquake shocks in Suruga Bay were much stronger than those during the Hoei earthquake. A "belt" of great seismic intensity occurred along the Fuji River, reaching northward into Nagano Prefecture. A tsunami caused heavy damage in Osaka Harbor after the Dec. 24 earthquake, and its aftershock activity was several times more severe than that which followed the earthquake of Dec. 21, 1946. Two months later, the aftershock zone had migrated to the vicinity of Kochi City. The land rose up approximately 1 m at Omaezaki, 1.2 to 1.5 m at Muroto, and 1 m at Kushimoto.

Table 2.5 Number of Destructive Earthquakes Classified by Magnitude

Years	$M < 5$	$5 \leq M < 6$	$6 \leq M < 7$	$7 \leq M < 8$	$8 \leq M$	Unknown	Total
500 ~ 599				1			1
600 ~ 699			1		1		2
700 ~ 799			2	4		1	7
800 ~ 899			6	10	2		18
900 ~ 999			3	1			4
1000 ~ 1099			6		2		8
1100 ~ 1199			1	1			2
1200 ~ 1299			4	3			7
1300 ~ 1399			4	2	1	1	8
1400 ~ 1499			4	3	1	3	11
1500 ~ 1599			9	3		3	15
1600 ~ 1649		1	12	8	1	4	26
1650 ~ 1699		6	16	8	1	8	39
1700 ~ 1749		3	21	3	2	6	35
1750 ~ 1799		3	13	11		8	35
1800 ~ 1849		5	17	6	1	4	33
1850 ~ 1867		5	10	2	2	10	29
1868 ~ 1899		15	21	13		6	55
1900 ~ 1924		34	42	17		9	102
1925 ~ 1949	2	17	34	18	3	7	81
1950 ~ 1973	8	23	33	19	6	2	91
Total	10	112	259	133	23	72	609

2.5.3 Zenkoji Earthquake

(May 8, 1847, at 21h00m. $\lambda = 138.2°$ E, $\phi = 36.7°$ N, M 7.4.)

An earthquake of similar magnitude hit this area approximately 1,000 years earlier, on Aug. 26, 887. Five days after the 1847 earthquake another earthquake of M 6.5 hit immediately north, namely, the Takada area and resulted in heavy damage. The Zenkoji earthquake is remembered for the damage it caused. There were more than 40,000 landslides in the area. The biggest landslide took place on Kokuzoyama Mountain and dammed up the Saikawa River, creating a lake some 30 km long and several km wide, with a maximum depth of 100 m on the upstream side. This dam broke on May 27 and flooded the area, causing great damage around the Zenkoji Plain. The Matsushiro Clan foresaw this flood and tried to prevent disaster, but 100 people drowned nevertheless.*

2.5.4 Ansei Edo Earthquake

(Nov. 11, 1855, at approximately 22h00m. $\lambda = 139.8°$E, $\phi = 35.8°$N, M 6.9.)

This is a typical example of an earthquake that occurred directly under a big city. Approximately 10,000 people died and some 30 fires occurred, burning a total area of 2.2 km² in Edo (now Tokyo). It was generally believed that damage from this earthquake was largely limited to Edo. Recently, however, it was found that damage also occurred along Tokyo Bay and the Shimofusa Plateau. Thus the latitude of the epicenter should be moved south some 10 to 30 km. Severe damage to storehouses in the Kisarazu area was reported, and the details of damage in the Jonan, Kawasaki, and Tsurumi areas are now known. Houses in what is now Setagaya-ku (southwestern Tokyo) did not collapse, and there were no injuries there. A few houses were damaged in other areas. This new historical information helps in estimating the effects of the earthquake.

---

* Other earthquakes which caused spectacular damage also occurred in the mid-nineteenth century. During the Hietsu earthquake of April 9, 1858, the Tateyama mountain range collapsed, damming up the Joganji River. The dam broke on April 23 and June 7, causing a great deal of flood damage downstream. A few days after the earthquake an evacuation of the Toyama area was apparently ordered. Some people complied and others didn't. There is no record of whether or not further evacuation measures were taken or enforced.

At the time of the Echigo-Sanjo earthquake of Dec. 18, 1828, a canal near Mitsuke-machi was dammed up and its waterlevel rose. There was fear of flooding, but, early the following year, the landlord ordered the accumulated snow and mud removed, and disaster was successfully prevented.

It is interesting to compare the measures taken at the time of these three earthquakes.

Table 2.6 Frequency of Tsunamis Classified by Magnitude (*m*: tsunami magnitude)

Year \ m	−1	0	1	2	3	4	Unknown	Total
500 ~ 599								
600 ~ 699				1				1
700 ~ 799			1				1	2
800 ~ 899		1	2	1	1			5
900 ~ 999		1						1
1000 ~ 1099			1					1
1100 ~ 1199								
1200 ~ 1299		2						2
1300 ~ 1399			1	1				2
1400 ~ 1499		3		1				4
1500 ~ 1599		2	1					3
1600 ~ 1649		2	1 + [1]	1	1			5
1650 ~ 1699		1	2	1			2	6
1700 ~ 1749		3		2	1		1	7
1750 ~ 1799		9	1	1	1			12
1800 ~ 1849		2	3					5
1850 ~ 1867			2	1	1			4
1868 ~ 1899	2	4	2	1		1		10
1900 ~ 1924	2	4	1	1				8
1925 ~ 1949	11	1	3	1	3			19
1950 ~ 1973	8	7	4	4		1	1	25
Total	23	16	36	22	13	7	5	122

2.5.5 Meiji Sanriku Earthquake and Tsunami

(June 15, 1896, at 19h32m. $\lambda = 144.2°\text{E}$, $\phi = 39.6°\text{N}$, $M\ 7.6$.)

On this occasion there was no earthquake damage. The earthquake's magnitude and seismic intensity along the coast was minimal—a fact which left people totally unprepared for the great tsunami that followed. In many areas the tsunami was preceded by a rumbling noise. In some coastal areas the water rose by as much as 20 m, and the death toll reached 26,360. This is a good example of the small-magnitude earthquake that often accompanies a big tsunami. This and similar cases will be the subject of further studies in disaster prevention. Some future countermeasures are already under consideration.

There are many more examples of large-scale earthquakes—far more than can be discussed here. Damaging earthquakes in Japan from the beginning of history through 1973 have been listed in the figures that accompany this chapter. There are 609 incidents in all, or one earthquake every 10 years, as is shown in Figure 2.1. Tables 2.5 and 2.6 demonstrate the frequency of earthquakes and tsunamis classified by

28 EARTHQUAKES REPEAT THEMSELVES

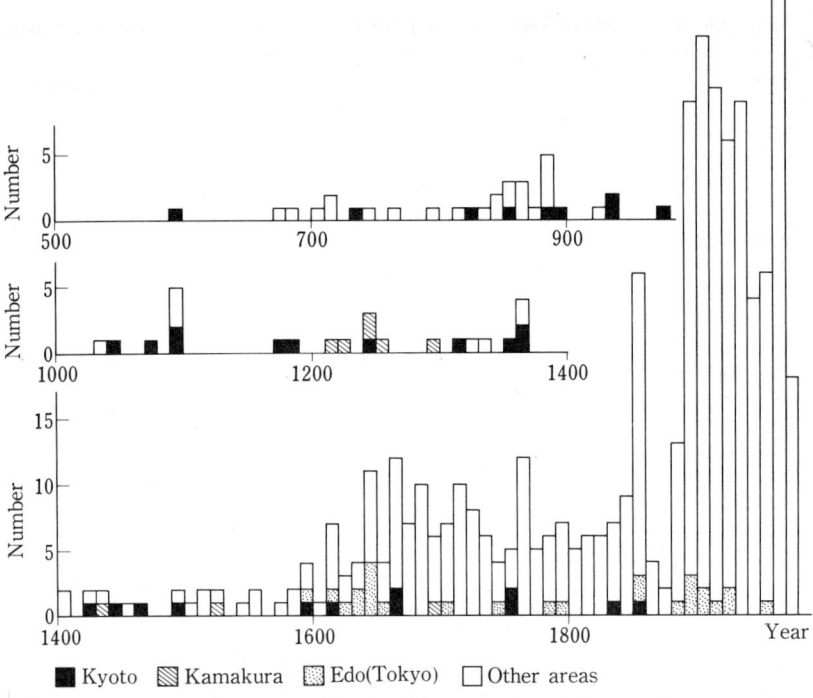

■ Kyoto ▨ Kamakura ▨ Edo(Tokyo) ☐ Other areas

Fig. 2.1 Number of recorded destructive earthquakes in each ten-year period of Japanese history.

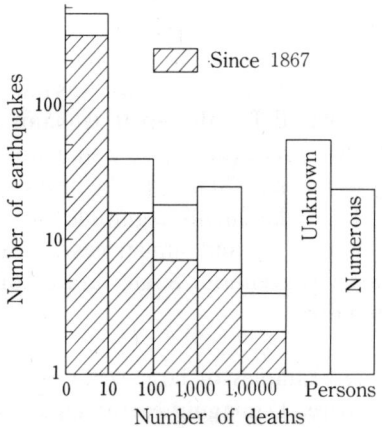

Fig. 2.2 Recorded destructive earthquakes classified by number of deaths.

magnitude. The degree of damage, as measured by the number of deaths, is shown in Fig. 2.2. There were only four earthquakes that resulted in more than 10,000 deaths. Seventy percent of the earthquakes, or 452, resulted in fewer than ten deaths.

References

Earthquake Prediction Research Group in Japan, 1962: Prediction of Earthquakes—Progress to Date and Plans for Further Development, 21 pp.

Imamura, A., 1927: On the seismic activity of the Kwanto district, *Jap. Jour. Astr. & Geophys.*, **5**, 127–135.

———, 1936: Seismicity of Japan in historical time (in Japanese), *Zisin*, **8**, 121–134, 600–606.

———, 1941: The great Hakuho earthquake (in Japanese), *Zisin*, **13**, 82–86.

Kanda, S., 1968: Large earthquake of Tosa in 1099 (in Japanese), *Zisin* (ii), **21**, 142–143.

Kawasumi, H., 1952: Distribution of earthquake danger in Japan (in Japanese), *Shigen Data Book*, **6**, 1–14.

———, 1963: On the expectation of earthquake intensity at Kamakura (in Japanese). Saigai Kagaku Kenkyukai, 1–24.

Musya, K., 1942–43: *Materials for the History of Japanese Earthquakes* (in Japanese), revised ed., Vol. 1–3, Shinsai Yobo Hyogikai.

———, 1949: *Materials for the History of Japanese Earthquakes* (in Japanese), Mainichi Press.

Nasu, N., ed., 1977: Data on Precursory Phenomena of Great Earthquakes. Posthumous manuscript by Dr. A. Imamura, Kokon Shoin, 13–139.

Press, F., H. Benioff, R. A. Forsch, D. T. Griggs, J. Handin, R. E. Hanson, H. H. Hess, G. W. Housner, W. H. Munk, E. Orowan, L. C. Pakiser Jr., G. Sutton and D. Tocher, 1965: Earthquake Prediction—A Proposal for a Ten Year Program of Research, Office Sci. Technol., Washington, D. C., 134 pp.

Rikitake, T., 1975: Earthquake precursors, *Bull. Seis. Soc. Amer.*, **65**, 1133–1162.

———, 1976: *Introduction to the Study of Earthquake Prediction* (in Japanese), Kyoritsu Press, 212 pp.

Scholz, C. H., L. R. Sykes and Y. P. Aggarwal, 1973: Earthquake prediction—A physical basis, *Science*, **181**, 803–810.

Seismological Society of Japan, 1967: Seismology in Japan (in Japanese), *Zisin* (ii), Special issue, Seismol. Soc. Japan, 326 pp.

Suyehiro, S., T. Asada and M. Ohtake, 1964: Foreshocks and aftershocks accompanying a predictable earthquake in Central Japan—On a peculiar nature of foreshocks, *Papers in Meteor. and Geophys.*, **15**, 71–88.

Tsubokawa, I., 1973: On relation between duration of precursory geophysical phenomena and duration of crustal movement before earthquake, *Jour. Geophys. Soc. Japan*, **19**, 116–119.

Usami, T., 1979: Study of historical earthquakes in Japan, *Bull. Earthq. Res. Inst.*, **54**, 399–439.

Utsu, T., 1972: Large earthquakes near Hokkaido and the likelihood of the occurrence of a large earthquake off Nemuro, *Rep. Coordinat. Comm. Earthq. Pred.*, **7**, 4–6.

Chapter 3 Earthquake Scars

Tokihiko Matsuda

A knowledge of the past is necessary in order to predict the future. But how much "past" is needed to predict the earthquakes to come? Recorded history in Japan spans some 1,000 years; is that enough? In some cases, yes, but in others even this time period is inadequate.

To predict worldwide seismic activity over the next few years, data from the past several decades are sufficient. It is just about certain, for instance that within the next few months there will be an earthquake of 7+ magnitude somewhere on the face of the earth. It is just about impossible, however, to predict that a great earthquake will occur in a particular city or area based on information from the last several years or even decades about that area. This is because the past as we know it is far too short to take into account the intervals between great earthquakes in specific areas. The more limited the area and/or the greater the earthquake, the harder it is to collect sufficient evidence of prior incidents on which to base a scientific prediction.

To minimize this difficulty, the "past" must be extended and more information mined from it. The discovery and analysis of ancient documents is one way to do this (see Chapter 2). But they are not the only records of the past. Landforms and strata—materials that precede human history—must contain some imprint of the past. What imprints do great earthquakes leave? Finding and decoding these imprints will be helpful in providing the information needed for earthquake prediction. I would like to discuss methods of prediction based on such decoding, with a view to preventing future disasters.

3.1 Earthquake Scars Left on Landforms

What do earthquakes leave behind? Large-scale earthquakes leave cracks, offset, uplift, subsidence, tilting, and landslides. The greater the earthquake, the more noticeable are the traces of terrestrial movement. In other words, the earthquakes that are most important to learn about are the ones that leave the most traces.

Cracks are the most common form of earthquake trace. Most of them appear on the slopes of mountainous areas, on manmade piles of earth, or on soft alluvium on level land. The erosion that takes place over the years, however, erases most of these from the surface. Those that remain are not on the surface layer but are deep down in the earth and result from such crustal movements as (1) vertical or horizontal slips or faults (fault displacement), (2) a change in the earth's height relative to sea level (uplift or subsidence), or (3) the deformation of the earth (tilting or flexure). When an earthquake fault appears at an earthquake's epicenter it is in fact an underground fault that has surfaced, and the movement of that fault is what causes the earthquake. Such a fault is the most direct and significant of all earthquake traces.

Earthquake faults cannot be observed directly if the earthquake occurs offshore on the ocean floor, but the crustal movement within the ocean floor may be transmitted to land via a tsunami, or the movement of the ocean floor may extend to the adjacent land area, resulting in uplift or subsidence in the coastal areas which is easily recognizable. Uplift or subsidence also occurs in the areas adjacent to the focal area of an inland earthquake, but it is harder to detect since there is no tandem plane like the ocean surface with which to compare it.

Crustal movements sometimes continue to advance in the same direction, or even in the opposite direction, after the earthquake is over. But in either case the land's original position can never be restored. As earthquakes recur, the displacement accumulates and is recorded within the landforms. By studying these landforms one can trace the incidence and nature of prehistoric earthquakes.

To put it differently, a new landform is created every time there is crustal movement due to an earthquake. Fault scarps, uplifted coasts, tilted mountains—all are landforms created by crustal movement. The landforms created by fault movements are particularly useful when trying to identify the location of a fault.

In the case of shallow earthquakes—those whose focus is in the upper crust—the size of the "scar" the earthquake leaves is in direct proportion to the earthquake's magnitude. Understanding this relationship quantitatively is very helpful in estimating the size of an earthquake from a past scar. According to our experience, these relationships are as follows:

(1) The minimum size (M^*) of Japanese earthquakes that are accompanied by crustal deformations on the earth's surface such as slippage, uplift, or subsidence can be represented by the following equation [Kasahara, 1957]:

$$M^* = 6 + 0.22H^{1/2} \qquad (3.1)$$

where H represents the focal depth of the earthquake. According to this equation, the smaller, or shallower, H is, the smaller the earthquake that accompanies crustal deformation will be; when $H = 0$, $M^* = 6$. In other words, earthquakes with a magnitude of less than 6 usually leave no trace of crustal movement. The value of M^*, however, has tended to decrease in recent years as the precision of observation techniques has improved and the number of observation points has increased. According to Yonekura [1975], for instance, $M^* = 5.6 + 0.0175H$. During the Matsushiro earthquake of 1966, the greatest shock did not even reach a magnitude of 5.5, yet a group of fault cracks appeared on the earth's surface due to the nature of the earthquake, which is known as a swarm earthquake. The Matsushiro earthquake is the smallest earthquake that is known so far to have left behind visible earthquake faults on the land surface of Japan. Of the 23 damaging earthquakes with a magnitude of more than 6.5 that have occurred in inland Japan since the Meiji Era, 10 (or approximately 40%) have left behind surface faults.

(2) The relationship between the maximum amount of slip that accompanies an earthquake (D) and the magnitude of the earthquake (M) can be represented by the following equation insofar as historical earthquakes in inland Japan are concerned [Matsuda, 1977]:

$$\log D \text{ (m)} = 0.6M - 4.0. \qquad (3.2)$$

In brief, in an earthquake with a magnitude of 8, a slip of approximately 6 m is usually observed, while in an earthquake with a magnitude of 7 the displacement is usually 1.5 m. Since there is a great deal of local variation in this relationship, this equation is just an approximation.

(3) The relationship between the length L (km) of the fault (earthquake fault line or group of lines) that appears on the earth's surface at the time of an earthquake and the magnitude M of the earthquake can be represented as follows for earthquakes in inland Japan (in most cases $M > 6.5$) [Matsuda, 1977]:

$$\log L \text{ (km)} = 0.6M - 2.9. \qquad (3.3)$$

In the case of an earthquake with a magnitude of 7, the length of the fault will be about 20 km; an earthquake with a magnitude of 8 will produce a fault about 80 km long.

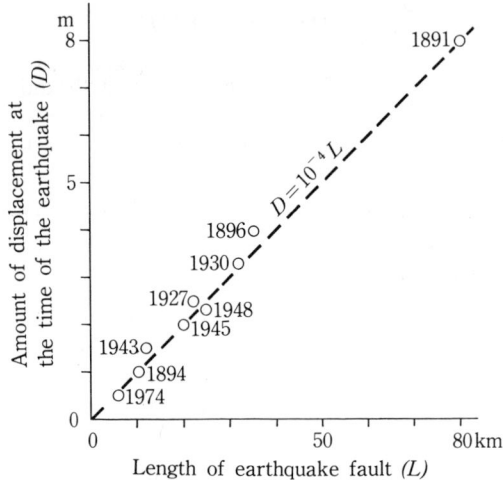

Fig. 3.1 Relationship between length of earthquake fault and amount of displacement in major inland earthquakes [Matsuda et al., 1980]

(4) Based on the relationships in equations (3.2) and (3.3), the relationship between the length of fault L and the amount of fault slip D can be represented approximately as follows (see Fig. 3.1):

$$L \approx 10^4 D$$

or

$$L \text{ (unit: km)} \approx 10D \text{ (unit: m)}. \qquad (3.4)$$

If the amount of fault slip during an earthquake is a maximum of 2m, then the earthquake fault itself will be about 20 km long.

(5) A similar relationship exists between the magnitude of an earthquake and the horizontal extent of uplift. Postulating the extent of surface deformation (found by remeasuring the level line) caused by an earthquake with a magnitude of M as $2r$ km (diameter), the relationship can be represented as follows for earthquakes in inland Japan [Dambara, 1966]:

$$\log 2r = 0.5M - 1.97. \qquad (3.5)$$

Compared with Equation (3.3), this shows that the length L of the earthquake fault resulting from an inland earthquake corresponds rather well to the diameter $2r$ of the deformation area on the surface. The length of the fault seems to correspond to the extent of crustal deformation the earthquake causes.

3.2 Reading Crustal Movement—Methods and Their Accuracy

3.2.1 Methods of Reconstructing Past Crustal Movement

When the topography is flat or there are topographical lines such as ridges or channels, a surface deformation or fault slip can be noticed more easily. Landforms such as these are helpful in detecting crustal movement and are called reference planes (or lines).

One common topographic surface that records crustal movement clearly is the terrace. This topography consists of a near-horizontal and a steep cliff-like surface, and usually was formed originally near a shoreline or riverbed. If the abrasion platform on the coast or the bed of a river is raised above the water level by a lowering of sea level or accelerated undercutting of a river, a terrace forms.

Terraces are the result of a sudden uplifting of the ground relative to its former position. This uplift is usually triggered by causes quite unrelated to crustal movement—a drop in sea level caused by the beginning of a universal glacial period, for instance, or a change in the climate such as an increase in rainfall that accelerates the undercutting forces of a river. The number of terraces, then, does not necessarily reflect the number of times the earth has been uplifted by crustal movement. On the other hand, an earthquake can form a terrace. These cases will be discussed in detail later in this chapter.

Even if an active fault did form a displacement along a fault line on an even part of the earth's surface that was exposed to the action of sea or river water, the even surface will not begin to preserve the evidence of the fault activity until it ceases to be exposed to water. Then the terrace formation finally begins.

When using a terrace landform to analyze the direction and amount of fault activity, it is best to examine the intersection of a flat terrace surface and the steep former river- or sea-cliff behind the terrace. This intersection is the base of the terrace cliff bordering two flat terrace surfaces. If the intersection is staggered at the fault line, the direction and length of a straight line drawn from either end of the intersection will show the displacement vector of fault activity from the time the terrace was formed to the present. The average rate of displacement of the fault during that time can be obtained by dividing the length of the vector by the age of the terrace. The amount of uplift and tilting that occurred after the terrace was formed can also be determined from the present state of the terrace. Examination of the landform, then, can reveal the direction, amount, and velocity of crustal movement during a certain period in the past or from a point of time in the past to the present. If several reference landforms of different ages are available in an area, the locus of the crustal movement over time can be examined.

In general, for restoration of the movement, location of the reference landform must be determined at various points of time. This raises the following questions: (1) How closely can the time intervals be measured—or, in other words, using the landform as a clue, how fine are the gradations on the chronological measuring stick? (2) How precisely can the number of years since the formation of the reference landform be measured—or, how accurate are the chronological gradations? (3) How accurately can the amount of displacement or deformation of the landform be measured—or, in how much detail does the landform record and preserve displacement?

First, let us examine question (1).

3.2.2 Chronological Gradations within Landforms

The more reference topographies there are from different periods, the more fully changes in crustal movement can be traced. The most widespread topographical features of the Japanese coastline are the terraces that were formed during the last interglacial period, some 120,000 years ago, and the landforms that emerged approximately 6,000 years ago during the early part of the Jomon era. In the Kanto district, the former is called the Shimosueyoshi Terrace; this terrace constitutes the even surface of downtown Tokyo, a few meters above sea level. This same surface is more than 20 m above sea level in the Boso Peninsula, and there it is called the Numa Terrace. In addition to these two, there are other marine terraces formed by changes in the level of seawater some 100,000, 80,000, and 60,000 years ago. In the Miura Peninsula, where such terraces can be readily examined, they are called the Hikihashi, Obaradai, and Misaki Surfaces, respectively.

The chronological gradation over the past 120,000 years in southern Kanto, then, is approximately 20,000 years for marine terraces.

As for the terraced surfaces that form along rivers, they have several steps that were formed during the past 40,000 to 50,000 years. Along the Tama River, for example, the major ones are the Musashino Terrace (approximately 50,000 years old), the Tachikawa Terrace (20,000 years old), and the Aoyagi Terrace (15,000 years old). Along the Kiso River, near Sakashita, Gifu Prefecture, where the river is crossed by the Atera Fault, there is a terrace about 27,000 years old. Between this terrace and the Kiso River's present river bed, there are four terraces, each several hundred meters wide with an average time interval of several thousand years, which provide a record of the movements of the Atera Fault since these terraces were formed.

The chronological gradations imprinted by the major marine and river terraces are in general a few tens of thousands of years apart during the past 100,000 years. Under favorable conditions gradations of several thousand years may be recorded. There are not many examples of reference topographies that record changes during the last 10,000 years, with the exception of the 6,000-year-old terrace surfaces discussed above. Finer degrees of gradation can sometimes be obtained, however, during a specific period or in certain localities. In areas where volcanic activity has recurred frequently, depositing unique layers of ash each time, these layers form chronological gradations. If there are ruins from various periods, such as the ruins of prehistoric multiple dwellings, buried in soil on a fault, they can form gradations that are several hundred years apart, providing a helpful guide to fault displacement.

In the last 100,000 to 500,000 years the gradations are approximately 100,000 years apart, as in the case of Tama Hill which was formed over the last 100,000 to 400,000 years and contains several steps of marine terrace surface.

3.2.3 The Accuracy of Chronological Gradations

Let us now consider the second question—the accuracy of chronological gradations. Radiocarbon dating can determine age for periods ranging from several hundred to almost 40,000 years with a precision of 10% or better when there are materials available that can withstand the method. The F.T. (fission track) method, using the track of radioactive particles contained within volcanic glass, can be used to determine the age of materials that are between 40,000 and several hundred thousand years old, although, in general, it is very difficult to make these direct chronological determinations. Usually the age of landforms older than ca. 40,000 years must be estimated on the basis of indirect evidence, such as the vicissitudes of glaciers or changes in sea level or climate. Extensive changes such as these have occurred more than once or twice during this period of time, however, and each time they have created similar landforms, such as marine terraces. Thus it becomes difficult to determine exactly which cycle in climatic change or change in sea level does correspond to a certain landform. During the period, for example, between 120,000 years and 60,000 years ago, periods of high sea levels (when marine terrace might be formed) occurred about 120,000, 100,000, 80,000 and 60,000 years ago, with intervals of about 20,000 years. Consequently there are various opinions about when, for example, Omaezaki Terrace (on which the Omaezaki Meteorological Observation Station is located) was formed. Estimates range from 60,000 years ago to 120,000 years, but no one is able to offer conclusive proof.

In determining the absolute age of landforms it is difficult to generalize, but at least ± 10,000 to 20,000 years must be allowed where landforms from 20,000 to 150,000 years old are concerned, and ±50,000 years must be allowed when determining the age of landforms between 200,000 and 500,000 years old, even in cases when the correlations of the terrace are deemed to be qualitatively accurate.

3.2.4 Accuracy in Measuring the Amount of Displacement

The next topic of discussion is the accuracy and detail with which we can determine the amount of crustal deformation from a given landform.

In general, old topographic surfaces are severely eroded and have lost their original forms. As a result, restoration of the minute details of crustal movement within ancient landforms is extremely difficult.

Based on experience, it can be assumed that a terrace surface that is flat enough to be an airstrip is not more than 30,000 to 50,000 years old. A surface suitable for a golf course, with its gentle undulations, is likely to be about 100,000 years old. The Tachikawa Airfield outside Tokyo, for example, is built on a 15,000-year-old terrace, and the Osaka International Airport in Itami on a 30,000-year-old terrace. Tama New Town in the Tama Hills, however, was built on a river terrace surface that is 200,000 to 400,000 years old, and extensive landscaping was required.

The highland plateau (also called an elevated low relief surface) carved out of a deep valley, such as the Hida or Kibi Heights, is a landform that has been smoothed by erosion over several million years, then thrust upward during the past million years to form a highland. Although it is relatively flat, such a low relief surface will vary in elevation by 100 meters or so; it is hilly or gently mountainous land but can still be useful in determining displacement.

The fine traces of displacement are no longer visible on the surface of old landforms. If erosion has caused an unevenness of 100 meters or more, it becomes difficult to read a smaller slip in the land. And yet, for the purpose of analyzing crustal movement, a younger landform surface is not necessarily superior to an older one, even though its surface is smooth. This is because an old landform surface has simply experienced more crustal movement over time and a younger surface suffers a smaller amount of displacement. For example, it is difficult to detect scars of movement in an extensively developed alluvial plain—the surface is simply too young.

The relationship between the landform's age and the smallest vertical displacement detectable is outlined in Table 3.1. From this table it is apparent that the limitations of detection per unit year are approximately 10^{-1} mm/y over various spans up to one million years. This relationship indicates that, in measuring crustal movement that is proceeding uniformly, the degree of accuracy is similar to that in precise leveling

Table 3.1 The Accuracy with Which Crustal Movement Can Be Detected Using Landforms (modified from Kaizuka, 1968)

Age of landform (T)	Limit of detectable crustal movement by height variation (H)	Accuracy of detection per unit period (H/T)
4,000–6,000 years (Numa Terrace and other low river terraces)	± 1 m	$n \times 10^{-1}$ (mm/y)
40,000–50,000 years (Musashino Terrace, etc.)	1–5 m	10^{-1}
100,000–150,000 years (Shimosueyoshi Terrace, etc.)	5–10 m	$n \times 10^{-1}$
200,000–500,000 years (Tama Hills, etc.)	10–50 m	10^{-1}
1 million years (elevated low relief surface)	100 m +	10^{-1}

which measures topographical changes at several-year intervals. Thus the reconstruction of crustal movement through the use of landforms is not such a crude method after all.

3.3 From the Past to the Present—Prediction and Its Problems

3.3.1 The Present Viewed From the Past

In many cases past activity can be reconstructed by using landforms as a guide, as we have seen. A graph in which the age of the landform (T) as the x-axis is plotted against the amount of slip in that landform (D) as the y-axis elucidates the history of fault movement. Figure 3.2, for instance, shows the Median Tectonic Line, the most active fault in southwestern Japan [Okada, 1980]. Even though the dots on the graph do not fall in a strictly straight line, the gradient of the approximately straight line that connects them through 0 point represents the average rate of slip of the Median Tectonic Line during this period. (This value is called the average rate of displacement of the fault.) From materials such as these the average rate of displacement of the Median Tectonic Line during the past 50,000 years is determined to be 5 to 10mm per year.

The average rate of displacement over the last several tens of thousands to several hundred thousand years has been determined in this way for many other active faults. Active faults with an average rate of displacement in the range of mm per year, like the Median Tectonic Line, are called Class A faults. Ten or so such faults are centered in the Chubu district—among them the Atera and Atotsugawa Faults in Gifu Prefecture, the Tanna Fault in northern Izu, and the Kozu–Matsuda Fault in Kanagawa Prefecture as well as the Median Tectonic Line. Many Class B

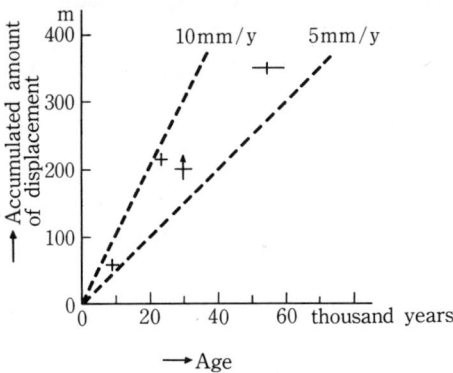

Fig. 3.2 Relationship between the age of a landform or stratum and the amount of slip caused by an active fault—an example from the Median Tectonic Line in eastern Shikoku [after Okada, 1980].

The older the landform or stratum is, the greater the amount of displacement. The gradient of the broken lines represents the average rate of displacement.

faults (one order smaller in slip rate than the above, with an average rate of displacement ranging from 0.1 to 1 mm/y) exist in the area between the Tohoku district and Kyushu, as well as numerous Class C faults (0.01 to 0.1 mm/y in slip rate). Generally speaking, the lower-class or less active faults are greater in number.

Table 3.2 represents the degree of activity of the major active faults (or active fault systems) in inland Japan. The degrees A, B, and C denote the range of average rate of displacement, as explained above.

All of the examples mentioned concern the amount of slip on the fault. It is also possible, however, to obtain the rate of uplift of the land compared to sea level—the average rate of uplift. Figure 3.3 demonstrates the average rate of uplift at Chikura-machi at the southern tip of the Boso Peninsula. The y-axis represents the height above the present sea level at which the dated materials were collected. The shellfish whose shells were used for the dating lived near sea level during their lifetimes, and the x-axis shows the age of death of the shellfish according to radiocarbon dating. As is apparent from the figure, the shoreline topography 5,500 years ago was 25 m above the present sea level. Since it is known that sea level at that time was perhaps 2 m higher throughout the world than it is at present, that much can be subtracted from the 25 m. It can then be concluded that

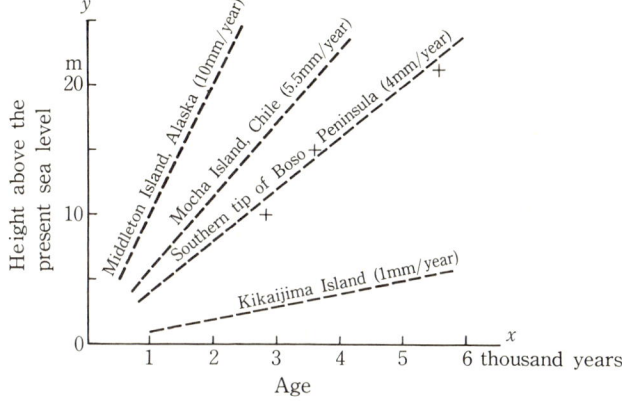

Fig. 3.3 Average rate of uplift of the southern tip of the Boso Peninsula compared with that in other areas [from Yonekura, 1975].
+ denotes age determined by radiocarbon dating of seashells and other materials gathered near Chikura (Boso Peninsula).

Table 3.2 Major Active Faults in Japan [Matsuda, 1976; partially revised]

Site (prefecture)	Name of fault or area of distribution	Type of fault	Length (km)	Activity
Aomori	Western Aomori	Reverse fault	40	B
Aomori–Iwate	Orizume Fault	Reverse fault	40	C
Akita	Hanawa	Reverse fault	20	B
Akita	Senya Fault	Reverse fault	40	B
Iwate	West of Shizukuishi	Reverse fault	20	B
Iwate	Morioka Faults	Reverse fault	40	B
Iwate	Isawa–Aburashima	Reverse fault	40	B
Miyagi	Nagamachi–Rifu Line	Reverse fault	30	B
Miyagi–Fukushima	Shiroishi–West of Fukushima	Reverse fault	45	B
Yamagata	West of Nagai Basin	Reverse fault	10	B
Yamagata	Vicinity of Kaminoyama	Reverse fault	10	B
Fukushima	West of Aizu Basin	Reverse fault	55	B
Fukushima	Futaba Fault	Reverse fault	80	B
Niigata	West of Nagaoka Plain	Reverse fault	30+	B
Niigata	Southwest of Kuninaka Plain	Reverse fault	10	B
Tochigi	West of Nasuno	Reverse fault	30	B
Gumma–Saitama	Hirai Fault	Reverse fault	20	B
Saitama	Fukaya Fault	Reverse fault	10	B
Saitama–Tokyo	Arakawa Fault	Reverse fault	30	B
Tokyo	Tachikawa Fault	Reverse fault	20	B
Chiba	Iwai–Kamogawa	Dip-slip fault	20	B
Kanagawa	Takeyama–Kitatake Fault, etc.	Right-slip fault	10	A
Kanagawa	Kozu–Matsuda Fault	Right-slip fault	15	A
Kanagawa	Kannawa Fault	Reverse fault	15	A

Table 3.2 Continued

Site (prefecture)	Name of fault or area of distribution	Type of fault	Length (km)	Activity
Sagami Bay	Sagami Trough Fault	Right-slip fault	300	A
Shizuoka	Tanna Fault	Left-slip fault	35	A
Shizuoka	Irozaki Fault	Right-slip fault	10	B
Suruga Bay	Suruga Trough Fault	Reverse fault	100+	A
Yamanashi–Nagano	Itoigawa–Shizuoka Line	Left-slip fault	80	A
Nagano	Zenkojidaira–Iiyama	Reverse fault	70	A
Nagano	Kisodani	Reverse fault	70	A
Nagano	Inadani	Reverse fault	60	A
Toyama–Gifu	Atotsugawa Fault	Right-slip fault	60	A
Gifu	Atera Fault	Left-slip fault	60	A
Gifu	Neodani Fault	Left-slip fault	80	A
Ishikawa	Ouchi Lowland	Reverse fault	20	B
Fukui	Nosaka Fault	Left-slip fault	10	B
Fukui	Mikata Fault	Dip-slip fault	20	B
Fukui–Shiga	Yanagase Fault	Left-slip fault	50	B
Gifu–Mie	Yoro Fault	Dip-slip fault	30	B
Mie	Ichishi Fault	Reverse fault	60	B
Shiga	West of Suzuka Mountain Range	Dip-slip fault	60	B
Shiga	Hira Fault	Reverse fault	50	B
Mie–Nara	Nabari Fault	Dip-slip fault	40	B
Mie–Kyoto	Kizugawa Fault	Reverse fault	30	B
Shiga–Kyoto	Hanaore Fault	Right-slip fault	60	B
Kyoto	Kameoka Fault	Dip-slip fault	20	B
Hyogo–Osaka	Rokko Fault, etc.	Dip-slip fault	50	A
Osaka	Ikoma Fault	Dip-slip fault	20	A
Kyoto-Nara	East of Uji–Nara	Dip-slip fault	40	B
Kii–Shikoku	Median Tectonic Line	Right-slip fault	500+	A
Hyogo	Yamasaki Fault	Left-slip fault	50	A
Okayama	North of Tsuyama Basin	Reverse fault	30	B
Oita	Kuju–Haneyama	Normal fault	40	B
Kumamoto	Hinagu Fault	Right-slip	60	B
Nagasaki	Unzen	Normal fault	20	B
Kagoshima	Izumi Fault	Right-slip fault	20	B

Length here is approximate. Some figures include the length of the entire fault system.

the earth has been elevated more than 20 m within the last 5,500 years. As was discussed in the case of the Median Tectonic Line, older landforms show greater displacement than younger ones. The crustal movement in the southern tip of the Boso Peninsula can therefore be assumed to be cumulative. This region was elevated about 1.5 m during the Kanto earthquake of 1923. This particular seismic uplift is only one scene from a long drama of movement in geologic history.

The average rate of uplift of the southern tip of the Boso Peninsula is approximately 4 mm per year. Figure 3.3 also shows the uplift at several other sites along the line of the Pacific Ocean. The average rate of uplift of the Muroto Cape (uplifted during the Nankai earthquake of 1946) is approximately 1.5 mm per year; the average rate of uplift of Mocha Island off the coast of Chile (uplifted during the Chile earthquake of 1960) is approximately 5.5 mm per year; and the average rate of uplift of Middleton Island (uplifted during the Alaska earthquake of 1964) is approximately 10 mm per year. All of the above are near the subduction zone of the oceanic plate.

Crustal movement that is accompanied by earthquakes also results in the tilting of the earth. The Ogi Peninsula on Sado Island, for instance, uplifted and tilted during an 1802 earthquake [Ota et al., 1976]. The old uplifted shoreline (which must have been horizontal then) is apparent in the terrace landform and tilts northward. The degree of tilt is greater in the older shorelines than the younger ones. (The tilt angle is approximately 2 sec for the 6,000-year-old shoreline, 15 sec for the 80,000-year-old shoreline, and 38 sec for the 120,000-year-old shoreline.)

Apparent from these examples is the fact that crustal movements, both uplift and faulting, have tended to occur in roughly the same direction over the past 120,000 years. The velocity of such movement is also almost the same within each region, whether it is measured during the past several decades, the past several thousand years, or the past 120,000 years. From this fact it can be concluded that crustal movement has been progressing in these directions with almost equal velocity. As far as the nature of the movement is concerned, there is no break between the past (at least since the late Quaternary Period) and the present. This knowledge of the past can be projected into the future.

3.3.2 Predicting Earthquake Recurrence Intervals

Reading crustal movement from the record provided by landforms is almost like viewing the image of time-lapse photography recorded at intervals of several thousand years or longer. If each image were lined up in a row, crustal movement over the past one hundred thousand years would be seen to be progressing linearly at an almost uniform speed toward the present. The average velocity of the movement during this period is represented in the average rate of displacement of faults and in the average rate of land uplift, as was discussed in the preceding section. These velocities are averaged over a long period of geological time, however, so at any given moment they may not be occurring at that speed. The average rate of displacement along the Median Tectonic Line, for instance, is 5 to 10 mm/year (see Fig. 3.2), and yet, during the past several decades, nothing has happened to the houses or the railway that rest on this line. In other words, there has been no movement. Putting together this fact of daily life with the realities of geology, it is safe to conclude that although no movement is taking place right now, a sudden movement will occur some day and the geological average velocity will thus be maintained. Such sudden and intermittent displacements take the form of earthquakes. The question is how to predict their size and scale and when they will occur.

Seismically active faults in Japan are usually at rest but move suddenly when the time comes. The average rate of displacement S is, in this case, made up of the sum total of displacements at the time of earthquakes, ΣD, divided by the age of the reference topography, T years:

$$S = \Sigma D/T. \qquad (3.6)$$

If the number of earthquakes which occurred during T years is n,

$$\Sigma D = D \times n.$$

Given equation (3.6), T/n, or the average recurrence interval between earthquakes R, can be represented as

$$R = D/S. \qquad (3.7)$$

If there is an ongoing displacement (as in the case of creep displacement) moving at an average velocity C during the nonseismic period between earthquakes, the denominator S in the above equation can be replaced by $S - C$. (C is positive if its direction increases ΣD.)

When the faults in Japan are currently said to be at rest, this only means that there is no daily slip movement along them. The ground near the faults, however, is constantly and gradually changing shape. There comes a time when the resulting deformation rebounds elastically, causing a slip to form along the fault. This is the fault movement. When this occurs, the strain within the rocks in the area dissolves, as does the deformation. The continuous deformation of the ground is most noticeable in the coastal areas. The Muroto Cape and the southern Boso Peninsula, for instance, are known to be subsiding. They uplifted suddenly, however, during the Nankai earthquake of 1946 and the Kanto earthquake of 1923, respectively. After that they returned to their usual state of subsidence and have been quiet ever since. Given this background, the average rate of uplift v of an uplifted landform in the coastal areas can be represented as

$$v = (\Sigma h + \Sigma h')/T. \qquad (3.8)$$

Σh represents the sum total of the amount of uplift during n earthquakes in T years. $\Sigma h'$ represents the sum total of the amount of vertical movement during a quiet period (when its subsidence is a negative value). R, the average interval between earthquakes during T years, then, is

$$R = T/n = \frac{h}{(v - v')} \tag{3.9}$$

where v' equals $\Sigma h'/T$, or the uplift rate in the absence of earthquakes; it is a negative value that represents subsidence.

R can also be obtained in cases where the ground has tilted:

$$R = \theta/(\alpha - \alpha') \tag{3.10}$$

where θ represents the amount of tilting during an earthquake, α is the average rate of tilting during a period of T years, and α' represents the average rate of non-seismic tilting (α' is negative when the sense of tilting is reversed).

The denominators S, v, or α used to obtain R in equations (3.7), (3.8), (3.9), and (3.10) are topographically and geologically obtainable quantities. C, v', and α' can also be estimated using precise measurements that have been developed in recent years. Given the numerator—i.e., the amount of displacement at the time of the earthquake—the recurrence interval, R, of earthquakes of this size can be obtained. Since the fault displacement at the time of an earthquake, D, is related to the magnitude, M, of the earthquake, as defined in equation (3.2), R in (3.7) can be defined as follows:

$$\log R = 0.6M - (\log S + 1.0). \tag{3.11}$$

S, or the average rate of displacement, is in mm per year. Table 3.3 represents the earthquake interval R when S and M have several different values, and Fig. 3.4 is its nomograph. Roughly speaking, the interval between earthquakes of M 8 or so caused by Class A faults is about 1,000 years. The intervals between earthquakes of M 8 caused by the less active (Class B or C) faults are longer. The same applies to earthquakes with smaller magnitudes, except that these smaller events are more frequent than M 8 events.

Table 3.3 The Approximate Recurrence Interval of Earthquakes along Class A, B, and C Active Faults

Class (average rate of displacement S)*		A (1~10 mm/y)*	B (0.1~1 mm/y)*	C (0.01~0.1 mm/y)*
Earthquake magnitude assumed	M 8	1,300 years	13,000	130,000
	M 7	300	3,000	30,000
	M 6	80	800	8,000

* For purposes of approximation S is assumed for calculation to be 5, 0.5, and 0.05 mm/y in Class A, B, and C active faults, respectively.

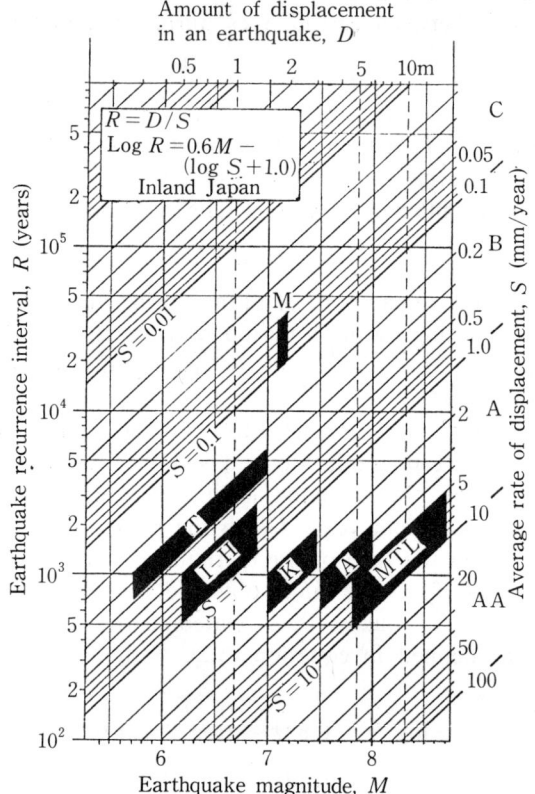

Fig. 3.4 A nomograph to obtain earthquake recurrence intervals (R) from the average displacement of faults (oblique lines, S).

Shaded areas represent examples of R obtained from the value of S and the value of the x-axis (earthquake magnitude M or earthquake displacement during a single event). MTL: Median Tectonic Line; A: Atera Fault; K: Northern Izu Fault System; I-H: Irozaki Fault; T: Tachikawa Fault; M: Fukozu Fault.

3.3.3 Estimating the Sizes of Earthquakes

The points discussed so far are rough criteria for determining the relationship between active fault activity and the intervals at which great earthquakes occur. But if they are to be applied in a practical way, further questions must be taken into consideration:

(1) How do we infer the size of an earthquake or the amount of displacement caused by an earthquake (M in equation (3.11) or D in (3.7))?

(2) The earthquake intervals arrived at by using the equations in the

preceding section are averages. What are the actual intervals like?
(3) Even though the sizes of and the intervals between earthquakes that occur in a specific fault can be determined, how can the overall degree of earthquake risk in an area be evaluated when multiple faults exist?

First, let us consider question (1). The records contained in landforms are usually not precise enough to accurately estimate the amount of displacement, D, that occurred at the time an earthquake took place in an active fault. M, or the size of an earthquake, has been inferred on the basis of the empirical rule that great earthquakes occur in long fault systems (the reverse is not necessarily true, however). Since the relationship between earthquake magnitude, M, and the length of a fault, L, can be expressed for Japanese inland historical earthquakes by equation (3.3), the maximum magnitude, M_L, expected from an active fault system L (km) long would be

$$M_L = (\log L + 2.9)/0.6. \tag{3.12}$$

Another way of estimating the size of an earthquake is based on the idea that an active fault on which an earthquake has not occurred for a long time has stored up great seismic energy. The amount of seismic energy stored around an active fault for as long as t years is considered to be proportionate to the strain velocity represented by the average rate of displacement of the fault, S. Using equation (3.11), M_t—that is, the stored energy converted to magnitude—can be expressed as

$$M_t = \{\log (t \cdot S) + 1.0\}/0.6. \tag{3.13}$$

In most cases, t represents the years for which there is no historical record of an earthquake; the energy actually stored, therefore, is likely to be greater than M_t. Hence $M_{\text{prob.}}$, or the most probable size of an earthquake on that particular active fault, is considered to be between M_t and M_L.

In the central part of the Median Tectonic Line (in eastern Shikoku), for example, $L = 200$ (km), $t =$ approximately 1,000 (y), $S = 5 \sim 10$ (mm/y). The expected magnitude, therefore, is

$$M_t = 7.8 \sim 8.3 < M_{\text{prob.}} \leq M_L = 8.7.$$

Thus, a great earthquake of M 8, or thereabouts, is likely. The intervals between M 8 earthquakes, R, can be obtained, through (3.11), as approximately 1,000 years ($S = 10$ mm/y) $\sim$ 3,300 years ($S = 5$ mm/y).

3.3.4 The Uneven Distribution of Earthquakes

The estimated intervals between earthquakes are only averages. The intervals between great earthquakes and the magnitudes of those earthquakes are constant to some extent, but predictions cannot be based on such broad figures. The regularity (or, rather, irregularity) of earthquakes must be examined using historical records as well as landforms.

Great earthquakes have occurred along the same fault during recorded history in the Nankai Trough off the Nankaido coast. Since the great earthquake of 1498, this area has been hit by M 8-class earthquakes in 1605, 1707, 1854, and 1946. The intervals between these earthquakes were 107, 102, 147, and 92 years, respectively, or 120 ± 30 years. This example gives the impression that nature behaves more or less regularly where crustal movement is concerned. The 120-year interval almost agrees with the figure calculated from the average rate of uplift over the past 120,000-odd years of the Muroto Cape: i.e., when $h = 1.2$ (m), $v = 190$ (m/120,000 years), $v' = 8.5$ (mm/y) is applied to equation (3.9), $R \approx 120$ (y). Hence it can be concluded that the 120-year interval obtained from the historical records is constant for the Nankai earthquakes. The range of fluctuation is ± 30 years and the rate of irregularity is within 30% for the average interval.

The incidence of repeated earthquakes along the same fault during recorded history is far less frequent where inland earthquakes are concerned. This could be predicted from the fact that the average rate of displacement of inland faults is an order of magnitude smaller than that in the Nankai Trough. There are, however, a few examples of inland earthquakes that seem to have occurred twice on the same fault during recorded history. The Northern Izu earthquakes of 841 and 1930, both M 7 according to the Science Almanac, took place in the Northern Izu Fault System; the Nobi earthquakes of 745 and 1891, both M 7.9, took place in the Nobi Fault System; the Zenkoji earthquakes of 887 and 1847, both M 7.4, took place in Shinshu. The Shinanogawa Fault Zone that caused these Zenkoji earthquakes is a much more extensive fault system than the others, so it is hard to ascertain whether both earthquakes were caused by the same segment of the fault or by neighboring segments.

Regularity cannot be discussed with any certainty given so few examples. If, however, the actual intervals between earthquakes in recorded history agree with the calculated average intervals, it can be assumed that actual earthquakes are occurring at intervals close to the average figure. Taking the Northern Izu Fault System as an example, the average recurrence interval R calculated on the basis of equation (3.7) would be 800 to 1,500 years since $S = 1,000$ (m/400,000–500,000 years) and $D = 2$ to 3 (m).

There is agreement between this figure and the actual interval between the two great Northern Izu earthquakes ($1930 - 841 = 1089$ years); great inland earthquakes also appear to occur at intervals close to the average estimated. From the above examples it can be surmised that, in the case of the Northern Izu Fault System, the actual intervals between earthquakes with a magnitude in the range of $M\,7$ are $1,000 \pm 500$ years.

Historical records concerning specific faults are much scarcer. Traces of each earthquakes can be found in landforms, however, and their regularity traced. The easiest way to do this is through the marine terraces caused by great earthquakes. When a great earthquake occurs, the even surface created near sea level by the action of the water is suddenly thrust above sea level. It becomes a terrace, in other words. In a terrace formation in which this process has been repeated, the height differences between the terraces represent the magnitude of seismic uplift, which is roughly proportional to earthquake magnitude. The gaps between the formation of the terraces are the lengths of the intervals between earthquakes. Although the influence of changes in sea level should actually be considered, it will be ignored here since change in sea level has been extremely slow since the early part of the Jomon period (approximately 6,000 years ago). The unevenness of the heights of terrace surfaces seems to correspond to the unevenness in the sizes of the earthquakes. Let us examine this unevenness.

Four terraces with a width of more than 100 m have been formed within the past 5,000 years in the area around Chikura near the southern tip of the Boso Peninsula, so there are traces of four seismic uplifts during this period [Matsuda et al., 1978]. The heights of these terraces (strictly speaking, the intersection of the terrace surface with the terrace cliff on the mountainward side = the height of the old shoreline) are 25, 16, 12, and 6 m above sea level. The highest terrace (25 m) was formed some 5,000 years ago, and the lowest (6 m) was formed during the Genroku earthquake (1703) some 270 years ago. The differences in height between these terraces are 9, 4, 6, and 6 m, respectively. The extent of difference in this case is ± 3 m from the average 6 m. The actual difference, however, may be a little smaller since the highest one (9 m) may have been influenced by a change in sea level.

In determining the time of prehistoric uplifts caused by great earthquakes, the radiocarbon dating method, using shellfish buried in the sediments of each terrace, is useful, although some speculation is necessary since the period when these shellfish lived does not necessarily coincide with the time of seismic uplift. It is generally estimated, however, that seismic uplift occurred approximately 6,150, 4,350, 2,850, and 270 years ago [Nakata et al., 1979]. The intervals between earthquakes range from 1,500 to 2,800 years. These are the intervals between the great earthquakes that cause extensive terrace surfaces to uplift (such as the Genroku earthquake). There may be smaller earthquakes in between (such as the Kanto earthquake of 1923) that cause only a quarter to a fifth of the amount of uplift caused by great earthquakes.

Fluctuations in seismic uplift can be traced from the relative heights of these Holocene marine terraces. The group of terraces on Middleton Island that were uplifted by the Alaska earthquake of 1964 provide a good example. There are six terraces that have been forming over the past 4,500 years. The sixth and lowest one was formed by the seismic uplift in 1964. The present heights of the old shorelines are (from the bottom) 4, 14, 23, 30, 40, and 46 m, respectively [Plafker, 1972]. The height differences are 4, 10, 9, 7, 10, and 6 m, respectively, or 7 ± 3 m. The intervals between seismic uplifts range from 500 to 1,400 years (an average of 800 years).

At beach ridges on Cape Turakirae east of Wellington on the North Island of New Zealand, six seismic uplifts within the past 6,000 years can be traced. Their height differences range from 2.5 to 9 m [Wellman, 1967]. There is no proof that all these coastal elevations were caused by the same fault activity, but the fluctuation is limited, in this case also, to 6 ± 3 m.

From the above examples it can be concluded that the range of fluctuation in earthquake intervals or in the displacement caused by a particular fault is 1/2 to 2 times that of the average figure. Thus this is the fluctuation that can be expected between the average figure and actual earthquake incidence.

If the movement of a particular fault (the length of its inactive period and its displacement) is completely arbitrary, then earthquake prediction is, indeed, difficult. There is a working hypothesis, however, derived from the above information, that "each fault plane moves with its own characteristic seismic displacement and time intervals."

3.4 Active Faults and Disaster Prevention

3.4.1 Definition of Active Faults

Faults that have been active repeatedly in the recent geological past are considered likely to repeat their activity in the future. Such faults are called active faults. Many geological textbooks and dictionaries define active faults as "faults that were active during the Quaternary period" [*e.g.* Research Group For Active Faults, 1980]. This is because there is proof of the repeated activity of faults that have been active during this period (from approximately 2 million years ago to the present), and it is assumed that such activity will continue.

There are some faults that were active during the Quaternary period but that will not be active again. Some faults formed as a result of volcanic activity during this period, for instance, ceased to be active when the volcanoes themselves became extinct. Such faults, active only as a result of certain specific conditions, cannot rightly be called active faults. Ordinary faults formed as a result of extensive crustal stress, however, are considered to be active since such stress fields still exist today.

"Active," in short, refers to the possibility of activity in the future. Consequently such faults are sometimes called potentially active. The definition of an active fault often includes such phrases as "active in late geological time" or "active during the Quarternary period," but such phrases are far too inexact. They certainly are not suitable for legal purposes where an exact definition is required. In California special permission from the state is required to build directly above or near faults designated as active by law (Alquist-Priolo Geologic Hazard Zones Act). Such faults are defined in a regulation as "potentially active faults—those that have been significantly active during the Holocene period (approximately the past 10,000 years) with well-defined positions, and those that caused slips on the earth's surface during the Quaternary period." The regulation defines which active faults are covered and the extent of the areas involved, and makes this information public. The United States Atomic Energy Commission, in a regulation regarding nuclear power plant safety, defines the faults subject to the regulation much more broadly, as "those active during the past 500,000 years." It also defines the length of the active faults to be taken into consideration when a nuclear power plant is proposed. For example, faults that are longer than one mile are to be taken into consideration if they are within 30 miles of the proposed site; faults that are 10 miles long or more are to be considered if they are within 50 to 100 miles of the site.

In New Zealand, the New Zealand Geological Survey defines active faults as those that have been active repeatedly during the last 500,000 years and classifies them in three categories (I through III), which are shown on geological maps in different colors.

Active faults have thus been variously defined according to the purpose of the definition (disaster prevention, for instance) and its use in the real world. There are all kinds and scales of active faults, and it is unwise to consider earthquake countermeasures without taking these differences into consideration. Countermeasures are determined by the possibility of future activity and by considerations such as the strength of seismic shaking and the slip or deformation likely to result. In addition, the nature and uses of the land adjacent to a fault are a factor. Based on these considerations, a fault can be either totally ignored or examined with the utmost care.

3.4.2 The Location of Active Faults

Since active faults mean earthquakes in Japan, it is important to know where they are located. Usually they can be recognized by the traces they have left in present-day landforms. Those which cannot be seen either were active a very long time ago and are active no longer, or have not been very active. The traces left by fault activity become less and less clear with time, due to the effects of erosion or sedimentation. What can be seen in present-day topography is a result of the competing forces of fault activity that leaves traces behind, on the one hand, and the actions that erase those traces on the other. Consequently, traces of highly active faults (those with high average rate of displacement) that leave a strong impression on the land are easy to find, but the traces of less active faults are sometimes obscured or overlooked. Sometimes the question of whether they are active or not is the subject of considerable debate.

Class A and B active faults which are more than 10 km long are visible in aerial photographs of 1/40,000 scale. Class C active faults, on the other hand, are harder to locate on aerial photographs—especially the shorter ones. And determining whether they are active or not is even harder. On regional maps of active faults, therefore, Class C faults are often omitted.

Figure 3.5 is a distribution map of the major active faults known to

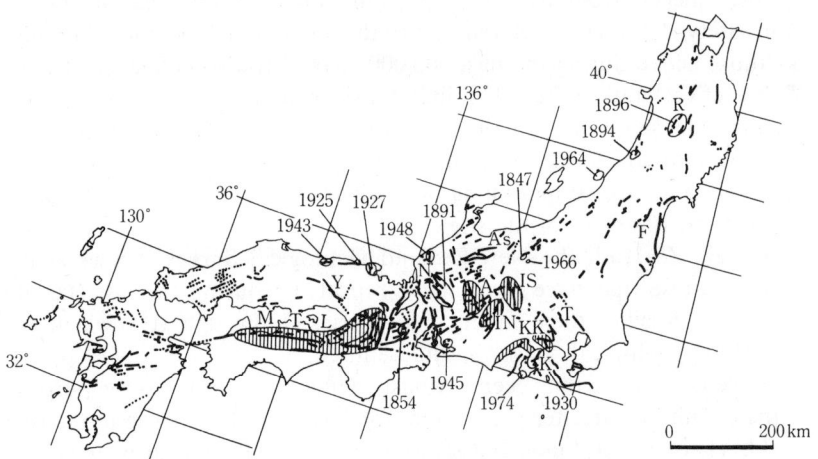

Fig. 3.5. Active faults in Japan [Matsuda, 1977].

Shaded areas are those that warrant special attention because they contain major active faults, with (t/R) 0.5. Dates are the years of great earthquakes which accompanied the appearance of seismic faults originating from these active faults. A: Atera Fault; As: Atotsugawa Fault; IS: Itoigawa–Shizuoka Tectonic Line; K: Northern Izu Fault System; KK: Kozu–Matsuda and Kannawa Faults; M: Median Tectonic Line; N: Nobi Fault System; T: Tachikawa Fault; Y: Yamasaki Fault.

Table 3.4 Major Inland Earthquakes, Active Faults Involved, and Felt Foreshocks

Earthquakes (Classified by magnitude)	Major existing active faults systems involved	Major surface faults and their activity classes	Foreshocks felt several days before earthquake
I. $8.0 > M \geq 7.5$			
1891 Nobi earthquake	Nobi Fault System	Neodani Fault (A)*	yes
1927 Northern Tango earthquake	Yamada Fault (B)	Gomura Fault (C)	no
II. $7.5 > M \geq 7.0$			
1894 Northern Tokyo Bay earthquake	Unknown	None	no
1896 Rikuu earthquake	Rikuu Fault System	Senya Fault (B)	yes
1899 Mie–Nara earthquake	Unknown	None	no
1930 Northern Izu earthquake	Northern Izu Fault System	Tanna Fault (A)	yes
1931 Western Saitama earthquake	?	None	no
1939 Oga earthquake	Unknown	?	no
1943 Tottori earthquake	Shikano Fault	Shikano Fault (C)	no
1945 Mikawa earthquake	Fukozu Fault	Fukozu Fault (C)	yes
1948 Fukui earthquake	Fukui Fault	Fukui Fault (C?)	no
1948 Kii earthquake	None	None	no
1961 Northern Mino earthquake	Hatogayu–Koike Fault (B?)	None	no
III. $7.0 > M \geq 6.5$			
1894 Shonai earthquake	Shonai Fault System (B)	Yadarezawa Fault	no
1895 Tonegawa earthquake	None	None	no
1900 Northern Miyagi earthquake	Unknown	None	no
1902 Sannohe earthquake	Unknown	None	no
1924 Tanzawa earthquake	Unknown	None	no
1925 Tajima earthquake	Unknown	Tai Fault	no
1949 Imaichi earthquake	Unknown	None	(rumbling)
1962 Northern Miyagi earthquake	Unknown	None	no
1969 Central Gifu earthquake	Hatasa Fault (C?)	None	no
1974 Izu Peninsula offshore earthquake	Irozaki Fault	Irozaki Fault (B)	no

Earthquakes listed here are $M \geq 6.5$ destructive inland earthquakes between 1868 and 1975 according to the *Science Yearbook*.

* (A), (B), and (C) after the faults represent activity based on long-term average rate of displacement.

The earthquakes near Sannohe (1902) and Tanzawa (1924) are considered to be the aftershocks of offshore earthquakes which took place in the preceding years.

exist today. Most of them are Class A and B faults. Looking at those earthquakes with magnitudes of more than 6.5 that have caused damage, that have had hypocenters in the upper crust, and that have occurred since 1868 (Table 3.4), it can be seen that in most cases the active faults corresponding to those earthquakes are topographically visible. About half of these faults are Class A and B, and the other half are Class C faults. Extrapolating from the experience of the past 100 years, about one-third of the damaging earthquakes (M 6.5+) in the future will originate along the active faults or fault systems shown on the Active Fault Distribution Map. More thorough examination will be necessary before the prediction of earthquake location becomes possible for the remaining two-thirds of all destructive earthquakes.

3.4.3 Active Faults That Warrant Special Attention

The great earthquakes that originate in Class A active faults occur, roughly speaking, once every 1,000 years, while those originating in Class B faults occur once every 10,000 years, as has been mentioned. The latter are rare indeed. There is a Japanese saying that "A crane lives a thousand years and a tortoise ten thousand." If one buys a tortoise, however, there is no guarantee that it will live 10,000 years. It might die the next day. To avoid such a mishap, a knowledge of averages is not enough. One must know the age of the tortoise and how long individual tortoises live. Predicting the time at which an active fault will move also requires additional information.

The average earthquake recurrence interval for an active fault corresponds to the average life span of the tortoise. It can be obtained geologically for each individual fault. The range of variation is from 1/2 to 2 times the average figure, as we have seen. The other piece of information needed is the length of time since the fault was last active (the tortoise's birthdate). If such activity is a part of recorded history, then the figure is known. If not, the number of years not known may number more than several hundred in Japan. The closer the number of years, t, is to the recurrence interval, R, of the active fault, the greater the possibility of a great earthquake in the near future. The areas that include active fault systems that have passed the halfway mark in the average recurrence interval are shown in Fig. 3.5 marked with straight lines. These are the fault systems that warrant special attention.

These faults are all Class A faults that have not been active during recorded history. There are two kinds of active faults that are not marked with straight lines. One are active faults for which t is known and for which t/R is clearly less than 0.5. These are the less worrisome faults. The other kind are the Class B and C faults for which there is no record of activity.

Since t is the number of years without known earthquakes, t cannot exceed 1,000 since that is the period covered by historical records. Consequently, in active faults where R exceeds 2,000 years (mostly Class B and C faults) t/R never reaches 0.5. To obtain a more meaningful number for t it is necessary to go beyond recorded history and study the prehistoric strata or landforms. This can be done by examining the records left by deposits of alluvium and volcanic ash. For example, if a buried ruin of the latest Jomon period (the Jomon period is about 10,000~2,200 years B.P.) on an active fault is found to have been displaced by the fault and have been covered with a ruin of the youngest Yayoi period (the Yayoi period is about 2,200~1,500 years B.P.) which do not show such a slip, it can be inferred that the most recent fault activity occurred between those two periods ($t \doteqdot 2,000$ years). As is obvious from this example, it is critical to know, for each fault, the age of the newest sediment that was severed and of the oldest sediment that was not severed by the fault. The most efficient way to obtain this information is by digging at the most convenient location and then examining a cross-section of the strata above the fault.

3.4.4 Successive Seismic Activity

In areas that include more than one active fault, the average interval between great earthquakes, R_m, can be calculated by using the average recurrence intervals, R_1, R_2 . . . etc., of each of the faults mentioned in the preceding section. (The sum of the inverted number of each of the average recurrence intervals equals the inverted number of R_m.) In actuality, however, will the activity of each fault recur independently of the activity of nearby faults? There are examples that would indicate otherwise.

Historically, when a part of an extensive fault system becomes active, it sometimes seems to trigger other sections of the fault system in a sort of chain reaction. In the Anatolia Fault System in Turkey, for instance, more than 10 great earthquakes of M 7 to M 8 occurred between 1930 and 1970 after a non-seismic period of approximately 1,000 years. These earthquakes eventually spread throughout the length of the fault system (about 1,000 km), and now most of the seismic gaps seem to have been filled. In Japan, the following earthquakes occurred in the Shinanogawa Fault System, which stretches from Nagano to the bottom of the Japan Sea off the coast of Niigata: 1828, downstream from the Shinano River, M 6.9; 1847, near Zenkoji, M 7.4; 1964, the Niigata earthquake, M 7.5. The following earthquakes occurred in the Sagami Trough Fault System which extends from the southern edge of the Tanzawa Mountains of Southern Kanto to the Sagami Trough: 1605, off the coast of Boso, M 7.9; 1703, the Genroku earthquake, M 8.2; 1923, the Kanto earthquake, M 7.9 [Matsuda *et al.*,

1978]. In the Nankai Trough Fault System, which extends into Suruga Bay, the Nankai earthquake, the Tonankai earthquake, and the Tokai earthquake have occurred in succession [Ando, 1977]. It is obvious that earthquake gaps in these fault systems must be closely watched.

The earthquakes above are examples of successive earthquakes occurring within relatively extensive fault systems. In areas that contain comparatively short and scattered active faults, successive earthquakes also seem to occur. Good examples of such earthquakes in inland Japan are hard to find, but the seismic activity in the Izu Peninsula region within the past several years (since the Izu Peninsula offshore earthquake of 1974) and the last several decades (since the Northern Izu earthquake of 1930) is one example. The intervals between successive earthquakes are completely different from the recurrence interval R in the same area; further research is necessary before the mechanism and natural laws pertaining to successive earthquakes can be understood.

3.4.5 The Basis of Evaluating Earthquake Danger

The size of the greatest seismic shaking that can take place in any one area in any given period can be estimated from past seismic activity (see the Kawasumi [1951] map, for example). The probability that there will be an earthquake of more than a certain size in a given area during a given time can also be estimated.

These figures are all based on the records of past seismic activity, however, and these records are incomplete—whether because it has only been several decades since the nationwide earthquake observation program began, or because there are only 1,000 years or so of recorded history in Japan. Such records are inadequate as a base from which to predict future great earthquakes. The following postulates, for instance, can all be correct in certain cases:

(1) Since there have been great earthquakes in Area A during recorded history, the possibility that there will be a great earthquake in the near future is relatively
        (1a) great   (1b) small

(2) Since there have been no great earthquakes in Area B during recorded history, the possibility that there will be a great earthquake in the near future is
        (2a) small   (2b) great

In the past, earthquake risk has been seen as a function of [1a + 2a]; in some cases, however, the direct opposite (*i.e.,* [1b + 2b]) seems to be more appropriate.

(1a) is correct, for example, in the case of the Nankaido region, in which great earthquakes have occurred repeatedly during recorded history.

(1b) is more accurate, however, where inland active faults (faults with a great t/R) are concerned—faults in which the degree of activity is greater, but which have not been active during recorded history. (2a) is appropriate for areas when there are neither historical records of great earthquakes nor any remarkable active faults. In the absence of these conditions, however, (2b) is more likely to be correct.

Thus predictions based solely on the historical past may lead to totally inaccurate conclusions, especially in the case of active fault regions in inland Japan (Fig. 3.6). What is necessary, then, is an earthquake risk map that takes into account both the geological past and the geological locality. This work has at last begun recently.

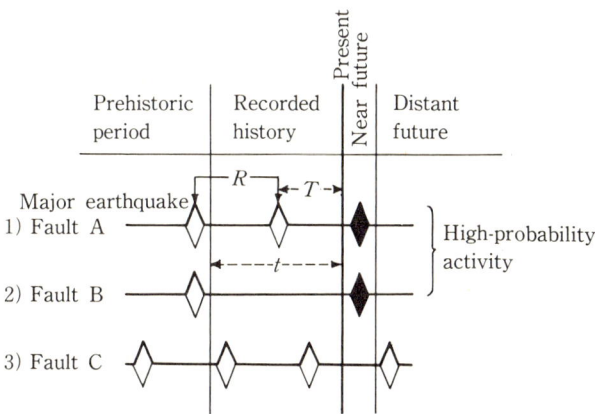

Fig. 3.6 Three examples illustrating the relationship between recorded movement along an active fault and the probability of earthquake activity in the near future.
Case 1: Because movement has been recorded along Fault A within recorded history, there is a good probability of activity in the near future.
Case 2: Because Fault B has *not* been active within recorded history, there is a good chance that activity will occur in the near future.
Case 3: Because movement along Fault C has been recorded very recently, the probability of activity in the *near* future is low.
In calculating the probability of movement along an active fault, the historical record is examined, and the periodicity of fault movement (R), if known, is compared with the time period up to the present (t) during which the fault has been quiescent (t/R).

3.4.6 Active Faults and Felt Foreshocks

Active faults in inland Japan are generally characterized by intervals of several hundred years or longer between great earthquakes. But there is still no way to tell—even with the topographic or geological means already discussed, or with precise observation methods—whether the fault activity will occur in several years or several decades. Consequently, at this time the observations and experiences of the people living in these active fault areas are regarded as critical.

In the fall of 1930, when foreshocks began in the northern Izu area, an old woman went out into her yard and began cooking rice. Her bewildered family thought she had gone mad. Early the next day, the Northern Izu earthquake struck. The woman had somehow sensed the earthquake coming and was simply preparing emergency food. I heard this story in the community of Ukihashi, which is on the Tanna Fault, where that particular earthquake occurred. Other examples of people who lived near an active fault sensing an earthquake several days before it occurred were reported at the times of the Nobi, Rikuu, and Mikawa earthquakes.

In Table 3.4 the right-hand column indicates whether foreshocks were felt within several days of the major inland earthquakes in Japan since 1865 (the beginning of the Meiji Era). About 10 of these earthquakes of M 6.5 and more can be attributed to the movement of segments of active faults. Four of these earthquakes were accompanied by pronounced felt foreshocks. Five of the six remaining earthquakes occurred along active faults on the coast of the Japan Sea. It is not clear why earthquakes on these particular faults are not accompanied by felt foreshocks. If foreshocks not felt by humans could also be detected, fault activity could be anticipated much sooner. Earthquake observation networks covering each of the major inland active faults would contribute greatly to understanding precursory foreshock activity.

References

Ando, M., 1977: Possibility of a major earthquake in the Tokai district, Japan, and its pre-estimated seismotectonic effects, *Tectonophys.*, **25**, 69–85.

Dambara, T., 1966: Vertical movements of the earth's crust in relation to the Matsushiro earthquake, *J. Geodet. Soc. Japan,* **18**, 18–45.

Kaizuka, S., 1968: On study of Quaternary crustal movement from geomorphological point of view, *Memoir Geol. Soc. Japan*, **2**, 75–76.

Kasahara, K., 1957: The nature of seismic origins as inferred from seismological and geodetic observations (1), *Bull. Earthq. Res. Inst.*, **35**, 473–532.

Kawasumi, H., 1951: Measures of earthquake danger and expectancy of maximum intensity throughout Japan as inferred from the seismic activity in historical time, *Bull. Earthq. Res. Inst.*, **29**, 469–482.

Matsuda, T., 1976: Empirical rules on sense and rate of recent curstal movements, *Jour. Geodet. Soc. Japan*, **22**, 252–263.

———,1976: Earthquakes and active faults. *Earthquakes*. Univ. Tokyo Public Lecture Series, No. 24, Univ. Tokyo Press.

———, 1977: Estimation of future destructive earthquakes from active faults on land in Japan, *J. Phys. Earth*, **25**, *Suppl.*, S251–S260.

Matsuda, T., Y. Ota, M. Ando and N. Yonekura, 1978: Fault mechanism and recurrence time of major earthquakes in southern Kanto district, Japan, as deduced from coastal terrace data, *Geol. Soc. Am. Bull.*, **89**, 1610–1618.

Matsuda, T., H. Yamazaki, T. Nakata and T. Imaizumi, 1980: The surface faults associated with the Rikuu earthquake of 1896, *Bull. Earthq. Res. Inst*, **55**, 795–855.

Nakata, T., M. Koba, W. Jo, T. Imaizumi, H. Matsumoto and T. Suganuma, 1979: Holocene marine terraces and seismic crustal movement, *Sci. Rept., Tohoku Univ.*, 7th Ser., **29**, 195–204.

Okada, A., 1980: Quaternary faulting along the Median Tectonic Line of southwest Japan, *Memoir Geol. Soc. Japan*, **18**, 79–108.

Ota, Y., T. Matsuda and K. Naganuma, 1976: Tilted marine terraces of the Ogi Peninsula, Sado Island, central Japan, related to the Ogi earthquake of 1802, *J. Seismol. Soc. Japan*, Ser. **2**, 55–70.

Plafker, G., 1972: Alaskan earthquake of 1964 and Chilean earthquake of 1960: Implications for arc tectonics, *J. Geophys. Res.*, **77**, 901–925.

Research Group for Active Faults (ed.), 1980: *Active Faults in Japan: Sheet Maps and Inventories*, Univ. Tokyo Press, 1–363.

Wellman, H. W., 1967: Tilted marine beach ridges at Cape Turakirae, N. Z., *Jour. Geosciences, Osaka City Univ.*, **10**, 123–129.

Yonekura, N., 1975: Quaternary tectonic movements in the outer arc of southwest Japan with special reference to seismic crustal deformations, *Bull. Dept. Geogr., Univ. Tokyo*, **7**, 19–71.

PART II LONG-RANGE PRECURSORS

The earth's crust is under stress; as time goes by, strain accumulates. When this strain goes beyond a certain critical point, certain changes are thought to take place in the earth's crust. The time-span of maximum accumulation can be as great as 1,000 years where earthquakes in inland Japan are concerned or less than 100 years for Pacific coast earthquakes.

These changes take various forms, such as an increase in the crust's volume and variations in seismic wave velocity. They affect the way small earthquakes occur, the nature of electromagnetic phenomena, and, in some cases, the water content within the crust or the migration of radon and helium. From the fragments of information gathered so far it seems that the period during which such "anomalous conditions" exist—that is, from the onset of the anomalies to the moment of the earthquake—is related to the magnitude of that earthquake.

The greater the magnitude of the earthquake, the longer the precursory period will be. This relationship is different for inland earthquakes and Pacific coast earthquakes in Japan. In the case of inland earthquakes the precursory period is long despite the lesser magnitude of the earthquakes. Pacific coast earthquakes have a shorter precursory period even when their magnitude is as great as 8. This is a very important point.

Although the study of the physical aspects of long-range earthquake precursors is critical to better earthquake prediction, the amount of time and space required for such studies makes them very difficult to do. In this part, examples of physical research being done in this area will be discussed.

PART II: LONG-RANGE PRECURSORS

Chapter 4   How Small Earthquakes Occur

Akio Takagi

4.1   Smaller Earthquakes Occur More Often

The study of how and where earthquakes occur is one of the most basic areas of earthquake research. Before pursuing the subject further, however, it is necessary to look at the standard that determines the scale of earthquake size.

Richter [1935] observed Southern California earthquakes with numerous Wood-Anderson seismographs and studied the data he obtained. As a result of these studies he came to the conclusion that although the decrease in the amplitude of ground motion during an earthquake is quite varied, it does seem to follow a uniform rule with regard to the distance from the hypocenter. This means that if two earthquakes of different sizes were measured at equal distances from their hypocenters, the ratio of one amplitude to the other would remain uniform no matter what the distance. This discovery made the determination of earthquake scales possible. There is no particular reason to choose one standard over another; yet, if earthquakes are to be interpreted uniformly, a standard is definitely necessary. Richter called an earthquake that recorded a maximum amplitude of one micron on the Wood–Anderson seismograph (natural period: $T_0 = 0.8$ sec; attenuation constant: $h = 0.8$; magnification: $V = 2,800$), positioned at an epicentral distance (the distance between the epicenter and the observation site) of 100 km, the standard earthquake; he set its magnitude at zero. He represented the magnitude of any earthquake as M_L, defined as the common logarithm of the ratio of the maximum amplitude of a particular earthquake to that of the standard earthquake.

This, of course, does not mean that the same method can be used to obtain the attenuation of seismic waves in parts of the world other than Southern California. Furthermore, it is almost impossible for a seismograph of this sensitivity to observe body waves from a distance of more than 600 km. Richter's method is used for shallow earthquakes but does not work for deep-focus earthquakes. That is why various other methods for determining magnitude have been proposed. All of them, of course, include newly created standards that have been adjusted to the Richter scale so as to minimize the differences among the various methods.

Observation records in Japan were gathered during the early days by the Japan Meteorological Agency using Wiechert seismographs. These were replaced in 1959 by modern electromagnetic seismographs. These modern seismographs have amplitude characteristics that are identical to the old ones. The magnitude for earthquakes close to Japan was selected from those obtained by Richter and Gutenberg to fit an equation that would agree with the value of the selected magnitude as much as possible, with A as the maximum amplitude (the composite of two horizontal components; unit μ) and Δ km as the epicentral distance:

$$M = 1.73 \log \Delta + \log A - 0.83.$$

This is called Tsuboi's formula.

Noting that earthquake frequency decreases exponentially as the magnitude M (which defines the earthquake's scale) becomes greater, Gutenberg and Richter came up with an empirical formula for Southern California earthquakes ($3.5 \leq M_L \leq 6.5$):

$$\log n(M) = a - bM.$$

This formula is now called the Gutenberg-Richter formula, and represents the relationship between $n(M)$ and M, when $n(M) \, dM$ is the frequency of earthquakes that occurred in certain times and places, the magnitude of which is from M to $M + dM$.

The total number of earthquakes with magnitude above M is:

$$N(M) = \int_{M}^{\infty} n(M) \, dM.$$

Hence, the relationship between $N(M)$ and M can be represented as:

$$\log N(M) = A - bM.$$

In these formulas a, A, and b are constants. A value close to one is obtained for b.

In the meantime, in Japan, Ishimoto and Iida [1939] came up with an empirical formula (Ishimoto and Iida's formula) that gives the relationship between $n(A)$ and A when the frequency of earthquakes with a maximum seismic wave amplitude recorded during a certain period of time—of A to $A + dA$, is set at $n(A) \, dA$:

$$n(A) = kA^{-m}.$$

Where k and m are constants. 1.7 to 2.0 is the value generally obtained as m. Later, Asada et al. [1950] demonstrated that, as far as magnitude distribution is concerned, the above formula and the Gutenberg–Richter formula are identical, and that the following relationship can be established:

$$m = b + 1.$$

The fact that b is close to 1 means that $(m - 1)$ earthquakes occur about ten times more frequently than magnitude M earthquakes during a fixed period of time in any one area, and that the number of earthquakes with about one-tenth the amplitude of a certain earthquake will be as many as 100. These equations seem fairly accurate for great earthquakes that take place in shallow areas; but for smaller earthquakes, those with less than $M3.5$, there is less certainty.

The observation of smaller earthquakes must be carried out in places where there is little artificial noise as the instruments used are highly sensitive and rely on a method where S/N (signal/noise ratio) is as great as possible. Generally speaking, these places are not conveniently located.

Compared to the observation of great earthquakes, the observation of smaller earthquakes is a recent phenomenon; it began in Japan, for instance, in the late 1940s. While observing the aftershocks of the Fukui earthquake of 1948 and the Imaichi earthquake of 1949, using newly developed and more sensitive seismographs, Asada and Suzuki recorded innumerable minor earthquakes which were much smaller than any ever recorded before. There has been continuing research and observation in this area ever since, and several significant facts have emerged. One of them is that the mechanism of an earthquake is almost the same regardless of its scale, and so are the characteristics of its activity. In short, it became clear that the Gutenberg–Richter and the Ishimoto–Iida formulas apply to both small and large earthquakes. This points to the fact that by observing frequent small earthquakes in an area, one can learn how greater earthquakes occur. Thus, even very small earthquakes have meaning and should not be overlooked. Another significant fact that emerged is that when earthquakes take place in a given area, they do so in a series of all sizes ranging from very small to very large.

As the importance of small earthquake research was recognized, observation of these earthquakes began throughout Japan. The observations of the Research Group for Explosion Seismology in Japan have greatly contributed to research on crustal structure throughout the country since the group began its work in 1950. Its research, based first on a quarry explosion using 56 tons of dynamite during the construction of the Ishibuchi Dam (Iwate Prefecture), was applicable to microearthquakes. Thus the knowledge acquired by the Research Group for Explosion Seismology supplemented the information acquired during the observation of the Fukui earthquake aftershocks, resulting in rapid progress in what has become a major field of research in Japanese seismology.

In January 1962, the Earthquake Prediction Research Group published an article entitled "The Present Condition of Earthquake Prediction and a Plan for the Future." The article established a clear standard for earthquake scale so that future seismic investigations could be planned that would lead to better earthquake prediction. According to this scale:

Great earthquakes	$M \geq 7$
Intermediate earthquakes	$7 > M \geq 5$
Small earthquakes	$5 > M \geq 3$
Microearthquakes	$3 > M \geq 1$
Ultra-microearthquakes	$1 > M$.

Great earthquakes, as defined here, are all destructive unless their foci are quite deep. Intermediate earthquakes can be destructive if their foci are shallow, but the area of damage will not be as extensive. Small and micro earthquakes seldom cause damage. In this chapter smaller earthquakes, as opposed to larger ones, will be examined and discussed, but the above standard will not be adhered to strictly. The foci of over 100 such earthquakes are recorded daily throughout Japan.

4.2 Where Do Small Earthquakes Occur?

One of the basic tasks necessary for accurate earthquake prediction is the construction of a map that clearly presents the information gathered through the detailed investigations of seimsic activity done by scientists in the various fields of seismic research—information on which areas are prone to great and intermediate earthquakes, and which areas are not. This task has been undertaken by the Japan Meteorological Agency (JMA) and university-affiliated observation stations.

In recent years almost all the earthquakes with a magnitude of 5 and above have been observed and their foci determined. Put simply, these earthquakes are not evenly distributed throughout the world. They are concentrated in limited metamorphic belts: shallow earthquakes occur along the oceanic ridges, some forming a belt on the outer side of the trench; deep-focus earthquakes are distributed along the planes that incline from the oceanic trenches toward the continents or island arcs; and other earthquakes take place in the new orogenic belts such as folded mountain ranges. The earthquakes included on the epicenter distribution map in Fig. 4.1 are those of *M* 5 and above. There is not much uniformity to be found in the distribution of smaller earthquakes due to the uneven regional density of observation networks. In Japan, however, where networks affiliated with the universities and the JMA are densely distributed, it is possible to determine the foci of any small earthquake. The results indicate that the distribution pattern of these earthquakes is in excellent agreement with that of the greater earthquakes. The detailed distribution pattern of smaller earthquakes provides an important clue to earthquake phenomena.

Figure 4.1 is an earthquake frequency distribution map for the 14-year period between 1961 and 1974. The epicenters of the earthquakes on this map were determined by the JMA. Each number on the map represents the number of earthquakes that occurred within an area bounded by 5 minutes of latitude and longitude and at a depth of 0 to 39 km. In places where 10 or more earthquakes occurred, the occurrences are represented by A for 10 times, B for 11 times, C for 12 times, and so forth.

During this period (1961–74), seismic activity in Northern Japan was pronounced in the oceanic regions; in Southwest Japan, the coastal and inland areas were most active. According to the JMA, the characteristics of seismic activity during this period were: (1) active aftershocks in the wake of the Nemuro Peninsula offshore earthquake of 1973, the Tokachi offshore earthquake of 1968, and the Niigata earthquake of 1964; (2) the existence of seismic gaps in the Enshu Sea and the southern sea off the Boso Peninsula; (3) the long-range aftershock of the Nankai earthquake of 1946 which continued well beyond 1961; and (4) marked seismic activity in the Hyuga Sea and the Bungo Channel. The inland regions were characterized by small-scale seismic activity and swarm earthquakes in various locations. Aftershock activity following the Northern Miyagi earthquake of 1962 and the Southeastern Akita earthquake of 1970 is visible on the map, as are the aftershocks following the Mikawa earthquake of 1945, which continued well past 1961. The distribution map in Fig. 4.1 includes earthquakes with minimum magnitudes of 3.5 in the inland regions and slightly greater magnitudes in the oceanic regions. What would a distribution map of smaller earthquakes look like?

In 1963 the Microearthquake Research Group began to observe ultra-microearthquakes in the Neodani Fault region and adjacent areas. Since then routine or temporary observations have been undertaken by universities. By November 1972, approximately 20,000 foci had been determined by such teams. Figure 4.2 is a micro and ultra-microearthquake epicenter distribution map of Japan. The materials used to make this map were not all collected during the same period, nor were the methods of observation and of hypocenter determination (including methods using under-

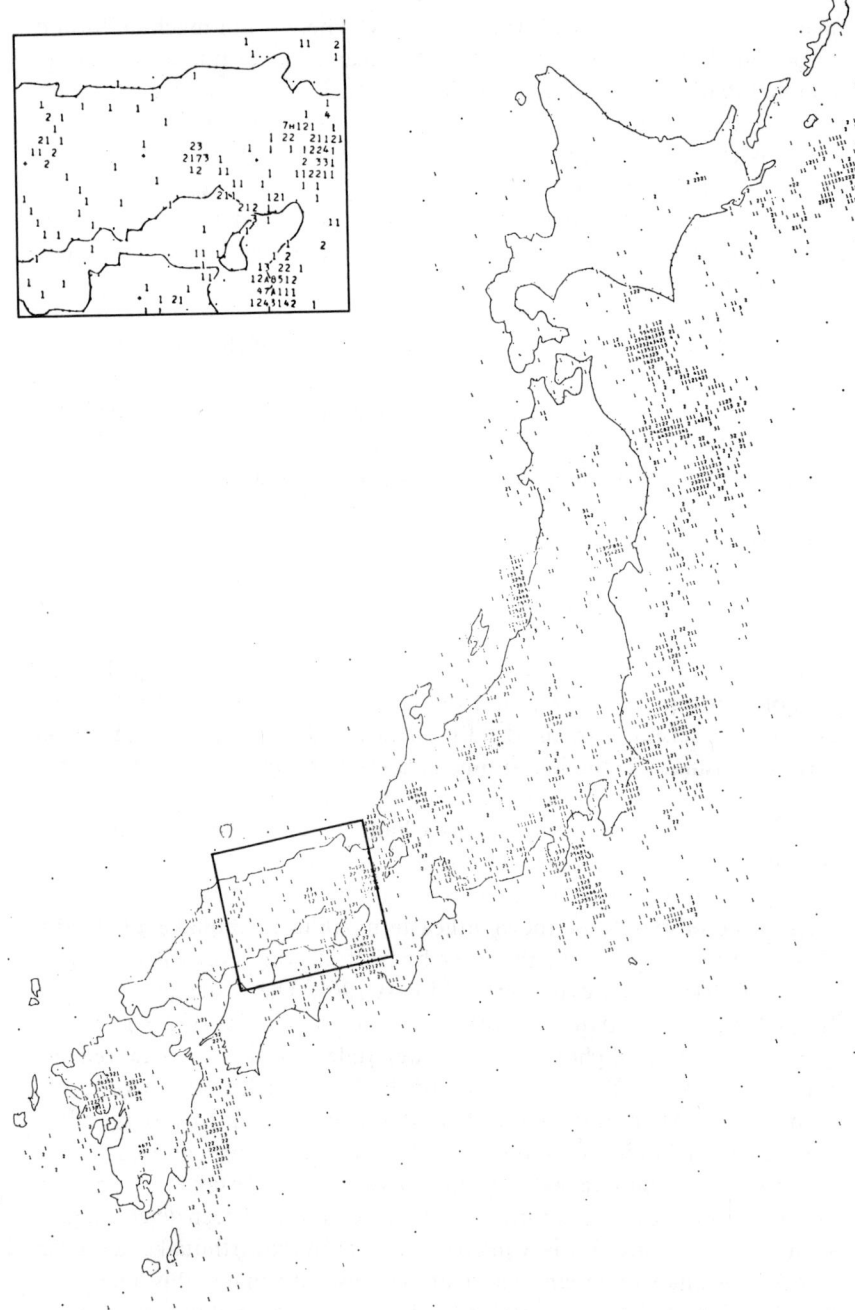

Fig. 4.1 Epicenter distribution in and around Japan, 1961–1974 ($h = 0$–39 km).

HOW SMALL EARTHQUAKES OCCUR 69

Fig. 4.2 Epicenter distribution of microearthquakes.

ground structure) uniform. Some small earthquakes are also included. Some areas such as Kyushu and the Southern Tohoku district are excluded because there had been no observation of microearthquakes there. Consequently the seismic activity represented by the light and dark distribution dots is not necessarily comparable. Despite these inconsistencies, however, and despite the fact that the data on smaller earthquakes are based on short-range or temporary observations, a comparison of these two distribution maps (Figs. 4.1 and 4.2) reveals that they are in excellent agreement.

The agreement between the two maps points to the fact that because smaller earthquakes occur more frequently than large ones, a special distribution pattern for a certain area can be obtained by observations made during a limited period of time. It also shows that the activity of smaller earthquakes can provide information that is directly relevant to a better knowledge of crustal activity. On the other hand, the correspondence between seismic activities and crustal movement is affected by the unique structure of each individual area. Thus it is necessary to take into account the spatial and chronological features affecting seismic activity, especially where smaller earthquakes are concerned. Two of the indispensable aspects of effective smaller earthquake studies are (1) an observation network of appropriate density to insure accurate focus determination and (2) an observation system that can function effectively for an extended period of time over a large area so that uniform data can be obtained.

Based on the findings of the third Five-year Earthquake Prediction Project, in 1974 the intensive telemeter observation of microearthquakes began. As facilities throughout the country (with the exception of Kyushu) became fully equipped, great progress was made in both the quantity and quality of small earthquake observations and research, providing an increasing number of new and significant findings. Figure 4.3 shows the new observation network compiled by Utsu [1978]. After this network is completed it will be possible to record an average of 100 small earthquakes throughout the nation in one day. A comparison of these results with other geophysical data is expected to reveal some fundamental facts about the conditions of crustal activity.

As is apparent in Fig. 4.3, observation sites are densely distributed in the Chubu and Chugoku districts, south-central Honshu, and in Shikoku. These areas contain many regions with exposed ancient bedrock and are quite different from the northern Tohoku district, for instance, which is thickly covered with volcanic sediment. For this reason, detailed maps of the area's active faults and geological structure could be made from the surface topography and geology. These maps make it possible to elucidate the relationship between the area's active faults and geological structure, and to obtain a highly accurate and homogeneous picture of the distribution of its small earthquakes. Observation at these sites has continued over a long period of time, and massive amounts of information have been accumulated. Focal distribution, as determined by the network, therefore, is spatially and chronologically quite homogeneous.

Figure 4.4 is a microearthquake epicenter distribution map prepared by the Microearthquake Research Group at Kyoto University. Figure 4.5 is a map of active faults and active tectonic lines prepared by Fujita (1977). From this data Oike [1975] obtained intriguing results. They found that many of the foci distributed in belt or linear patterns are closely

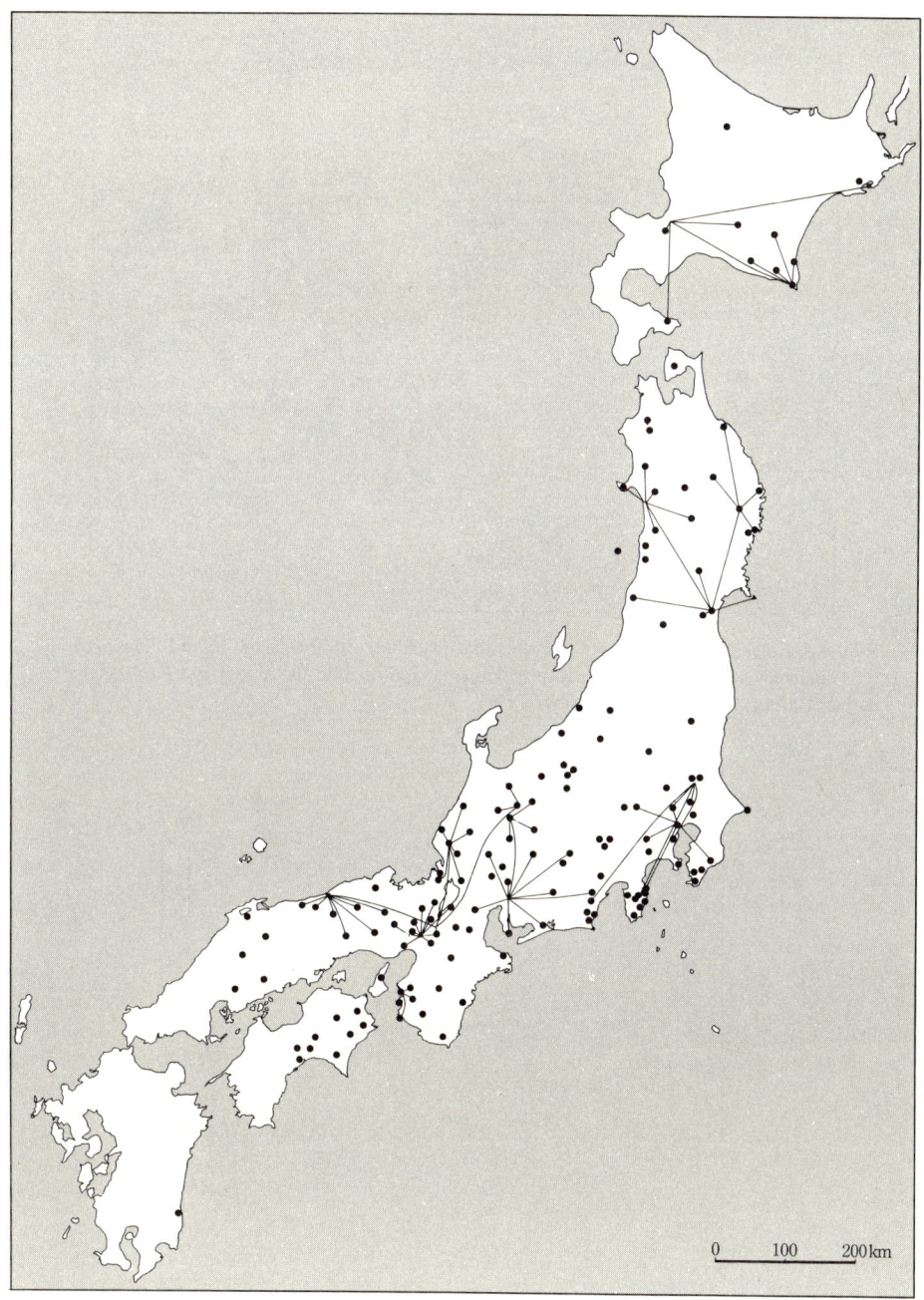

Fig. 4.3 Distribution of microearthquake observation sites as of November 1977.

72 LONG-RANGE PRECURSORS

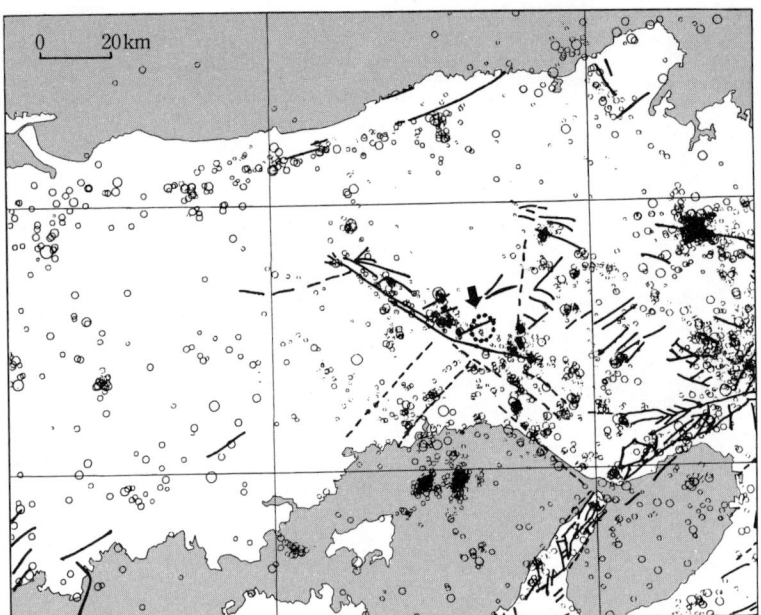

Fig. 4.4 Epicenter distribution of microearthquakes in the Seto Inland Sea area. The arrow points to a seismic gap which occurred around the Yamasaki Fault at the time of an earthquake in September 1977.

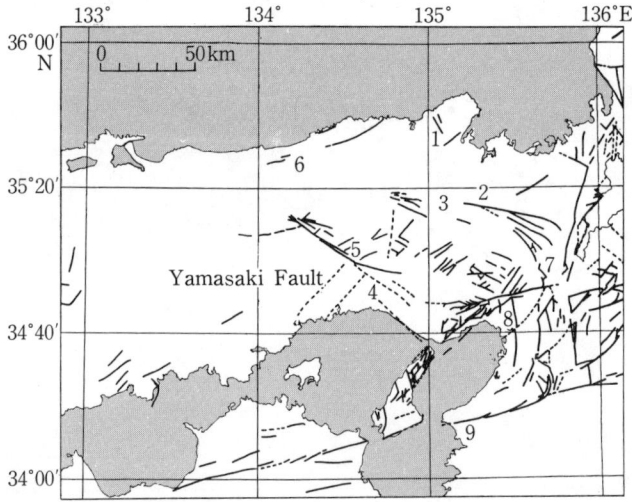

Fig. 4.5 Active faults and tectonic lines in the Seto Inland Sea area.

related to active faults. Areas where there is a lot of microearthquake activity and areas of comparatively little activity are clearly divided by large-scale faults or tectonic lines. The map also shows that active aftershock activity still continues along the faults involved in the Tottori earthquake (1943) and the Northern Tango earthquake (1927).

The relationship between active faults and smaller earthquakes is the subject of research throughout the world. California's San Andreas Fault, for which the term "active fault" was first coined, has been the subject of extensive and intensive investigations of crustal movement and the characteristics of microearthquakes. According to these studies, smaller earthquakes were found to be concentrated in a straight line 100 km long south of San Francisco, where seismic activity is currently most vigorous; in the epicentral zone of the San Francisco earthquake of 1906, seismic activity is low.

Future research on active faults and earthquake phenomena should focus on: (1) the discovery of latent active faults from detailed earthquake distribution studies; (2) determination of the maximum earthquake potentiality of various areas through an examination of how earthquakes occur in active faults; (3) study of the characteristics of the focal process and the temporal and spatial distribution of the seismic activities corresponding to each stage of the process—ranging from the accumulation of strain energy to its release. Some of the research findings in these areas will be discussed in the following section.

Detailed research into small earthquake activity in Japan is also under way in the Tokai district, the Kii Peninsula, and eastern Shikoku Island. The general thinking here is that intra-crustal and subcrustal seismic activities are clearly distinguishable, and are especially remarkable on the low-angle plane that dips from the Pacific Ocean toward the inland region. Figure 4.6 (a) shows the distribution of subcrustal earthquakes compiled by Mizoue [1971] and others in the Kii Peninsula area classified according to depth. Figure 4.6 (b) shows similar distribution for the north–south cross-section of eastern Shikoku compiled by Kimura. It is notable that seismic activity in these figures inclines approximately 14 degrees to the north down to a depth of 45 km. Such a tendency is also found in the Tokai district. The Wakayama Microearthquake Observation Group of the Earthquake Research Institute of the University of Tokyo, in their study of these unique seismic activities, analyzed the reflected seismic waves on the upper plate of the inclined subcrustal earthquake distribution zone, using precise long-range observation techniques. From this analysis they made the important discovery that the subcrustal earthquake distribution zone comes into contact with the crust above at exactly the point of velocity discontinuity. In other words, this zone corresponds to the

74 LONG-RANGE PRECURSORS

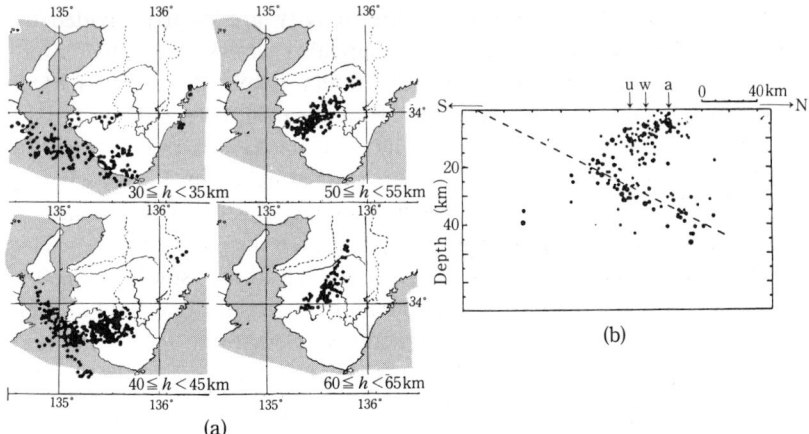

Fig. 4.6 (a) Epicenter distribution of crust-bottom earthquakes in the Kii Peninsula and its vicinity; (b) Vertical cross-section of the distribution of microearthquakes in the Kii region.

subduction of the plate. This layer is characterized by greater seismic wave velocity. Based on this finding, the group arrived at the significant conclusion that the great earthquakes off the coast of Nankai and Tonankai and their aftershock activities are closely related to the Philippine Sea Plate which subducts into inland Japan from the Nankai Trough.

Small earthquakes in the Tohoku district, which is located on a typical island arc—an oceanic trench system roughly parallel to the Japanese Trench—are distributed as represented in Figs. 4.7 through 4.9. Figures 4.7 and 4.8 represent the epicentral distribution of microearthquakes that occurred at depths of 0 to 20 km and 20 to 60 km, respectively. Figure 4.9 is a stereographic distribution diagram of microearthquakes that occurred at depths of 10 to 20 km. The epicentral distribution can be roughly divided into three areas: seismic activity off the coast of Sanriku; that along the volcanic front that runs through the Tohoku district; and that along the coast of the Japan Sea. As is apparent in Fig. 4.9, inland earthquakes 10 to 20 km deep occur only on an arcuate line in the center of the Tohoku district. There are hardly any earthquakes here deeper than 20 km. The epicenters of all the earthquakes along the volcanic front are shallower than 10 km (with the exception of the central areas), which perfectly agree with the areas of earthquake swarms. In these areas, no earthquake of more than M 6 has ever been recorded. On the other hand, the central area, where epicenters go down to a depth of 20 km and are distributed in an arcuate pattern, is where destructive earthquakes of M 6

HOW SMALL EARTHQUAKES OCCUR 75

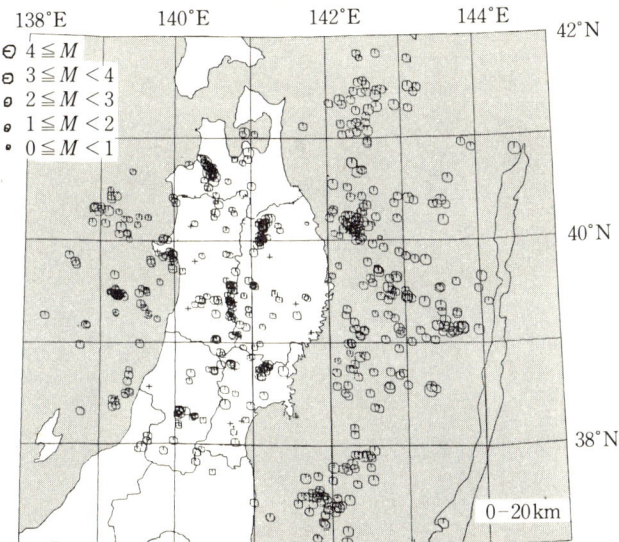

Fig. 4.7 Epicentral distribution of microearthquakes, depth 0–20 km, in the Tohoku district from July 1 to December 31, 1977.

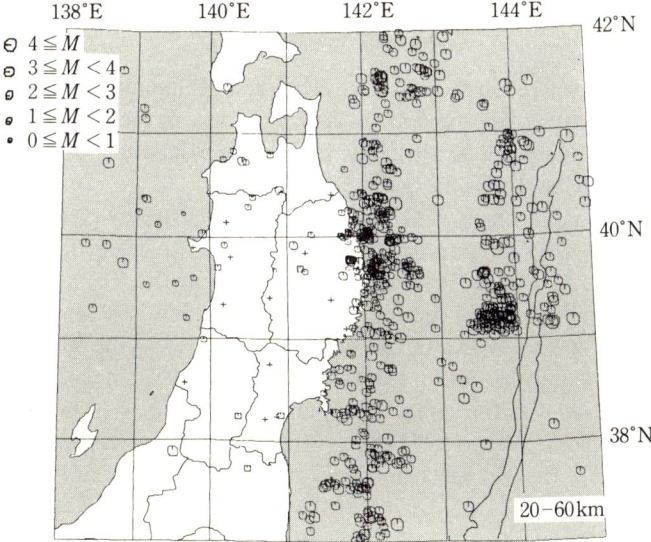

Fig. 4.8 Epicentral distribution of micro earthquakes, depth 20–60 km, in the Tohoku district from July 1 to December 31, 1977.

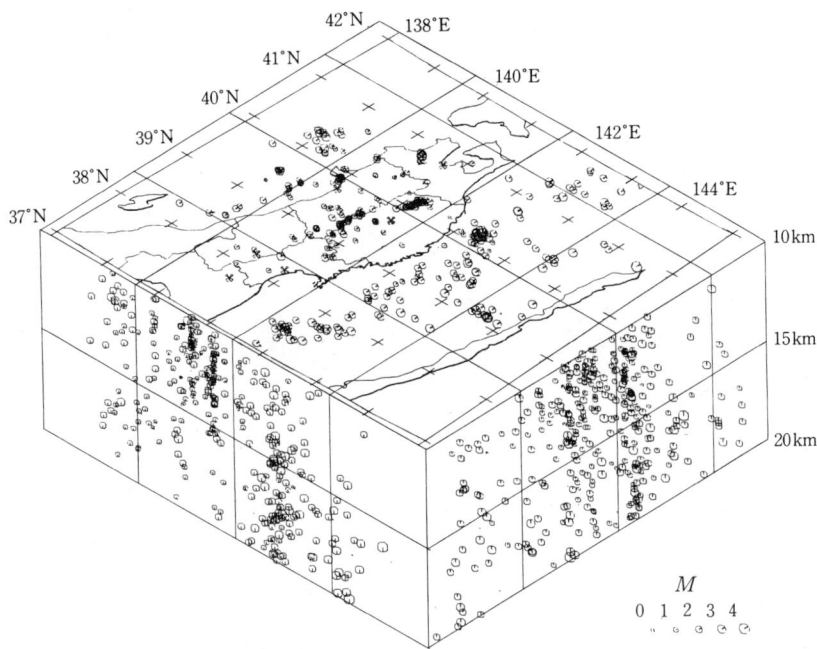

Fig. 4.9 Epicentral distribution of micro earthquakes in the Tohoku district from July 1 to December 31, 1977.

to M 7 have occurred in the past. The Kitakami Mountains have this arcuate line as their western margin and the Sanriku coast as their eastern margin. Small earthquakes are remarkably few in the Kitakami Mountains; the area seems to be a seismic gap.

Figure 4.10 is the result of GDP (Geodynamics Project) traverse surveys carried out by the Geographic Research Institute. It shows the variations in lateralation and angle from the Pacific coast of the Tohoku district to Tobishima on the coast of the Japan Sea from 1894 to 1973 and during the years from 1962 to 1973. As the figure demonstrates, the changes from Goyozan to Taneyama and from Taneyama to Yakeishidake, which span the Pacific coast and the central mountain range, were +3 cm and −2 cm respectively from 1894 to 1973 (but −1cm and +16cm respectively from 1962 to 1973); the change from Yakeishidake to Mitsumori was −147 cm (−24 cm for 1962–1973), from Mitsumori to Oguni, +48 cm, and from Oguni to Tobishima, −3 cm. The Rikuu earthquake (1896) and the Southeastern Akita earthquake (1970) occurred in the central mountain area during this period of observation. The changes in lateralation in an east–west direction in the central Tohoku district were great in the

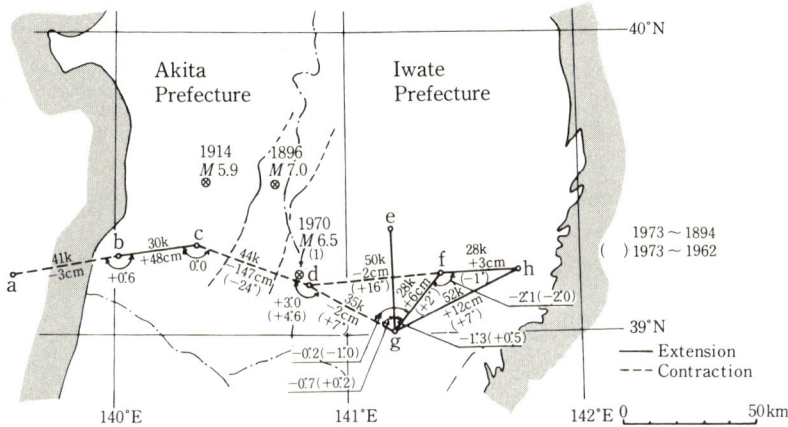

Fig. 4.10 Angle variations and extension and contraction distances in the Tohoku district according to a GDP high-precision traverse survey.
a: Tobishima   b: Okunibayama   c: Mitsumori   d: Yakeishidake   e: Raijintoge
f: Taneyama   g: Tabashineyama   h: Goyozan

central mountain ranges and along the Japan Sea coast. The Kitakami Mountain Range remained virtually unchanged, however. This phenomenon bears witness to the fact that the Kitakami block, compared with the adjacent areas, had not accumulated strain energy within itself. Rather, it acted as a rigid body. Movement relative to the block, then, is considered to have occurred at its boundary, resulting in the arcuate pattern of seismic activity. This idea seems to be in agreement with the displacement diagram of the Tohoku district made by the Geographic Research Institute. Can a block within a plate have a different character from the rest of the plate, though? The clue to this question's answer seems to lie in the physical properties of the lower part of the earth's crust, which will be discussed later.

Seismic activity off the coast of Sanriku apparently occurs at the plate boundary and in the upper plate. As is seen in Fig. 4.11, small earthquake activity occurring at a depth of 90 to 100 km is distributed in a pattern that is almost linear and that parallels the oceanic trench. The pattern becomes slightly irregular for those earthquakes that occur at a depth of 60 km or thereabouts. Earthquakes that take place in the upper plate (0 to 20 km) have a very complex distribution pattern. These various patterns are thought to reflect the relative movement of the subducting oceanic plate and the continental crust structure above it.

There is a great deal of micro earthquake activity at a depth of 0 to 20 km in the coastal area of the Japan Sea. This is important from the standpoint of earthquake prediction since this area is the very place where

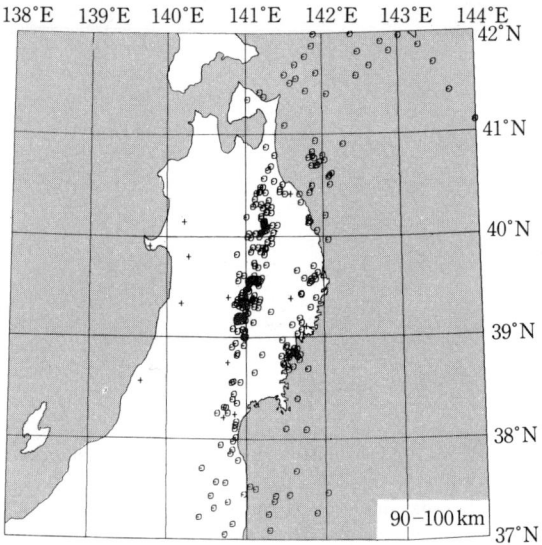

Fig. 4.11 Epicentral distribution of intermediate-focus earthquakes, 90–100 km deep, in the Tohoku district between April 19, 1975, and July 31, 1977.

many destructive earthquake have historically occurred—most of them directly below big cities.

The causes of these earthquakes may not be understood completely for some time, but they can be attributed in part to the concentration of stress created by an inhomogeneity in the crustal structure.

The Research Group for Explosion Seismology (RGES) has been conducting study of crustal structure in the island arc area of Northeast Japan, and recently studied crustal structure along the survey line between Oga and Kesennuma. The results are shown in Fig. 4.12 (a) and (b) [RGES, 1977]. The microearthquakes that occurred within 60 km of this survey line are demonstrated in a vertical cross-section in this figure. These observations were made over a period of 18 months, as were the observations in Fig. 4.11. This work led to an extremely interesting discovery: the foci of almost all inland earthquakes are in the 5.9 km/sec velocity layer; there are hardly any earthquakes in the 6.6 km/sec layer, which is more than 18 km deep. The few earthquakes that do occur in the lower crust and immediately below the Moho discontinuity are concentrated directly below active volcanoes such as those in the Iwate and Kurikoma Mountains (see Fig. 4.13). The Microearthquake Research Group of the Disaster Prevention Research Institute, as well as the Science Department

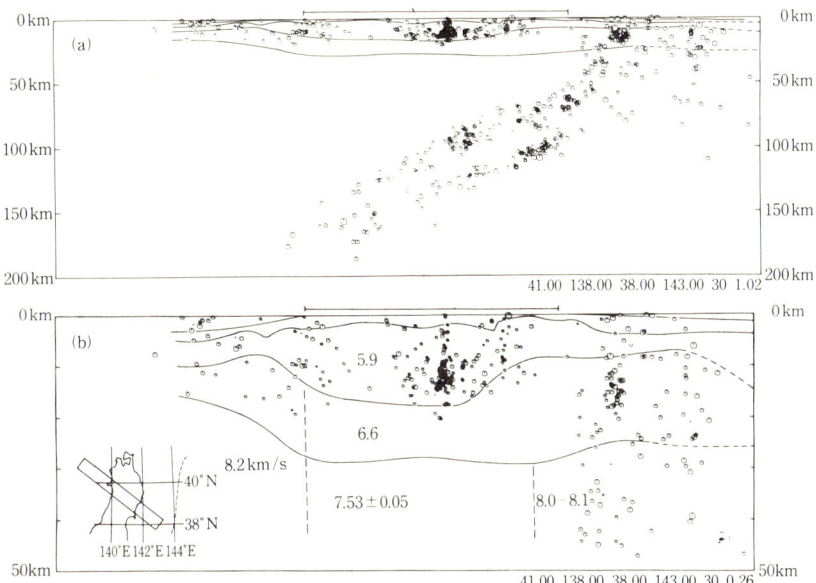

Fig. 4.12 Vertical distribution of the foci of microearthquakes and crust and upper mantle structure along the Oga–Kesennuma Line, between April 19, 1975, and October 31, 1976.

of Kyoto University, made a similar discovery in Southwest Japan. Most of these earthquakes are concentrated in the 6.0 km/sec layer, while very few occur in the 6.6 km/sec layer [Oike, 1975].

These findings point to the need for a detailed investigation of how granitic or basaltic layers respond to the crustal movement caused by the relative motion of the oceanic and continental plates. This study will provide the key to inland earthquake prediction. Since research is already being done that indicates that the P wave velocity within the lower crust is 6.6 km/sec or less, the question of the physical properties of the lower crust is expected to be resolved in the near future with the assistance of the research being done in experimental petrology.

The region along the Japan Sea coast where microearthquakes are active is a transitional area. As the oceanic crust meets the continental crust, not only does the crust's thickness change, but the seismic wave velocity within the crust may change from 6.6 km/sec to 7.0 km/sec. Stress concentration, therefore, is probably intense in this particular area.

Reports that intermediate-focus earthquakes directly below the Northeastern Japan island arc have a double-planed structure have been made by Tsumura [1973], Umino and Hasegawa [1975], and Hasegawa et al.

80 LONG-RANGE PRECURSORS

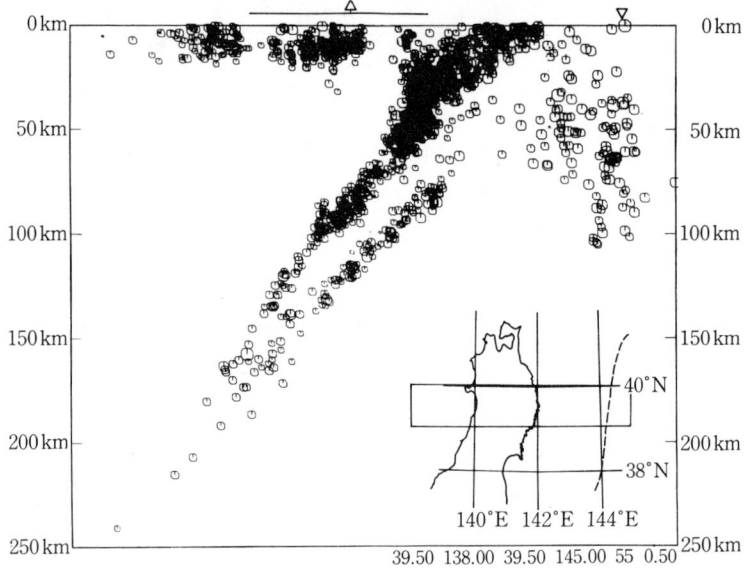

Fig. 4.13 East–west vertical cross-section (39° N to 40° N) of the distribution of microearthquakes between April 19, 1975, and October 31, 1976.

[1978]. As the result of an intense telemeter survey, a highly accurate determination of earthquake foci has recently been made, as demonstrated in the east–west vertical cross-section in Fig. 4.13. The figure is an earthquake foci distribution diagram of intermediate-focus earthquakes that occurred within 39° N to 40° N. It can be clearly seen in this diagram that microearthquakes are concentrated on two planes. The distance between the two planes is 30 to 40 km, and the microearthquakes occur down to a depth of 200 km. Hasegawa et al. [1978a] sought the common earthquake source mechanism for intermediate earthquakes occurring on two planes using P waves. The result revealed the characteristics of the earthquake source mechanism in the Northeast Japan Island arc: earthquakes occurring in the upper crust are characterized by reverse fault movement or down-dip compression (i.e., compression axis parallel to the dip of the deep seismic zone), while those occurring in the lower plane are characterized by down-dip extension (i.e., tension axis parallel to the dip of the zone). Microearthquake activities on the two planes are neither chronologically nor spatially uniform—a direct reflection of the relative movement of the subducting oceanic plate and the continental plate. Further investigation of the dynamic roles of these two planes is essential.

On June 29, 1974, a deep-focus earthquake of M 7.7 struck in the western part of the Japan Sea. An examination of the records revealed that approximately 13 minutes after the initial motion was recorded, ScS waves and ScSp waves (which precede the ScS waves) were clearly recorded in both the horizontal and vertical components. The ScS wave propagates from the focus as an S wave, then is reflected from the surface of the earth's core, and reaches the surface of the earth as an ScS wave. The ScSp wave is an ScS wave that is reflected from the surface of the earth's core and, on the way to the earth's surface, is converted to a refracted P wave when it hits the plate boundary before finally reaching the surface. Taking the time difference between the arrival of these two waves at an observation point, Hasegawa and Umino sought to determine the position of the plate boundary, using both the earthquake ray tracing method and the new information available on seismic velocity—that the seismic wave velocity of the oceanic plate is 6% greater than that of the continental plate at the same depth. They found that the boundary between the oceanic and the continental plates coincided with the distribution of deep focus microearthquakes and that the earthquake-generating stress followed a reverse fault pattern along the distribution plane. These findings, then, confirmed the existence of a boundary between the two plates and defined some characteristics of the relative movement between them.

An examination of the activities of deep-focus earthquakes is one of the most important tasks of earthquake prediction research since the relative movement of the plate is directly related to the great interplate earthquakes beneath the Pacific Ocean as well as to the changes in crustal stress distribution in inland Japan.

The activity on the upper plane of the deep seismic zone near the earth's surface extends below the sea floor at 143.5° E. This area coincides with the topographic region in which the sea floor increases drastically in depth as it approaches the oceanic trench. As in the case of the distribution of shallow earthquakes off the coast of Sanriku, deep-focus seismic activity seems to be related to the structure of the continental plate. This is a very intriguing subject.

Since seismic activity around oceanic trenches is vitally related to deep-focus earthquakes, detailed research is necessary; the trench areas, however, are a great distance from the observation networks, and consequently no one can be sure of the depth distribution of seismic activity. For this reason, we have no detailed, conclusive research results. Microearthquake activity around the oceanic trenches is shown in Fig. 4.8, in which no earthquake exceeding 100 km in depth is listed. The distribution pattern of microearthquakes does not seem to differ radically from that of comparatively greater earthquakes.

4.3 How Smaller Earthquakes Occur

Great earthquakes occur when intra-crustal stress increases until a critical point is reached and the crust ruptures. To predict such earthquakes it is necessary to observe the various phenomena that reflect changes in crustal conditions and then perform a comprehensive analysis of the data. Research on smaller earthquake activity is an effective tool in monitoring the degree of crustal stress in specific regions. Earthquake prediction, in other words, hinges on knowing just how far along a certain area is in the process that culminates in a great earthquake. Asada (1972) advocated this standard for evaluating the seismic activity in a given area:

(1) Areas with continuing aftershocks: Once a great earthquake occurs, its aftershocks continue for a considerable period of time. As was discussed in connection with Fig. 4.4, Northern Tango and Tottori are two areas that continue to experience aftershocks as a result of the Northern Tango earthquake (1927) and the Tottori earthquake (1943). There are many other areas in which aftershock activities continue.

(2) Areas where aftershocks have ceased: In these areas the aftershocks of great earthquakes, which may well have continued for some time, have completely ceased. The Sagami Sea, for instance, used to be the focal zone of aftershocks of the Kanto earthquake (1923) but at present has very little seismic activity.

(3) Areas of constant earthquake activity: These are areas where comparatively minor earthquakes that are not considered to be aftershocks occur off and on. Examples are the Miyako area and the areas along the volcanic front.

(4) Innately aseismic areas: These are areas where there is little or no seismic activity.

This system of classification was created to facilitate microearthquake research and is based on comparatively short-range data. Thus it should only be used if its meaning is clearly understood. For instance, it is important from the standpoint of long-range earthquake prediction to know that a particular area is a seismic gap—*i.e.*, an area of little seismic activity compared with the adjacent areas. It is not easy, however, to distinguish between areas where aftershocks have ceased and innately aseismic areas, because some earthquakes have such long recurrence cycles that they cannot be determined from historical records. An effort is being made to determine which areas represent true seismic gaps by using the findings of active fault research. Due to its seismic inactivity, the Abukuma Mountain Range was thought to be an innate aseismic area. Recent studies of active faults, however, have led to its reclassification as a Class B active fault area, which means that it should probably be regarded as an area where aftershocks have ceased. The meaning of a seismic gap can differ depending on the kind of time scale one uses for seismic activity. Four types of seismic activity evaluation standard have been introduced. Of these, the relationship between seismic gaps and great earthquakes will be examined further.

As was mentioned earlier, the study of smaller earthquakes can provide information on great earthquake activity since smaller earthquakes are more numerous and since the proportional relationship between log N and M can be largely established for a specific area. This seems to apply to areas of constant earthquake activity—the Matsushiro earthquake, for example. It may seem contradictory that an aseismic zone, or seismic gap, can later host a great earthquake. There is a report, however, that seismic activities increase in an extensive area around a seismic gap which is undergoing an earthquake-causing process. There is no contradiction in these concepts, since it is expected that considerable seismic activity will be triggered by the increase in stress around a seismic gap where stress is accumulating.

In general seismic gaps are reported to be closely related to the great earthquakes that occur along the oceanic trenches. Seismic gaps are also reported to exist in the case of smaller earthquakes. According to Ohtake et al. [1978], seismic activity showed a marked decrease from 3 months to 3 years prior to the four intermediate shallow earthquakes that hit California in 1966, 1968, 1971, and 1973, with a temporary increase taking place during the last half of that period. The extents of the seismic gaps were generally greater than the rupture zones of the earthquakes. The difference in size between the seismic gap and the rupture zone depended on locality, but was greater in the case of smaller earthquakes. The results are shown in Table 4.1.

From this evidence it was concluded that the study of seismic gaps is useful not only for predicting great earthquakes along oceanic trenches, but also for predicting intermediate earthquakes in inland areas such as those in California. In inland Japan, seismic activities have also been reported to be occurring in seismic gaps. One recent report clearly indicates that seismic activity along a latent NE-SW fault near Wakayama has been invading the seismic gap near the fault's center with some regularity. Since April 1977, earthquakes have been occurring near Kumamoto City which have tended to "fill up" the seismic gap. These earthquakes were all around M 5 and seem to confirm that careful observation of such intermediate earthquakes can be an effective earthquake prediction tool.

Successful earthquake prediction has been based on a comprehensive analysis not only of seismic gaps but also of various other anomalous phenomena related to earthquakes. On September 30, 1977, an M 4 earth-

Table 4.1 Examples of Earthquakes Which Occurred in Seismic Gaps

Location	Date	M_s	T (days)	L (km)
Parkfield	28 June 1966	5.6	282	290
Borrego Mountains	9 Apr. 1968	6.4	752	280
San Fernando	9 Feb. 1971	6.5	597	240
Point Mugu	21 Feb. 1973	5.2	97	300

T: Time elapsed between first decrease in seismic activity and the actual earthquake.
L: Extent of anomalous zone.

quake occurred about 5 km north of the Yasutomi Observation Station on the Yamasaki Fault. The Yamasaki Fault Research Group, working out of the Kyoto University Disaster Prevention Research Institute, had been carrying out a comprehensive observation around the Yamasaki Fault, which was designated as a test field. To some extent the Group succeeded in predicting this particular earthquake. An account of their research follows.

Seismic research carried out over a long period of time had shown that earthquake activity around the Yamasaki Fault increased approximately every four years. The most recent earthquake had been in 1973, so heightened activity was predicted for 1977. Figure 4.14 shows the chronological change in seismic activity occurring around the area in which the principal earthquake took place. In May 1976, seismic activity dropped around the principal area but increased with the occurrence of the group of earthquakes at A. It was as if there had been a tendency to approach the position of the principal earthquake by filling the seismic gap from the outside. The concentration of chlorine ions in the underground water in two wells 10 km from the epicenter showed meaningful change two weeks before the

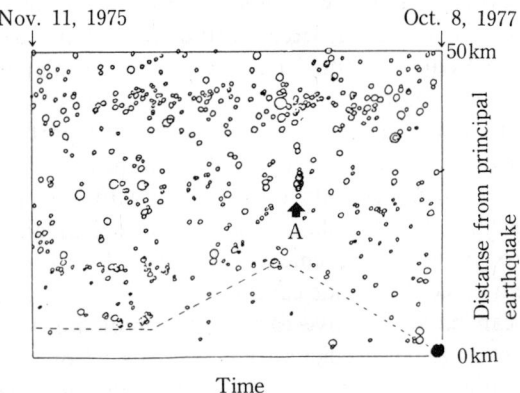

Fig. 4.14 Migration of epicenters around a principal earthquake in May 1976.

principal earthquake occurred. At the same time there were some intriguing anomalies in the radon concentration and the total magnetic force of the area. Then the rain that characteristically triggers earthquakes in the Yamasaki Fault began to fall, and the principal earthquake finally struck.

Thus the seismic gap in the Yamasaki Fault, combined with other phenomena, proved to be a reliable earthquake precursor. The earthquakes predicted in the Yamasaki Fault and the Wakayama and Kumamoto areas are not the results of hindsight but rather are predictions of principal earthquakes based on the discovery of seismic gaps. These results, therefore, merit attention. Recognizing the onset or the existence of anomalous seismic activity, including seismic gaps, is generally considered very difficult. In the past, such activities were usually discovered only after the principal earthquake occurred. Now it is clear that, with careful observation, even small earthquakes can be useful for short-range as well as long-range earthquake prediction.

Small earthquakes tend to occur in swarms within a limited time and space. The foreshocks of these earthquakes may prove useful in long-range earthquake prediction and need to be studied. Foreshocks are earthquakes that occur in the vicinity of the principal earthquake just before it takes place. Few earthquakes are preceded by foreshocks, however—probably just a small percentage of the total. Foreshocks often cannot be perceived, moreover, even by the most sensitive earthquake observation networks, even for earthquakes as large as $M\ 6$.

Figure 4.15 shows the distribution of destructive earthquakes preceded by foreshocks as compiled by Mogi [1963] for areas with high levels of volcanic activity which is thought to coincide with active swarm earthquake activity. Foreshocks, in the narrow sense of the term, originate near the

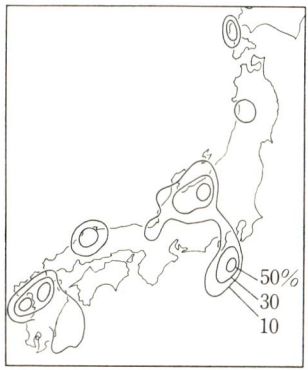

Fig. 4.15 Percentages of earthquakes accompanied by foreshocks.

principal earthquake and are the direct precursors of major destruction. Therefore they play an important role in short-term earthquake prediction. If a sensitive observation network could be established throughout Japan to collect data on the existence (or nonexistence) of foreshocks, the general characteristics of foreshocks would be better understood. It would be possible then to make a distribution map of crustal homogeneity throughout Japan since areas of foreshocks and active swarm earthquakes coincide and since Mogi's study indicates that swarm earthquakes are related to the heterogeneity of the earth's underground structure, or to stress distribution.

A detailed discussion of this matter will appear in Chapter 11. There are recorded incidents of earthquakes that are not considered to be foreshocks but do seem to be directly related to great earthquakes. These earthquakes become abnormally active at the focal zone prior to a great earthquake. Sekiya examined the relationship between T—the number of days from the onset of anomalous activities to the occurrence of a principal earthquake—and M, and arrived at the following equation:

$$\text{Log } T = 0.77M - 1.65.$$

This result may be useful in long-range earthquake prediction providing that a great deal more is known about the selection of the inferred focal zone and the definition of anomalous activities.

Figure 4.16 represents the number of small earthquakes that occur each season as recorded by seismographs stationed near the Kamafusa Dam. Observations at that location began 8 months before the dam was filled with water. On April 30, 1970, as the water rose to approximately 30 meters in depth, a sudden flurry of microearthquakes was recorded. Three months

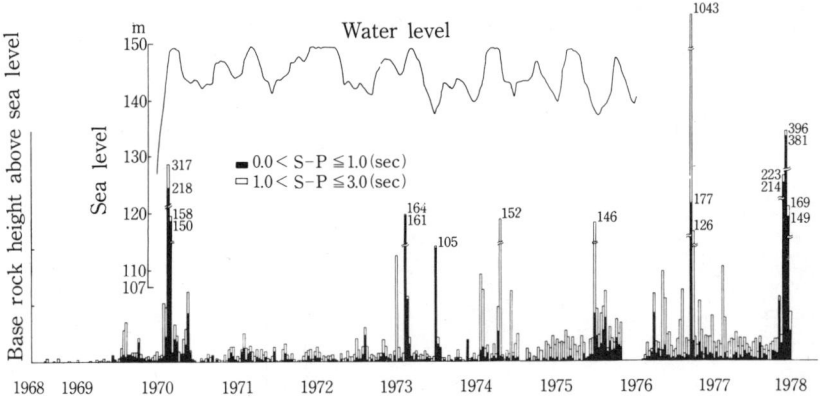

Fig. 4.16 Seasonal frequency distribution of earthquakes in Kamafusa, 1968–1978.

later seismic activity returned to the same level as it had been before the introduction of water. These induced earthquakes may be attributable to a series of events. There was increased water pressure at the bottom of the dam which forced water deeper into the ground. This, in turn, increased the water pressure in porous rocks, which made it easier for the shearing rupture of rocks to occur. Since 1973, microearthquakes have been occurring rather regularly in the area at fourteen-month intervals. If certain stresses cause strain to accumulate up to a certain limit, after which it is then relieved, it may be that the introduction of water into the dam triggered the alteration of the rocks in the surrounding area. This area is known to be prone to swarm microearthquakes induced by great natural earthquakes, but these swarm microearthquakes do not occur immediately after the regular release of strain. There are similar examples of earthquakes induced by other great earthquakes on the western border of the Kitakami block. The earthquakes induced in these areas seem to be intimately related to the concentration of crustal stress.

4.4 Problems in Smaller Earthquake Research

More than 100 small earthquakes (and their foci) are observed each day in the Japanese archipelago and vicinity. A careful examination of this enormous amount of spatially homogeneous data should yield some significant results. As is shown in Fig.11.3 of Chapter 11, with an increase in axial stress, b in the Gutenberg–Richter equation (or m in the Ishimoto–Iida equation) will decrease as the amount of stress drop increases and the high-frequency component within the spectrum of seismic waves is exceeded.

It should be possible to obtain the distribution of concentrated crustal stress by looking at such factors as high-frequency components in the enormous amounts of data already compiled and comparing them with the data available on crustal stress. Research in this area will prove to be just as significant to earthquake prediction as the research on seismic activity and foreshocks.

References

Asada, T., Z. Suzuki and Y. Tomoda, 1950: On energy and frequency of earthquakes (in Japanese), *Zisin* (J. Seismol. Soc. Jpn.), Ser. 2, **3**, 11–15.
Hasegawa, A., N. Umino and A. Takagi, 1978a: Double-planed structure of the deep seismic zone in the northeastern Japan arc, *Tectonophysics*, **47**, 43–58.
Hasegawa, A., N. Umino and A. Takagi, 1978b: Double-planed deep seismic zone and upper-mantle structure in the northeastern Japan arc, *Geophys. J. R. Astr. Soc.*, **54**, 281–296.

Ishimoto, M. and K. Iida, 1939: Observations sur les Seismes Eneregistrés par le microsismographe Construit Dernièrement (in Japanese), *Bull. Earthq. Res. Inst.*, Tokyo Univ.

Mizoue, M., 1971: Crustal structure from travel times of reflected and refracted seismic waves recorded at Wakayama Microearthquake Observatory and its substations, *Bull. Earthq. Res. Inst.*, **49**, 33–62.

Mogi, K., 1963: Some discussions on aftershocks, foreshocks and earthquake swarms— The fracture of a semi-infinite body caused by an inner stress origin and its relation to the earthquake phenomena (3rd Paper), *Bull. Earthq. Res. Inst.*, **41**, 615–658.

Ohtake, M. et al., 1978: Patterns of seismicity preceding earthquake in Central America, Mexico and California, *Proceedings of Methodology for Identifing Seismic Gaps and Soon-to-Break Gaps.* U.S.G.S., 585–610.

Oike, K., 1975: On a list of hypocenters compiled by the Tottori Microearthquake Observatory (in Japanese), *Zisin*, Ser. 2, **28**, 331–346.

Research Group for Explosion Seismology, 1977: Regionality of upper mantle around northeastern Japan as derived from explosion seismic observations and its seismological implications, *Tectonophysics*, **37**, 117–130.

Richter, C. F., 1935: An instrumental earthquake scale. *Bull. Seismol. Soc. Amer.*, **25**, 1–32.

Tsumura, K., 1973: *Microearthquake Activity in the Kanto District*, Publications for the 50th Anniversary of the Great Kanto Earthquake (in Japanese), 1923, 67–87.

Umino, N. and A. Hasegawa, 1975: On the two-layered structure of the deep seismic plane in the northeastern Japan arc (in Japanese), *Zisin*, Ser. 2, **28**, 125–139.

Utsu, T., 1978: An investigation into the discrimination of foreshock sequences from earthquake swarm (in Japanese), *Zisin*, Ser. 2, **31**, 129–135.

Chapter 5 Variations in Seismic Velocity

Toshikatsu Yoshii

5.1 Some Problems in Seismic Wave Variation

Seismologists are still talking about the meeting of the Seismological Society of Japan held at the Disaster Prevention Research Institute of Kyoto University in the fall of 1973. At this meeting nine research papers on variations in the velocity of seismic waves were presented. It is rare to have so many papers on such a specialized subject presented all at once. This phenomenon was triggered by a paper by Scholz *et al.* [1973] who explained variations in seismic velocity by using a dilatancy model.

Japanese research in this area appeared to be lagging behind that of seismologists in the U.S.S.R. and the U.S.A., but in fact pioneering studies on seismic wave velocity by K. Sassa, Masami Hayakawa, and others had already been done in the later 1940s and 1950s. They reported finding variations in seismic velocity before the Tottori earthquake of 1943, the Nankai earthquake of 1946, and the Fukui earthquake of 1948. The data available at that time, however, were not very accurate, and the analysis of the data was beset with errors as to the location of the hypocenter and the other parameters. In some cases, seemingly great variations in seismic wave velocity resulted from an analysis based on an incorrect hypocenter.

Before the appearance of the dilatancy model, variations in seismic velocity before a great earthquake were thought to be purely the result of variations in stress. In detecting the subtle velocity variations, the use of explosion seismology, in which the position and time of the event can be accurately determined, is clearly preferable to depending solely on natural earthquakes. Experiments in explosion seismology have been conducted in Izu Oshima and the Southern Izu Peninsula since 1968 under the auspices of the Geological Survey of Japan.

In the early 1960s Russian scientists reported pronounced changes in V_P/V_S (the velocity ratio between P waves and S waves) in the Garm district of the Republic of Tadzhik. Probably because these findings were published in Russian, seismologists in other parts of the world were not aware of them until Savarensky [1968] introduced them in English. Other papers by Nersesov, Semenov, and others followed in 1967.

American scientists were stimulated by the Russian research. Their work culminated in the discovery, by Aggarwal et al. [1973], of V_P/V_S variation and in the paper by Scholz et al. mentioned earlier. A number of papers on this subject have been published in Japan and elsewhere since 1973. Interested readers will find bibliographies in Rikitake [1976], Iizuka [1977], and Ohtake and Katsumata [1977].

At first, however, Japanese seismologists were suspicious of such a pronounced variation in V_P/V_S. At the 1973 meeting of the Seismological Society, researchers from the Japan Meteorological Agency (JMA) were very cautious. Since these were the people who supplied the others with vast amounts of data on seismic velocity variation, they were presumed to be in the best position to assess the accuracy of those data. Thus their opinion seemed much too authoritative to refute, and, under these circumstances, research on seismic velocity changes was not very popular. Rumors that the variation was minimal when the data were accurate, or that American researchers had found the discoveries in the Garm district of questionable value, circulated freely. Although I do not intend to question the credibility of all the findings made at that time, it is certainly true that the accuracy of some of the data left much to be desired.

This chapter is the result of my thoughts on this subject—what to watch for when reading papers on velocity variation, and what directions research in this area should take. It is not the usual exposition or outline of research. The most crucial element in research on velocity variation is the accumulation of accurate and dependable observational data; therefore, it would be dangerous to get bogged down in the mechanism of velocity variation at this point. It is commonly accepted that the observation of velocity variation is important in intermediate to long-term earthquake prediction—so important, in fact, that it merits a very careful approach.

5.2 Causes of Seismic Wave Velocity Variations

Before the dilatancy model was introduced, variations in seismic velocity were attributed directly to changes in underground stress. In laboratory experiments in which a rock is subjected to pressure while its elastic wave velocity is being measured, the velocity increases as the pressure increases, and gradually approaches a constant value. This is thought to result from the closing of minute pores within the rock. Placed under pressure equal to that several km below the earth's surface, the P wave velocity of such rocks as granite shows a change of 10^{-4} km/sec with a pressure change of one bar (approximately 1.02 kg/cm^2). Even if a great stress change of 100 bars occurs underground prior to an earthquake, the resulting variation in velocity will be only 0.01 km/sec, or approximately 0.2%. Within 100 km of an anomalous area, the travel-time anomaly will only be 0.03 sec or thereabouts. Obviously it would be impossible to detect such minute changes by analyzing natural earthquakes.

The dilatancy model was also based on the results of high-pressure laboratory experiments. As pressure on the rocks increases, cracks begin to appear once the stress is more than halfway to the breaking point of the rocks and the volume begins to increase. This phenomenon is called dilatancy, and scientists have been familiar with it since the 1960s. The velocities of P and S waves decrease in the presence of dilatancy; the effect of the phenomenon varies, however, depending on whether the rocks contain water or not.

The dilatancy model usually assumes that water flows into the cracks after dilatancy takes place: its full name is the dilatancy diffusion model or the wet model. A detailed explanation of the model can be found in papers by Nur [1972] and Scholz et al. [1973]. A simplified explanation of the model follows.

Dilatancy begins when pre-earthquake stress in water-saturated rocks builds up, causing desaturation and a drastic drop in the V_P. The V_S, on the other hand, is relatively unaffected by whether the rocks contain water or not, and thus does not decrease much. Thus the V_P/V_S ratio decreases. When the cracks created by dilatancy are refilled and the saturation point reached, the V_P/V_S ratio returns to normal. If the stress is further increased, rupture will finally occur, and an earthquake takes place. The dilatancy model has become extremely popular since it explains why the V_P/V_S ratio decreases before an earthquake and then returns to normal just before the earthquake strikes. Since the observed decrease in V_P or in the V_P/V_S ratio is usually more than 10%, only a special model such as this one can explain it.

There is a dry model as well as a wet model. In the former the diffusion of water is not taken into consideration. It is identical to the wet model insofar as the decrease in V_P and V_P/V_S after dilatancy takes place, but it differs in its view of the normalizing process. According to the dry model, as the rupture (or earthquake) approaches, small cracks concentrate near the fault plane where the rupture is about to occur. As a result, stress in other areas decreases. When the cracks in these areas close again, the V_P and V_P/V_S ratio return to normal. The dry model differs from the wet model in that it assumes that rocks which have expanded will contract before the earthquake. Detailed observation of crustal movement and V_S variation may provide the information needed to choose between these two models.

Dilatancy does not occur when the original confining pressure on the rocks is rather high. These models can only be applied, then, to earthquakes that occur in the shallow part of the earth's crust. Hayakawa and Iizuka [1976] suggested a new model that takes heat into consideration. They maintained that a great velocity variation was observed before an earthquake that occurred 30 to 40 km deep on the Pacific side of northeastern Japan. Their model failed to be convincing, however, due to some questionable points in the observational data on which the model was based.

There will undoubtedly be more new models in the future, but it is safe to say that finding an explanation for a reduction of 10% in the V_P is not an easy task. The fact that a seismic wave velocity of 6.0 km/sec can go down to 5.4 km/sec or one of 8.0 km/sec to 7.2 km/sec is a drastic event which can be compared with a great earthquake.

5.3 Various Methods for Detecting Velocity Variations

5.3.1 The V_P/V_S Ratio

The Russian and American research that insipired the Japanese work on variations in seismic velocity relied on the ratio of the velocity of P waves to S waves, as does most Japanese research. The "Wadati diagram" is one of the most common methods used; it plots the arrival time of the P wave as the x-axis and the time difference between P and S waves as the y-axis. Another common method compares the arrival times of P and S waves at two different observation sites. The following tests may be helpful in evaluating papers on V_P/V_S variation which are based on the above two methods:

(1) Has the accuracy of the data been fully tested? The weakest point of the V_P/V_S method is that the accuracy of the S wave reading can play a crucial role in the dependability of the result.

(2) Are the observation sites appropriately located? The V_P/V_S ratio obtained by these methods is that beneath the observation network. Thus one must be especially careful if the earthquake is outside such a network.

(3) Have other factors that may have a bearing on the V_P/V_S ratio been taken into consideration? Variations in the depth of the hypocenter, for instance, can result in apparent changes in the V_P/V_S ratio.

(4) Has the underground structure been considered? It is extremely difficult to postulate a seismic ray path through a three-dimensional underground structure as complex as the Japanese Islands, for example. It is also possible for P waves and S waves to travel totally different paths. These considerations are related to item 3.

(5) Has an unbelievable V_P/V_S ratio, such as 1 or less, been included?

If the V_P/V_S variation has been "amplified" for any reason, an explanation is necessary.

These criteria may be a little too strict; the research papers that can pass all these tests are very few. In the following section I would like to examine the two major methods for obtaining the V_P/V_S ratio more closely.

5.3.2 Using the Wadati Diagram

The primary examples of research using this method are the Russian research in the Garm district and that carried out by Aggarwal and others near Blue Mountain Lake in New York State. The "Wadati diagram" plots the arrival time of the P wave as the x-axis and the time difference between P and S waves as the y-axis, using information from several different observation sites. It was first used in an attempt to pinpoint the time at which an earthquake occurred—in other words, the origin time. Generally the plotted data form a more or less straight line. Assuming that the V_P/V_S ratio is constant at any part of the seismic wave propagation path, the ratio can be found by adding 1 to the inclination of this straight line. Although the assumption here is very important, it often tends to be ignored.

The accuracy of the S-wave reading is the most important factor in all the V_P/V_S ratio methods. The S wave arrives after the P waves, while the tremors created by the P wave are still present. In addition, its period is comparatively long, and thus the onset of the S wave is harder to read accurately. In the case of shallow earthquakes, it is sometimes difficult to detect the S wave at all. For these reasons, S wave readings may vary from one individual to another. It is also possible that the V_P/V_S ratio will vary according to who reads the observatory records. The S wave chosen may seem to be legitimate but in fact may not have been the first S wave to arrive. Such factors can all result in incorrect V_P/V_S ratio readings. There is evidence that the data collected by the JMA include some inaccurate S wave readings. Ohtake and Katsumata [1977] reported one such case. Whatever the method, the precision of the clock and the speed with which the recording paper is fed also need to be considered. Almost all of the observatories registered with the JMA or with the ISC (International Seismological Centre) feed their recording paper at a speed of one to two mm/sec, and only the minute marks are recorded on the paper. Obviously, if there is an unevenness in the feeding speed, the precision of the timing will suffer. Even with P waves, which usually have a clear beginning, one can only expect the reading to be accurate to within a tenth of a second. The accuracy of S waves is likely to be much poorer than that.

The observations of Aggarwal and others [1973], made near Blue Mountain Lake, are considered to be the most reliable research on V_P/V_S variations. The recording paper they used was fed at the rate of 2 mm/

94 LONG-RANGE PRECURSORS

sec. Despite the use of a high degree magnifier, their claim of 2/100 sec accuracy is questionable—especially where the S wave is concerned. Since the phenomena at Blue Mountain Lake are small in terms of time and space, highly accurate observation is essential.

The relationship between the earthquake and the position of the observation sites is also important. The V_P/V_S ratio obtained from observations is usually that under the observation sites (Fig. 5.1 (a), (b)). Let us assume that a V_P/V_S variation of 10% is observed in Hokkaido and Tohoku at the time of an earthquake off the Sanriku coast. This would mean that the V_P/V_S ratio changed 10% over a vast area on land. Considering the distance between the earthquake and the observation sites, this change must have occurred primarily within the upper mantle. If such a drastic change were to actually take place before a great earthquake occurred off the Sanriku coast, then earthquake prediction would be simple indeed.

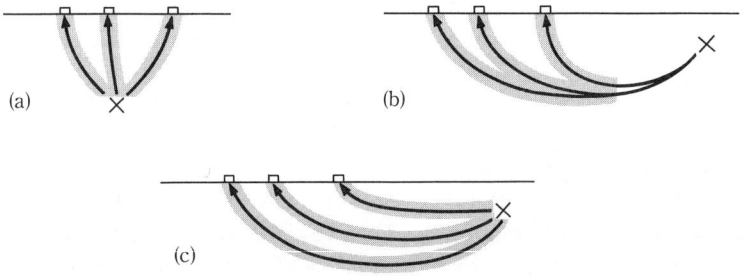

Fig. 5.1 Schematic diagram of seismic wave paths.
It is possible to detect anomalies if they occur in the shaded areas. In the case of (c), anomalies outside the observation network can also be detected.

Basically, while the V_P/V_S method obtains the V_P/V_S ratio under the observation network, there are cases that are much more complex. Figure 5.1 (c) shows one example. When the underground structure is special, the first arrivals have sometimes traveled by different paths, according to their distances from the observation sites. The wave to the closer observation site travels a shallower path, while that to the farther site travels a deeper path. In this case, anomalies that are not directly below the observation site may be detected, but the evaluation of such anomalies is extremely complex.

The accuracy of the researcher's picture of the area's underground structure also plays an important role. A region as complex as Japan is almost like a linked series of small distinct units, each with a different V_P/V_S ratio. This means that P waves and S waves may not travel the same path from the hypocenter to the observation sites. The condition of these paths will vary in subtle ways depending on the relative locations of the hypocenter and the observation sites. The apparent V_P/V_S ratio will change, therefore, depending on the location of the hypocenter.

If there is even one piece of questionable data mixed in with the rest of the information necessary to a Wadati diagram analysis, a change of 10% or so in V_P/V_S ratio can easily occur. Compared to the "two-station" method described below, however, the damage is minimal. It is wise, as was demonstrated by Ohtake and Katsumata [1977], to draw a travel time curve in order to insure dependable results, whether one is using the Wadati diagram method or the two-station method.

5.3.3 Obtaining the V_P/V_S Ratio Using Two Observation Sites

The results of the two-station method correspond to those obtained by the Wadati diagram method. Because of its simplicity, the two-station method is often used by Japanese scientists in their analyses. One simply obtains the ratio, $\delta T_S/\delta T_P$, by measuring the time difference between the arrival of the P wave (δT_P) and that of the S wave (δT_S) at two different observation sites. The objective of this method is identical to that of the Wadati diagram method. It attempts to obtain the apparent velocity of the P waves and S waves that pass two observation sites. The weakness of the method is the fact that it depends on only four pieces of data, two of which—the values of the S waves—are difficult to read. If any one piece of data is questionable, the effect on the calculation is obvious.

There is another potential trouble spot here too: the analyzer sometimes calculates the V_P/V_S ratio from the data mechanically, without taking the relevant travel time curve into account. This is obviously the case in the abnormally small V_P/V_S ratios that are often seen. When the V_P/V_S ratio is unexpectedly small, it is quite likely that one of the S wave readings was inaccurate (see Ohtake and Katsumata [1976]). If one part of the V_P/V_S data is questionable, then the other seemingly accurate data may also be unreliable. No technique such as averaging will help if this is the case.

The recent skepticism regarding seismic wave velocity variation is based largely on doubts about the V_P/V_S ratio—particularly about the accuracy of the S wave readings. A careful, fair analysis is necessary if this simple method is to work.

5.3.4 Variations in P Wave Velocity

The most critical issue in the direct detection of V_P or V_S variation is how to avoid the influence of the origin time and the hypocenter. These are estimated values, remember, based on observation. An error in hypocenter determination may result in an apparently great velocity variation.

On the other hand, a real anomaly over an extensive area may not be recognized since the determination of the hypocenter often tends to be skewed a little to one side, thus reducing the degree of anomaly.

At present two methods are usually used to obtain the V_P variation: in one the apparent velocity of the P wave as it passes between observation sites is examined, and in the other the travel time residual at a certain observation point is examined.

5.3.5 Method Using Apparent Velocity

Obtaining the apparent velocity of P waves between two observation sites is a reliable method only when the hypocenters are always located in a small area. When the positions of the hypocenters differ greatly, the result can be quite complicated to obtain.

Distance, for example, affects the apparent velocity. Since the underground structure is usually such that the velocity increases with depth, the travel time curve forms a convex shape. Thus, an earthquake that occurs farther away shows a greater apparent velocity as it travels through the observation sites than one that occurs nearer to the sites. This is because the farther away an earthquake is, the deeper its propagation path will be.

The influence of the depth of the earthquake cannot be ignored either. A deeper earthquake travels at a steep angle towards the observation site, and thus the apparent velocity will be much greater than that of a shallow earthquake. When the dimensions of the underground structure are extremely complex, the paths of the seismic waves will vary greatly with any change in the hypocenter, and the apparent velocity will then be affected.

In short, this method cannot be used unless many earthquakes are occurring in an area that is limited in size and depth. The meaning of "limited" can vary according to the size of the observation network or the complexity of the underground structure, but the important thing is to use earthquakes with similar propagation paths.

Another problem is the possibility that the onsets of P waves will be overlooked, as those of S waves often are. If the seismographs are not very sensitive, as in the case of the JMA observatory, it is not always clear that the real onset has been detected, especially if the hypocenter is far away. If such data are used in calculations, the apparent value of V_P can become very small. Seismicity around the region of a great earthquake usually decreases before its occurrence, and this can apparently lower the seismic wave velocity also.

5.3.6 The Travel-Time Residual Method

When determining a hypocenter from the arrival time data of the P wave at various observation sites, it is common to fit the data to a theoretical standard travel-time curve using the least-squares method. The difference between the arrival time actually observed and the theoretical value is called the travel-time residual. If a velocity anomaly were found under one observation site, it would influence the determination of the hypocenter. If there are a number of other normal observation sites, however, its influence is not going to be very great. The travel-time residual at an observation site where an anomaly is discovered will be quite different from the usual value at that site. In cases where the underground structure is complex, the travel-time residual may be different for each observation site, depending on the depth and position of the earthquake. In such cases it is desirable to examine the travel-time residuals of earthquakes occurring in a limited area.

The travel-time residual method has many variations. Utsu [1975] divided the central part of Japan into a grid of 0.2° squares and demonstrated a method for observing the travel-time residuals of waves as they pass through the grid. In order to actually use seismic wave velocity variation for earthquake prediction, the routine observation of V_P variation must be carried out over a wide area.

5.3.7 The Explosion Seismology Method

Artificial explosions are the best way of detecting an anomaly in P wave velocity because the position and time of the explosion are known accurately. It is sometimes said that velocity variation does not show up as readily in explosion seismology, but I regard this as a superstition. If the V_P does, indeed, decrease by 10% before a great earthquake, then the travel time must have slowed by at least 0.1 sec or more while passing through only 10 km of anomalous area. A modern-day explosion seismologist who missed an anomaly of this amount would be guilty of grave mismanagement. In the annual explosion experiments in the southern Kanto region conducted by the Geological Survey of Japan, attention is being paid to detecting anomalies as small as 0.01 sec. Such efforts are very important if anomalies, no matter how small, are to be distinguished from the mechanism that causes velocity variation. Maintaining precision of 0.01 sec is not easy, however, as we shall see.

A knowledge of underground structure is also important to the explosion seismic method. Without it, the focus of the analysis may well be confused. Several structures obtained in southern Kanto suggest that most of the seismic waves from Izu Oshima Island to the observation sites pass through the upper crust with a P wave velocity of 5.5 to 6.0 km/sec., and this seems a good area on which to focus an analysis.

Quarry blasting, in addition to specially planned explosions like those in Oshima, can also be used for analyses. The accuracy of data on an explosion's time and location is not as great, however, since the exact location of the blast and the time may not be accurately recorded. Experiments of this kind are being conducted in California by Kanamori and others, and, although there was some evidence of carelessness at the beginning of this project, their work is now heading in the right direction, with significant aspects, such as the underground structure, being taken into consideration [Kanamori and Fuis, 1976].

The effectiveness of explosion seismology is limited by the amount of energy that can be used to produce the explosion. Even if 500 kg of dynamite are used, precise observation is difficult from more than 100 km away. Since big explosions on land are increasingly undesirable, especially in a crowded country like Japan, the success of such experiments hinges on whether observation sites can be found where the noise level is minimal.

No matter how good the conditions are, however, the onset of the seismic wave is usually gentle and therefore not easy to read with an accuracy of 0.01 sec. Generally, seismologists want to read the onset as early as possible, particularly as the magnification is increased. If noise is present, it is more difficult to determine the actual onset of the wave. This noise can be quite troublesome, especially since it varies from experiment to experiment.

In order to avoid the problems of determining P wave onset, the Oshima researchers read the arrival times of peaks and troughs of the wave train around the first arrival and compute the average difference between this reading and that of past experiments. If this method of analysis is to be successful, there are several things that must be watched.

First of all, the transducer has to be placed in the same position at all times. Even a distance of several meters can change the profile of a wave pattern. Second, the same observation instruments should be used each time. Changes in amplifiers and recording machines, not to mention transducers, can shift the positions of the peaks and troughs of the wave pattern due to differences in their phase responses. If these instruments are not properly maintained, they can register different responses from one year to another.

The condition of the explosion site also must be constant. In explosion seismology, explosives are buried in a hole dug into the ground. In most cases, once the explosion takes place, the hole cannot be used again. Thus it becomes necessary to move the explosion site for any subsequent experiments. When this happens, a difference in the crustal structure near the earth's surface or in the depth at which the explosives are buried can change the wave pattern considerably. Furthermore, since explosives do not explode instantly—the shock propagates at a velocity of about 6 km/sec—even the position of the detonator can be crucial. In any case, matching the wave patterns is fraught with other difficulties, quite apart from that of reading the onset of seismic waves.

Moving the explosion site can sometimes cause a difficult problem of how to correct the shot time. Given 6 km/sec as the apparent velocity of the first arrival, a maximum correction of 0.005 sec becomes necessary once the explosion point is moved horizontally as little as 30 m. A vertical move has even greater consequences. Since the seismic velocity near the surface is very small, the same degree of correction becomes necessary for vertical moves of only 5 to 10 m.

In addition to the above factors, an error of 0.001 to 0.01 sec can be attributed to the accuracy of the clock and to the feeding speed of the data recorders and other recording machines. With all these factors affecting the result, the difficulty of maintaining an accuracy of 0.01 sec in explosion experiments becomes obvious.

Still, V_P variation observations using explosion seismology are clearly more accurate than those that rely on natural earthquakes. Future research in seismic wave velocity variation will probably center on such observations. If, in fact, a large V_P variation, such as the one that is expected from the dilatancy model, occurs in southern Kanto, the Oshima experiments or the Yumenoshima explosion experiment currently being conducted by Tokyo Metropolis will certainly detect it.

5.4 Future Problems

After this lengthy discussion of the difficulties inherent in analyzing seismic velocity variation, the reader may have been left with too great a distrust of such analyses. It is widely recognized, however, that an understanding of this variation is extremely critical to successful earthquake prediction. The timing of the decrease in the V_P and in the V_P/V_S ratio is thought to be related to the magnitude of the earthquake. The velocity appears to return to normal immediately before the earthquake. If seismic velocity can be monitored at appropriate time intervals, it may be possible to predict the position and size of an earthquake before it actually happens. A warning could be issued at the time the velocity returns to normal.

There are still many problems to be solved before velocity variation analysis can serve as an earthquake predictor. Let us examine some of them.

First of all, the reading of the arrival time of the seismic wave must be as accurate as possible. The feeding speed of the recording paper should be at least 10 mm/sec, although much depends on the degree of the anomaly one wishes to observe. Ideally, with this speed, a read-out accuracy of 1/100 sec should be achieved. If the onset of the seismic wave is dull to appear, however, there is nothing to be done. Since not many great earthquakes begin with a decisive initial motion, the sensitivity of the seismograph should be greatly improved. The clock must also be accurate. These conditions appear to have been met in the observations being done on microearthquakes by universities and other institutions in Japan.

If data on S waves were to be used, the accuracy of the readout would suffer even if all the above conditions had been met. Reliable results cannot be obtained if one relies on the S wave data as reported by each of the observatories. The analyzer must examine the original record itself in order to catch any questionable data that may have been included. Only then is it possible to evaluate the results objectively.

Second, it is most important to determine the objective of the observation more precisely. In other words, the path of the seismic wave must be recognized accurately. For this, a knowledge of the underground structure is indispensable. Even if the wave travels through an area on a flat plane, the depth of the path should be accurately determined by considering both the underground structure and the relationship between the hypocenter and the observation sites.

In a country like Japan, where the underground structure is as complex and varied as a mosaic, the shifting of a hypocenter can result in an apparent velocity variation. To avoid this it is desirable to focus the attention of an observation network on hypocenters that are concentrated in a limited area. The size of the area can vary depending on the anomalous zone to be studied and on the size of the observation network. The major consideration here is that the paths of the seismic wave always be similar.

The methods of analysis should also be reexamined. Given the fact that abnormal velocity changes have never been reported prior to any earthquake in Japan, it is obvious that keeping a daily watch over seismic velocities is extremely difficult using present-day methods. Routine surveillance of Japan in its entirety would clearly be an enormous task. Ideally, velocity would be watched by a method similar to that used for determining hypocenters, but, at this point, it is hard to imagine a practical way of doing it. The analysis of travel-time residuals may hold the key somehow. Utsu's analysis [1975] is heading in this direction. In any case there is no doubt that it will be quite an involved operation.

Methods of analysis based on totally different concepts may also propel velocity variation research way ahead. Bungum *et al.* [1977] succeeded in accurately determining seismic velocity by using the cyclic vibrations created by a power station generator. Reportedly, they could even detect velocity variations of 1% or less that were seemingly attributable to the earth tide. This method cannot be used everywhere, since power stations are far from abundant, but the possibility of using cyclic vibrations deserves attention. K. Aki and others at MIT are also examining this idea. Theoretically, a great many vibrations can be superimposed if the source of the cyclic vibration is known; thus it becomes possible to achieve precise velocity determination. It should be noted, however, that the most pronounced waves originating in this way are either surface waves or S waves. Surface waves merely provide information on S waves that travel near the earth's surface. There are additional complications in that surface waves change velocity with the period—a characteristic known as "dispersion."

In sum, the agenda for the future will be to increase the precision of data, to pinpoint scientists' objectives with the help of more information on areas' underground structures, and to achieve the routine surveillance of velocity variation. In my view, the widespread method of using V_P/V_S analyses based on read-out reports of P and S waves is not useful in Japan. If such data must be used then only the data on the P wave, which are more accurate, should be included. When the position of the earthquake relative to the observation site is properly chosen, higher accuracy in detecting velocity variation can be obtained by using the P wave alone than by the V_P/V_S method.

It would be helpful if explosion seismology were practiced throughout Japan in order to detect velocity variations. Although it is expensive in terms of both money and manpower, this method can produce data that are truly valuable. Its greatest shortcoming, as was mentioned, is the fact that its explosive energy is not as great as that of a natural earthquake. Experiments in urban areas such as Tokyo are difficult because of the presence of artificial noise. The time may have come for public administrators to consider controlling the noise of traffic, factories, etc., for a few hours a year in order to carry out such experiments.

5.5 Velocity Variation Research at the Starting Gate

There are a large number of books on earthquake prediction, many of which speak about the analysis of seismic velocity variation as if it were one of the established earthquake prediction methods. After examining many research papers, I don't think it is so simple. Velocity variation does seem to exist in medium-sized shallow earthquakes in inland Japan, and its detection does seem to be possible. And yet, one must be extremely lucky to find an anomaly before an earthquake since the anomalous area will be very limited.

Truly great earthquakes immediately to the landward side of the Japan Trench are important subjects of prediction-oriented study since they are potentially so damaging. Velocity anomalies cannot be detected easily, however, if the observation sites are located on land. For earthquakes 30 or 40 km deep, we do not even know whether there is great velocity variation.

While I was writing this chapter, an earthquake of magnitude 7.0 occurred in the sea near Izu Oshima Island on January 14, 1978. A hurried examination of the various research papers reporting precursory phenomena turned up no information on seismic wave velocity, although the explosion seismology experiments in the southern Kanto region completely covered the focal zone of this earthquake. This calls into question whether there is really a 10% decrease in seismic wave velocity—a question that should be studied immediately with repeated explosion seismology experiments.

Research on seismic velocity variation is still hovering at the starting gate—if it is not on the wrong track altogether. I look forward to the time when velocity variation, along with numerous other earthquake prediction methods, will come to be regarded as highly dependable. Earthquake prediction is an enormous problem that this generation alone cannot solve. It should be remembered, however, that dependable data are indispensable to the work of every generation.

References

Aggarwal, Y. P., L. R. Sykes, J. Armbruster and M. C. Sbar, 1973: Premonitory changes in seismic velocities and prediction of earthquakes, *Nature*, **241**, 101–104.

Bungum, H., T. Risbo and E. Hjortenberg, 1977: Precise continuous monitoring of seismic velocity variations and their possible connection to solid earth tides, *J. Geophys. Res.*, **82**, 5365–5373.

Hayakawa, M. and S. Iizuka, 1976: A mechanism to explain the changes in V_P/V_S (in Japanese), *Zisin*, Ser. 2, **29**, 339–353.

Iizuka, S., 1977: A review on the studies of seismic velocity changes as an earthquake precursor (in Japanese), *J. Fac. Marine Sci. Tech. Tokai Univ.*, No. 10, 213–242.

Kanamori, H. and G. Fuis, 1976: Variation of P-wave velocity before and after the Galway Lake Earthquake ($M_L = 5.2$) and the Goat Mountain Earthquakes ($M_L = 4.7$, 4.7), 1975, in the Mojava Desert, California, *Bull. Seismol. Soc. Amer.*, **66**, 2017–2037.

Nur, A., 1972: Dilatancy, pore fluids, and premonitory variations of t_S/t_P travel times, *Bull. Seismol. Soc. Amer.*, **62**, 1217–1222.

Ohtake, M. and M. Katsumata, 1977: Detection of premonitory change in seismic velocity, *Proceedings for Symposium on Earthquake Prediction Researches* (in Japanese), pp. 106–115.

Rikitake, T., 1976: *Earthquake Prediction*, Elsevier, Amsterdam, 295pp.

Savarensky, E. F., 1968: On the prediction of earthquakes, *Tectonophysics*, **6**, 17–27.

Scholz, C. H., L. R. Sykes and Y. P. Aggarwal, 1973: Earthquake prediction; A physical basis, *Science*, **181**, 803–810.

Utsu, T., 1975: Detection of a domain of decreased P-velocity prior to an earthquake (in Japanese), *Zisin*, Ser. 2, **28**, 435–448.

Chapter 6 Survey Repetition

Hiroshi Sato

6.1 Surveys and Earthquake Prediction

H. F. Reid, of the U.S.A., who developed the elastic rebound theory of earthquakes, made what was probably the world's first attempt at earthquake prediction using surveys of the earth's movement. He analyzed the San Francisco earthquake of 1906 (M 8.3) using the results of triangulation surveys made between 1851 and 1855 and surveys made immediately after the earthquake: these studies resulted in the elastic rebound theory. This theory maintains that elastic strain accumulates in the crust due to the relative movement of a fault, such as the San Andreas Fault. When the strain reaches a critical point, the fault rebounds, producing seismic waves. Reid estimated the San Francisco cycle to be (6.5m/3.2m) × 50 ≑ 100 years, noting that the relative displacement of the fault during the 50-year period before the earthquake was approximately 3.2m and that the greatest amount of slip during the earthquake was 6.5m*. Although Reid's long-term earthquake prediction method was somewhat naive, it was basically the same as the method as that used now, which also depends on observations of crustal strain to predict earthquakes.

In Japan, A. Imamura attempted to use surveys to predict earthquakes before World War II. Imamura thought that earthquakes were caused by the tilting of land mass blocks. He asked the Land Survey Department to conduct leveling surveys throughout Japan. Many of the level-route surveys that were repeated by the Land Survey Department before World War II were done at his request. The most significant of these was the survey made near Omaezaki in Shizuoka Prefecture immediately before the Tonankai earthquake of 1944. The land movement near Omaezaki associated with the Tonankai earthquake was discovered as a result of this survey, and is the foundation on which today's Tokai earthquake theory

---

* A recent excavation survey of the San Andreas Fault found that the average interval between great earthquakes in this area is between 150 and 200 years.

is based. It was also during this survey that an extraordinary tilting—which may be a short-term earthquake precursor—was observed. Thus repeated surveys have long been considered a very effective means of earthquake prediction in Japan.

Surveys detect land movement by comparing the results of past surveys with the results of present surveys. Where crustal movement surveys are concerned, there is no present unless there is past. Japan has been conducting such surveys nationwide for 80 years. This places Japan in an advantageous position vis-à-vis the rest of the world. This is precisely why the Blueprint of the Japanese Earthquake Prediction Project is committed to the importance of repeated surveys. The project is based on the assumption that a sure grasp of crustal movement and a knowledge of accumulated crustal strain, based on repeated surveys, is the most significant factor in successful earthquake prediction.

6.2 Surveys and Observation of Crustal Movement

In making an accurate topographical map, it is necessary to establish a base station, the position (both latitude and longitude) and the height of which have been accurately surveyed. A nationwide map such as the 1/50,000 topographical map requires the establishment of a network of base stations throughout the country. These stations must be based on accurate survey information that takes the ellipsoidal shape of the earth into consideration. The survey required to establish this network is called a geodetic survey and includes astronomical and gravity surveys.* The main survey methods used are triangulation (to determine the horizontal position of a point) and leveling (to determine the point's height above sea level). The position and height thus obtained are actually marked on the land as triangulation points and bench marks. Repeated surveys can then detect changes in height and position, and the earth's movement can be recorded.

Land movement can also be detected by the topographical methods mentioned in Chapter 3, or by using continuous observation instruments such as strain gauges and tiltmeters. The survey method is unique, however, in that movement over an extensive area can be charted. Table 6.1 summarizes the characteristics of the three methods of land movement detection.

Survey operations, which are the foundation for today's research on

---

* Gravity surveys are necessary in order to delineate the shape of the earth. Since gravity changes according to height, height variations can be found using the precise measurement of gravity variations. A gravity meter that can detect height variations of 2 to 3 cm is now being developed.

Table 6.1 Comparison of Crustal Movement Detection Methods

Methods	Continuous measurement	Geodetic survey	Geomorphological method
Order of time	Minutes, hours	1 ~ 10 years	1,000 years
Subject of research	Deformation of crustal rocks	Surface deformation	Deformation of topography
Method of detection	Extension meter, tiltmeter	Triangulation, leveling	Photogrametry, topogrametry
Unit of measured values	1 ~ 10 μm	mm ~ cm	1 ~ 10 m
Range of measurement	1 ~ 100 m	1 ~ 10 km	10 ~ 100 km

* 1μ (micron) = 1/1,000 mm

Table 6.2 Organization of Japan's Geodetic Triangulation Network

Classifications	First-degree triangulation points		Second-degree triangulation points	Third-degree triangulation points
	Principal points	Supplementary points		
Total number of points	390	559	5,046	32,662
Density of network	One point in every 1,600 km²	One point in every 800 km²	For those above third degree, one point in every 8 km²	
Average side-length	45 km	25 km	8 km	4 km
Error of closure	1 sec	2 sec	5 sec	10 sec

crustal movement, were begun in Japan during the early Meiji Era (1868–1911) by the Survey Department of the Ministry of Engineering, the Geology Department of the Ministry of Internal Affairs, and the Army. In 1884 these survey operations were combined under the General Headquarters of the Army, and the Land Survey Department was established to undertake a nationwide survey. The German geodetic survey method was chosen since it was highly regarded at that time and famous scholars, such as Gauss, had participated in its development.

The nationwide network of geodetic base stations was almost completed by the Army during the Meiji Era. Table 6.2 represents the triangulation network in Japan. To minimize the influence of errors, a high-precision survey was made throughout the country and a huge first-degree triangulation network established. This network was then filled in with smaller second and third-degree triangulation networks. The total number of triangulation points is approximately 39,000, or one point for every 8 km² throughout the country. The leveling network, on the other hand, consists of 10,000 first-degree bench marks buried at 2-km intervals along the nation's main highways. The height of these bench marks is based on the mean sea level. Tidal observation stations are positioned around the country to measure changes in sea level.

The Land Survey Department, which conducted the prewar survey operation in Japan, was run by the Army. It had a good understanding of the academic aspects of land surveys, however, and gave its full cooperation to the investigations of geology and seismology. It has made especially great contributions to Japanese seismology by re-surveying the triangulation and leveling networks in an earthquake area following each great earthquake, thus further elucidating the earth's movement.

The earth movements caused by great earthquakes have been recorded since ancient time. These records were based on visual observations, however, and their extent and accuracy were limited. Really accurate records of earthquake land movement only became available after the Land Survey Department began its work in the Meiji Era.

The first great earthquake that occurred after that time was the Nobi earthquake (M 8.4) in 1891. The geodetic survey network in Japan was not yet complete at that time, however, and only part of the focal zone had been surveyed. Therefore the re-survey could be only partially carried out after the earthquake.

The great Kanto earthquake of 1923 was significant in that it marked a new era both in seismological research and in the history of land survey operations. After the earthquake the Land Survey Department re-surveyed the triangulation and leveling networks from the Boso and Miura Peninsulas to the Southern Kanto district. The results were published as the *Kanto Earthquake Revision Survey Records*. A great many researchers have used these *Records* extensively since, to study the great Kanto earthquake. The Land Survey Department continued its re-survey operations in areas hit by such earthquakes as the Northern Tango earthquake (1927) and the Northern Izu earthquake (1930). These surveys are called revision surveys. Table 6.3 lists the major earthquakes in Japan for which such surveys were performed.

In August 1945, after Japan's defeat in World War II, the Army was disbanded and the General Headquarters closed. The survey operations of the Land Survey Department, however, were indispensable to the nation's well-being; thus the Geographical Survey Institute was founded on September 1 of that year, and it took over the operations of the Land Survey Department.

Due to postwar social conditions in Japan, survey operations were limited to the revision and publication of topographic maps. Actual surveys were minimal. The Nankai earthquake of 1946 (M 8.1), however,

Table 6.3 Major Japanese Earthquakes after Which Earthquake Revision Surveys Were Performed

Year	Earthquakes	M	Survey classification
1891	Nobi earthquake	8.4	Triangulation, leveling
1923	Kanto earthquake	7.9	Triangulation, leveling
1927	Northern Tango earthquake	7.5	Triangulation, leveling
1927	Sekihara earthquake	5.3	Leveling
1930	Northern Izu earthquake	7.0	Triangulation, leveling
1933	Sanriku offshore earthquake	8.3	Triangulation, leveling
1939	Oga Peninsula earthquake	7.0	Triangulation, leveling
1941	Hyuga Sea earthquake	7.4	Leveling
1943	Tottori earthquake	7.4	Triangulation, leveling
1945	Mikawa earthquake	7.1	Triangulation
1946	Nankai earthquake	8.1	Triangulation, leveling
1948	Fukui earthquake	7.3	Triangulation, leveling
1949	Imaichi earthquake	6.4	Leveling
1952	Tokachi offshore earthquake	8.1	Triangulation, leveling
1952	Yoshino earthquake	7.0	Leveling
1961	Northern Mino earthquake	7.0	Leveling
1964	Niigata earthquake	7.5	Leveling
1965~	Matsushiro swarm earthquakes	(6.3)	Triangulation, leveling
1973	Nemuro Peninsula offshore earthquake	7.4	Triangulation, leveling
1974	Izu Peninsula offshore earthquake	6.9	Triangulation, leveling

necessitated a large-scale revision survey of geodetic networks in the Kii and Shikoku districts. This operation marked the beginning of postwar Japanese geodetic survey operations.

The revision survey that followed the Nankai earthquake was very extensive. It spanned the Kii Peninsula, Shikoku Island, and the Inland Sea. It eventually disclosed the characteristic of the great earthquakes that occur along the Nankai Trough—a great contribution to the modern concept of earthquakes based on plate tectonics.

The Nankai Earthquake Revision Survey took about eight years to complete. As a result of the survey findings, the Geographical Survey Institute began a nationwide survey of the first-degree triangulation network, which had only been partially re-surveyed since its completion. This survey took some 20 years to complete. A nationwide re-survey of the first-degree bench mark network had been conducted before World War II on a limited scale, but a large-scale re-survey, which paralleled the re-survey of the triangulation network, was not performed until after the Nankai earthquake. The Earthquake Prediction Project was scheduled to repeat the survey of first-degree triangulation points (330) and first-degree leveling route (20,000 km worth) every five years, using the re-survey

108   LONG-RANGE PRECURSORS

prompted by the Nankai Revision Survey as a base. This schedule was revised, however, due to the progress made in light wave range finders. These instruments can measure the distance between triangulation points directly and with great precision. The third long-range project began in 1974 and will repeat the survey of 6,000 second-degree triangulation points by trilateration methods, instead of the first-degree triangulation.

6.3   Examples of Land Surface Movement Caused by Earthquakes

The surface anomalies that accompany earthquakes are determined by the size, shape, and amount of displacement of the faults involved. Consequently a fault's position, size, and displacement can be determined from the deformation of the earth's surface. Since these data are basic earthquake information, a post-earthquake survey of ground movement is of vital importance for earthquake prediction as well as for seismological research.

Japan has the most complete geodetic survey network of any earthquake-prone country, and the surface deformations caused by earthquakes have been studied thoroughly. The following three examples illustrate the information obtained from these studies.

6.3.1   Great Kanto Earthquake

Figure 6.1 represents the displacement of first-degree triangulation points caused by the Great Kanto earthquake. The tips of the Boso and Miura Peninsulas, located on the east side of the Sagami Trough (an oceanic valley), moved 3 to 4 m to the southeast—almost parallel to the trough—and rose about 1.5 m. Oshima Island, on the opposite side of the trough, moved 1m to the north. These displacements can be explained by postulating the formation of a particular fault in Sagami Bay, as shown in the rectangle in Fig. 6.1. According to calculations based on fault model theories —Ando [1971], for example—this fault would be 85 km long, 55 km wide, with a dip angle of 30°, a parallel displacement of 6 m, and a vertical displacement of 2 m. The area from the Tanzawa Mountains to Kofu subsided more than 60 cm after the earthquake. The fault postulated above explains this subsidence also.

6.3.2   Great Nankai Earthquake

Great earthquakes (M 8) occur along the Nankai Trough in the Nankai district every 100 to 150 years. The Nankai earthquake of 1946, and the crustal displacement it caused, was crucial to an understanding of great earthquakes along the Nankai Trough. The southern tip of the Kii Peninsula and the Muroto Cape were uplifted by 1 m and displaced 2

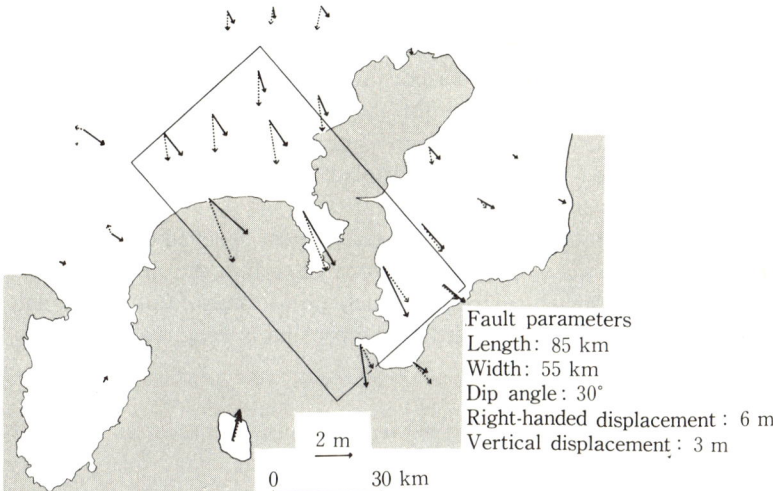

Fig. 6.1 Triangulation point anomalies caused by the great Kanto earthquake (1923). The bold lines show anomalies that were actually surveyed. The dotted lines represent those estimated from the fault model. (Source: Ando, 1973)

to 3 m to the southwest (perpendicular to the Nankai Trough). This displacement can be explained by postulating a reverse fault in which the continental side of the Nankai Trough thrusts up into the oceanic side. This is a good example of an earthquake caused by elastic rebound, which

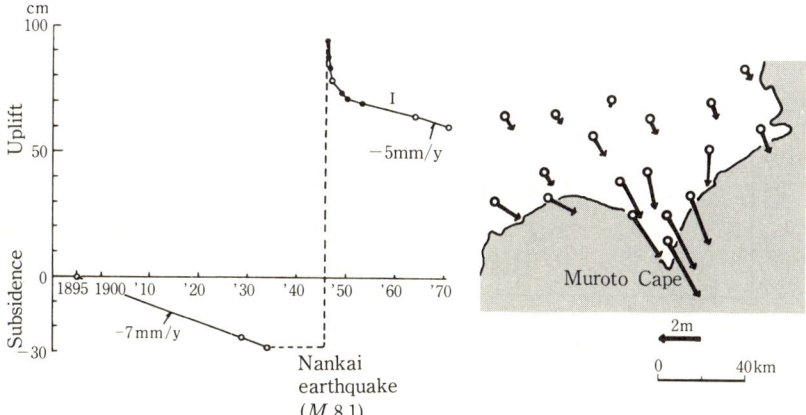

Fig. 6.2 Vertical (left) and horizontal (right) displacement of Muroto Cape as a result of the Nankai earthquake (1946).

occurs, according to the theory of plate tectonics, when the oceanic plate subsides beneath the continental plate (Fig. 6.2).

Crustal movement due to the Nankai earthquake covered a far greater area than the focal zone, based on seismological data such as the distribution of aftershocks, would seem to indicate. The coastline from Kochi to Tosa subsided and Ashizuri Cape uplifted about 40 cm as a result of the earthquake. Such extensive displacements cannot be explained unless the fault movement extended to off the coast of Tosa. Research done on tsunamis substantiated this idea, and it is now believed that there was a slow fault slip to the west of the Muroto Cape that did not generate any seismic waves. The elucidation of such non-seismic fault movement is thought to be important to earthquake prediction research.

6.3.3 Tottori Earthquake

The Tottori earthquake of 1943 was a typical transcurrent fault earthquake. It took place in inland Japan (intraplate). Since this earthquake occurred during World War II, the re-survey of triangulation points was not carried out until 1954, ten years later. Figure 6.3 analyzes these survey results. The broken line represents the fault as inferred from the displacement of the triangulation points. Exploration of the area led to reports that this earthquake had resulted in the formation of the Shikano Fault (8 km long) and the Yoshioka Fault (4.5 km long). The faults that can be observed

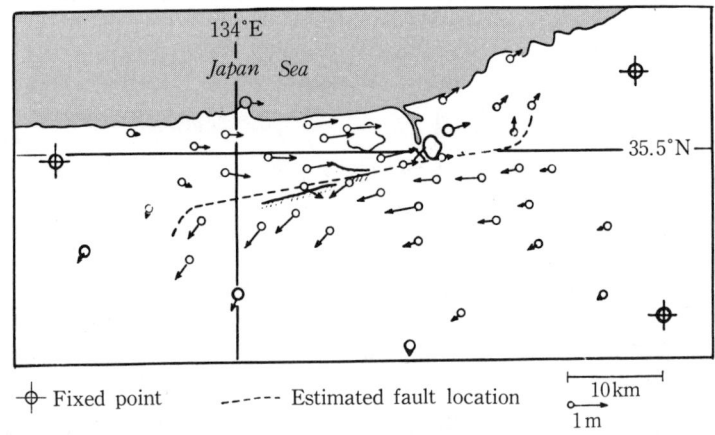

Fig. 6.3 The displacement of triangulation points caused by the Tottori earthquake (1943).

The dotted line shows the location of the seismic fault as postulated from the displacement of the triangulation points.

above ground, however, are merely a fraction of the entire fault. The fault as estimated by triangulation spans 34 km. Its relative displacement is 2 to 2.5 m. The depth of the fault, based on the fault model theory, is 10 to 15 km. Since the size of the fault is known fairly accurately from the survey results, the focal process of the earthquake can be analyzed using this information and data on seismic waves. Kanamori [1972] maintains that the initial rupture occurred in the center of the fault and traveled in both directions with a velocity of 2 to 3 km/sec, that the displacement velocity of the rupture was 80 cm/s, and that the stress that caused this movement (the effective stress) was approximately 30 atmospheric pressure. The Tottori earthquake is unique in that the tips of the fault are bent. This is characteristic of other transcurrent faults as well and was seen in earthquakes such as the Northern Tango earthquake and the Northern Izu earthquake.

6.4 Steady Crustal Movement in Japan

Detecting anomalies in crustal movement through repeated surveys requires a knowledge of how the crust moves under normal circumstances.

The bench mark network had been re-surveyed periodically since before World War II, regardless of whether great earthquakes had occurred or not. As a result, it was known that there was some steady vertical movement in the crust. For example:
(1) Mountainous areas tend to uplift.
(2) Under normal circumstances, the tips of peninsulas that protrude into the Pacific Ocean tend to subside (Fig. 6.4).
(3) In active fold zones, there is movement that corresponds to the fold structure.
(4) The crust is not a continuous structure, but is composed of blocks. Each block is about 10 km in diameter (Fig. 6.5).

Prewar re-surveys of the triangulation network were limited to the periods after great earthquakes, so knowledge of horizontal movement during normal times was not available until after the nationwide re-survey of the first-degree triangulation network that followed the Nankai earthquake revision survey.

Prior to this nationwide survey, the re-survey of the Aibano (Shiga Prefecture) and Tenjinno (Tottori Prefecture) base-line networks led to interesting results. Horizontal strain compressing in an east–west direction at a

---

* The block movement discovered by Captain Atami of the Land Survey Deptartment greatly influenced the study of earthquakes and crustal movement in Japan. It formed the focus, in fact, of prewar research in crustal movement. One reason that Reid's elastic rebound theory was not readily accepted in Japan may have been because of the theory that maintains that the crust moves in blocks.

112 LONG-RANGE PRECURSORS

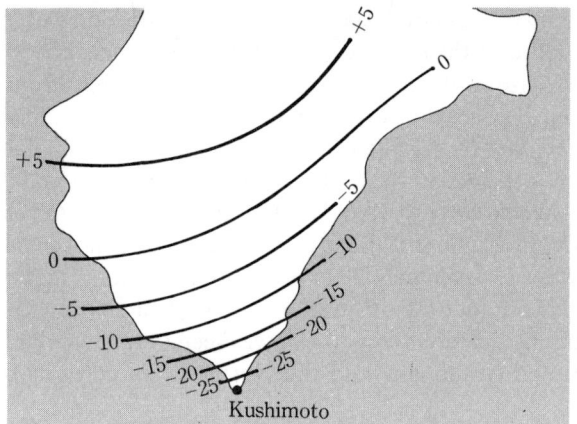

Fig. 6.4 Vertical movement (in centimeters) of the Kii Peninsula between 1890 and 1940, before the 1944 Tonankai earthquake. The tip of the peninsula dipped on the Pacific Ocean (right) side.

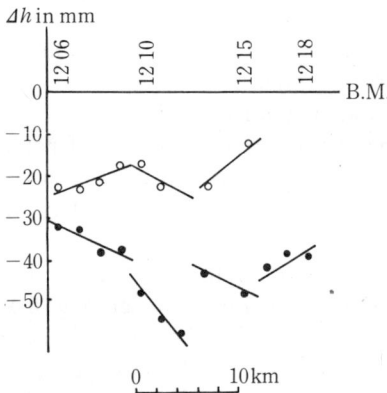

Fig. 6.5 Block movement as observed in a bench-mark survey. [by C. Tsuboi]
The ground is seen to be dipping in units of approximately 10 km.

rate of 1 to $2 \cdot 10^{-5}$ every 50 years was found in both base-line networks. This points to the fact that there is extensive crustal strain progressing in a fixed direction even during normal times—a very significant finding for crustal movement research. This was discovered, by the way, during the early 1950s, before the sea-floor spreading hypothesis or plate tectonics had yet been introduced to Japan. Figure 6.6 represents the distribution of

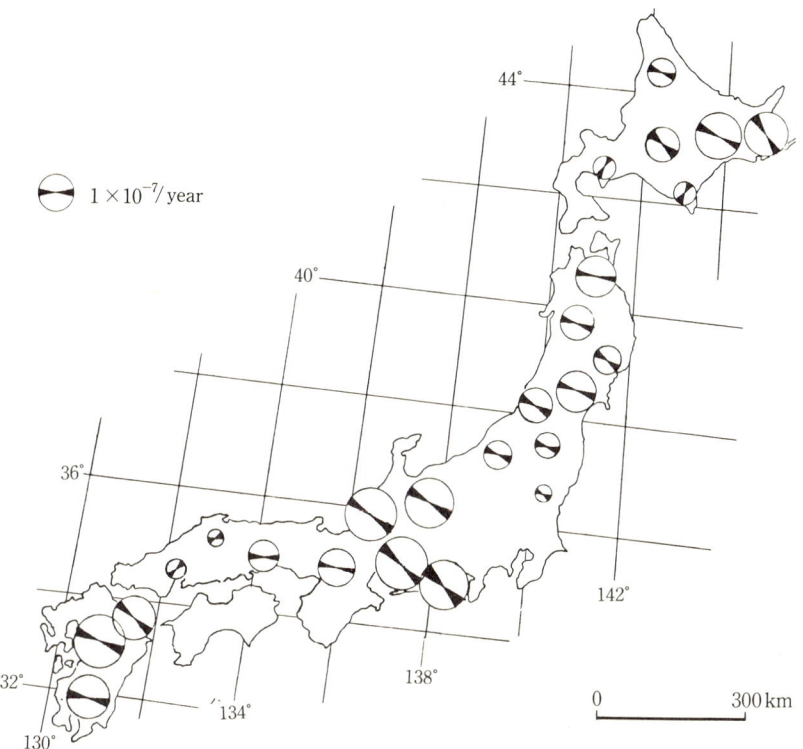

Fig. 6.6 Distribution of horizontal strain in the Japanese islands according to a first-degree triangulation survey.
The diameters of the circles indicate the relative amount of maximum shearing strain. The black lines indicate the direction of the maximum compression axis. (Indicators are missing for the Kanto and Kii areas and Shikoku because of the occurrence of recent earthquakes there.)

horizontal strain (maximum shearing strain) in the Japanese archipelago. I obtained this figure from the results of the nationwide re-survey of the first-degree triangulation network. It is based on the variations in horizontal angle that were directly observed during the triangulation survey. It is apparent from this figure that there is an east–west or southeast-northwest compression in progress throughout the entire Japanese archipelago. The average velocity of this deformation is about $2 \cdot 10^{-7}/y$.

The relative displacement of the first-degree triangulation points can be obtained by using the results of earlier and more recent first-degree triangulation surveys. Harada [1969] chose several fixed points along the strike of the Japanese islands and claculated the relative displacement of triangulation points over the past 60 years (Fig. 6.7). Such a calculation can be affected by apparent movement attributable to the choice of fixed points, but the dramatic inland displacement of the triangulation point along the Pacific coast of the Tokai and Sanriku districts, and that in southeastern Hokkaido, is clearly remarkable. This movement is one of the reasons that the area off the coast of Nemuro and Tokai was chosen as a candidate for a great earthquake in connection with Japanese earthquake prediction research.

Disregarding local movement, the movement within the Japanese archipelago can be seen as controlled by east–west or southeast–northwest tectonic forces. By analyzing the strike and the direction of displacement of

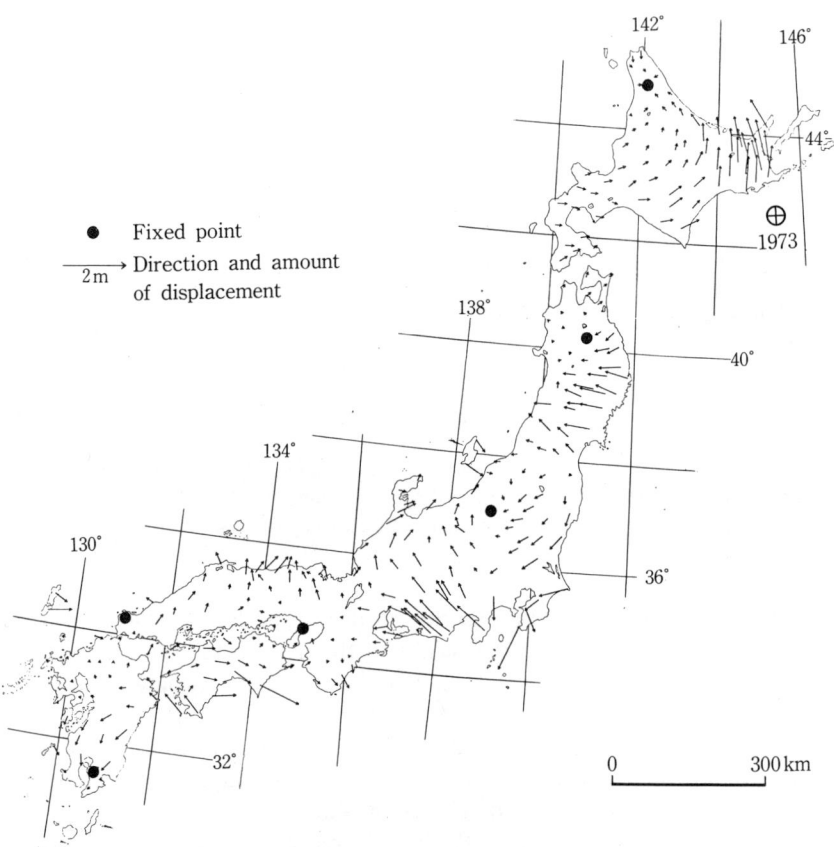

Fig. 6.7 Movement of first-degree triangulation points in the Japanese islands over the past 60 years [mapped by T. Harada].

active faults in the Chubu district, Matsuda [1969] discovered that this has been the direction of the tectonic or compressive force throughout the Quaternary Period (the past 2,000,000 years). Present-day crustal movement in Japan (referring to chronological units of 50 to 100 years), therefore, is basically a continuation of the Quaternary movement.

For purposes of long-range earthquake prediction, the steady crustal movement that should be receiving special attention is the pronounced subsidence of such Pacific coast peninsulas as the Nemuro, Ojika, Boso, Omaezaki, and Kii. (The Izu Peninsula is an exception.) These peninsulas are known to uplift when earthquakes occur off their coasts. Consequently, repeated bench mark surveys to observe their tilting and tide measurements to ascertain changes in sea level are important for the prediction of great earthquakes off the Pacific coast.

6.5 Earth Movement before and after Great Earthquakes

Topographical and historical research shows that great earthquakes occur repeatedly in the same locations. This means that the accumulation and release of crustal strain happens over and over again in the same place. Thus, understanding how crustal strain accumulates from the time of one earthquake to the next is very important.

A triangulation survey by light wave range finder was made of the entire Sagami Bay area, from the Boso to the Izu Peninsula, during 1970 and 1971. This was the first survey of this area since the revision triangulation survey following the great Kanto earthquake. The objective of the survey was to examine the movement of the triangulation points around Sagami Bay—which did move during the great Kanto earthquake—since 1923. This was the first organized survey of horizontal movement to use the light wave range finder for earthquake prediction.

The survey found that the triangulation points on the Boso and Miura Peninsulas which had moved 3 to 4 m to the southeast during the great Kanto earthquake (Fig. 6.1) had been shifting to the northwest since the earthquake (Fig. 6.8). The movement of these peninsulas, both during and after the earthquake, parallels the strike of the Sagami Trough. Clearly, activity in the Sagami Trough has a direct bearing on great earthquakes in the Sagami Bay.

From the survey results alone, it is not clear how the movement of the Boso and Miura Peninsulas has changed over time. Clues can be found, however, in the movement before and after earthquakes of the rhomboid base lines on the premises of the Tokyo Astronomical Observatory in Mitaka City. These base lines, with length of 100 m each, were established by the prewar Geodetic Committee and were surveyed repeatedly, even

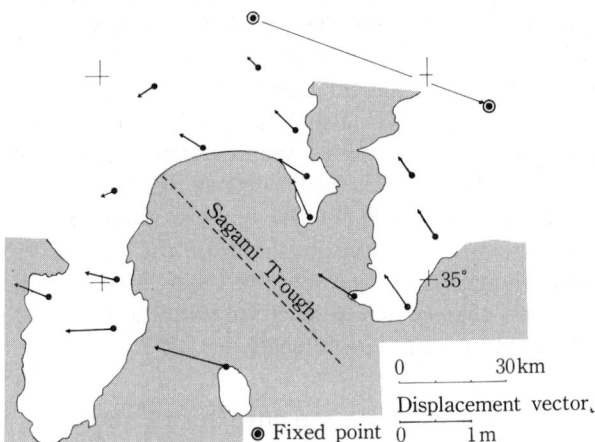

Fig. 6.8 Movement of triangulation points in southern Kanto since the great Kanto earthquake (1924–1971).

before the great Kanto earthquake. During the earthquake they showed a movement of $4 \cdot 10^{-5}$ in a north–south direction. As is demonstrated in Fig. 6.9, the relatively rapid movement began immediately after the earthquake and continued for 10 years. Seemingly, this was an aftereffect of the movement that accompanied the earthquake itself. Then the movement shifted and steadied.

Since the movement of the Mitaka rhomboid base lines is in excellent

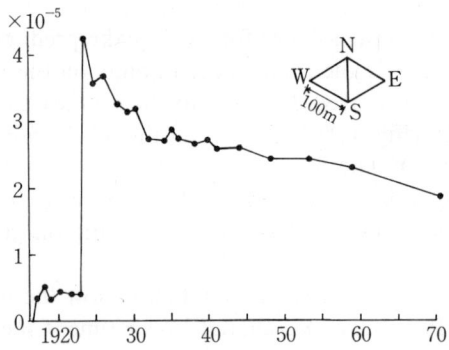

Fig. 6.9 Strain movement (maximum shearing strain) of the 100 m Mitaka rhomboid base line.

The axial direction of the strain at the time of the 1923 earthquake was almost directly north–south.

agreement with the extensive movement charted by the triangulation survey, it can be assumed that the time frame of the movement of the Boso and Miura Peninsulas also parallels that of the Mitaka rhomboid base lines.*

A similar pattern of aftereffect movement followed by steady movement can be seen in the vertical movement of the tide as recorded at Aburatsubo on the Miura Peninsula and Mera on the Boso Peninsula. At Aburatsubo there was a great rise in the level of the tide from the time of the earthquake to 1935 or thereabouts. After that time the tidal level continued to rise, but at an almost constant velocity.

Movement similar to the aftereffect movement that followed the great Kanto earthquake could be observed in the vertical movement of Muroto Cape (Fig. 6.2) after the Nankai earthquake of 1946. These movements, therefore, are thought to represent the movement that occurs in the plate boundary after a great earthquake.

For the purposes of earthquake prediction it is necessary to know how the steady movement that follows the aftereffect movement changes prior to the next great earthquake. There is little information available on such changes, however, because ground movement is usually surveyed after an earthquake strikes, and few surveys or observations are repeated before the next one. Aburatsubo on the Miura Peninsula and Kushimoto on the Kii Peninsula are exceptions in that continual tide level observations have been made there since the Meiji Era. The vertical movement of these two peninsulas before both the great Kanto earthquake and the Tonankai earthquake of 1944, therefore, can be estimated from these records. An examination of the tidal records of Aburatsubo reveals a nearly constant subsidence in this area, which stopped 10 years prior to the great Kanto earthquake (Fig. 6.10).

Such changes may simply be due to oceanographic phenomena. They cannot be disregarded as possible long-term precursory phenomena, however, since a similar tendency was found in the tidal records of Kushimoto prior to the Tonankai earthquake.

From Southern Kanto to the coast of Nankai, a change in the land's subsidence, whether it be in the velocity or the direction of the movement, appears to be a long-term precursory phenomenon, but this is not known for certain. Since such anomalous movements are not necessarily being observed as long-term precursory phenomena, their absence is no guarantee that a great earthquake will not occur.

---

* Roughly speaking, the amount of aftereffect movement following the great Kanto earthquake was a little less than one-third of the total amount of movement since the earthquake. Aftereffect movement is the process of adjustment that takes place after the abrupt movement of the earthquake itself.

118 LONG-RANGE PRECURSORS

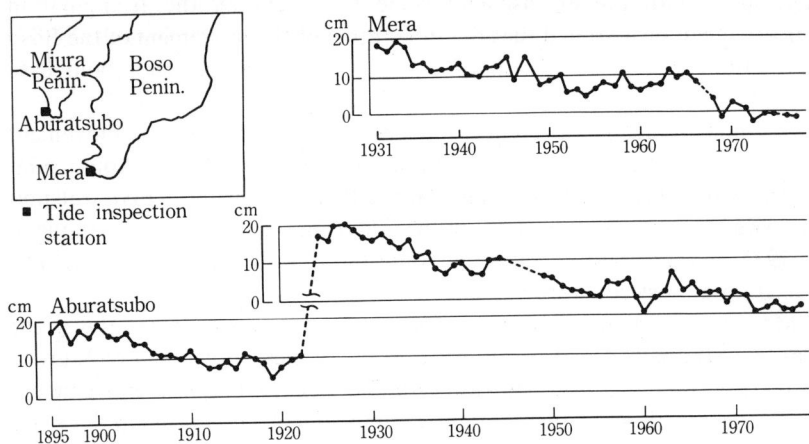

Fig. 6.10 Vertical movement of the Boso and Miura Peninsulas according to tidal records.
The movement velocity since 1935 has been 3 mm/y at both Aburatsubo (on the Miura Peninsula) and Mera (on the Boso Peninsula). At Aburatsubo there was no marked fluctuation in movement during the 10-year period prior to the 1923 earthquake.

How long before a great earthquake do precursory phenomena such as a change in the subsidence velocity of the land appear? This is a crucial point for long-term earthquake prediction. There are several equations that attempt to relate an earthquake's magnitude to the length of the precursory period, but they have drawbacks, especially where large-scale earthquakes are concerned. Some equations, for instance, posit several decades as the precursory period for M 8 earthquakes. And yet, as can be seen in Fig. 6.2, Muroto Cape continued to subside steadily until 1935, only 10 years prior to the Nankai earthquake. The tide level anomaly at Aburatsubo also began about 10 years before the great Kanto earthquake. Even if precursory phenomena do begin several decades before an earthquake, they are likely to be regarded as the norm if change within that period is fairly constant.

With regard to the movement that follows destructive inland earthquakes, some follow-up studies are being done on the Northern Tango earthquake (1927) and the Izu Peninsula offshore earthquake (1974). In the case of inland earthquakes, it seems that the aftereffect movement sometimes lingers on and takes the same direction as the movement during the earthquake. The pattern of aftereffect movement observed in Aburatsubo after the great Kanto earthquake and in Muroto Cape after the Nankai earthquake does not necessarily hold for destructive inland earthquakes.

6.6 Earthquake Prediction and Repeated Survey Operations

Some earthquakes are accompanied by remarkable foreshocks. There are cases where sudden anomalous ground movements have been observed immediately before an earthquake. Such cases would lead us to believe that the short-term prediction of earthquakes would soon be possible if seismic activity and crustal movement were observed with telemeters. Most short-term earthquake precursors, however, are recognized as such only after the earthquake strikes, not before. It is not easy to determine that such phenomena are, in fact, precursors of an earthquake. Even if it were possible to distinguish between foreshocks and swarm earthquakes, it would still be difficult to estimate the scale of a forthcoming earthquake from the foreshocks. This is obvious from the fact that some intermediate and minor earthquakes are accompanied by foreshocks, while some large-scale earthquakes occur with no noticeable foreshocks.

Accurate short-term prediction requires a knowledge of the estimated size and location of imminent earthquakes and some assurance that an earthquake is likely to occur in the near future. In other words, long-term prediction is necessary.

Long-term earthquake prediction begins by detecting the long-term precursors that can be expected to appear when the crust approaches the rupture point. Earthquakes occur when crustal strain reaches a critical point and is released through faulting within the crust. Since crustal strain appears as tilting of ground surface or as expansion or contraction of the ground, observation of ground movement is the foundation of long-term earthquake prediction.

When anomalous crustal movement is observed, it is extremely important to know how it is distributed throughout the region. Increasing the number of fixed observation stations may be helpful in achieving this goal, but observations from a fixed station are easily influenced by local conditions. Consequently, it is not easy to determine the extent of long-term anomalous crustal movement by continuous observations made from a station. This is exactly why repeated surveys are important for long-range earthquake prediction.

6.7 Critical Crustal Strain and Long-Range Earthquake Prediction

Crustal strain is limited in that when it reaches a critical point, a rupture, or earthquake, occurs. Thus, a knowledge of critical crustal strain is essential to earthquake prediction efforts. If the value of the critical strain and the present amount of crustal strain are both known, then it becomes possible to predict whether or not an earthquake is imminent.

Chuji Tsuboi, after analyzing the results of the triangulation survey made after the 1927 Northern Tango earthquake (M 7.5) and the 1930 Izu earthquake, postulated a critical strain of 1 to $2 \cdot 10^{-4}$. This was based on the fact that the maximum shearing strain near the ruptured fault was over $1 \cdot 10^{-4}$. At present, $1 \cdot 10^{-4}$ is widely accepted as the critical strain value. Earthquakes are rupture phenomena, however, and the magnitude of the strain necessary to cause a rupture is not constant. Inferring a variance of $1 \cdot 10^{-5}$ in the critical value and a strain progression velocity of 2 to $3 \cdot 10^{-7}$/y, for instance, a simple calculation will show the variance in the timing of earthquakes to be 30 to 50 years. (Actually, critical strain is not distributed evenly, but is thought to show certain patterns of distribution. In any case, it is clear that the variance in critical strain will make a great difference in predicting the timing of an earthquake.)

The greatest problem in basing earthquake prediction on critical strain is the difficulty in determining the amount of accumulated strain, except in regions where earthquakes struck fairly soon after surveying began. Repeated surveys can show only the amount of change in the strain from one survey to the next. If the amount of strain at the beginning of a series of surveys is unknown, there is no way of knowing the amount of strain in the present. Where destructive earthquakes in inland Japan are concerned, long-term prediction based on critical strain is almost impossible, since the intervals between them are thought to be more than 1,000 years.

If strain accumulated in short bursts rather than steadily, it could be measured even if an area were re-surveyed only once in several years or even several decades. In inland Japan, however, strain does not seem to accumulate rapidly: the actual average velocity of strain accumulation is estimated to be 2 to $3 \cdot 10^{-7}$/y, with earthquake intervals of 1,000 years and a critical strain of 1 to $2 \cdot 10^{-4}$. Actual observation also seems to eliminate the possibility of rapid strain accumulation before earthquakes. For example, a second-degree triangulation was made in March 1974, near the focal area of the Izu Peninsula offshore earthquake (M 6.9), which occurred two months later. This area was the subject of surveys following the Northern Izu earthquake (M 7.0, 1930), so approximately 40 years' worth of strain variation information was available. The variation, however, was less than $1 \cdot 10^{-5}$ near the focal zone.

A similar conclusion can be drawn from the Tottori earthquake of 1943. Kanamori [1972] analyzed the seismic wave of this earthquake and postulated a fault slip of 2.5 m. The amount of fault displacement obtained by triangulation survey was 2.3 m, which is in excellent agreement with the above figure. Since the survey detects the total land movement from the time of the Meiji Era until after the earthquake, the fact that these amounts agree proves the absence of a great strain change between the time of the first survey and the time of the actual earthquake.

Surveys can, at most, only reveal the crustal movement of the past several decades. Thus it is dangerous to directly relate the amount of crustal strain over several decades to the possibility of an earthquake. The exceptions are in regions where great earthquakes occurred shortly before or after the Meiji Era survey began and which are known to have relatively short intervals of 100 years or so. This is the case in the Sagami and Suruga Bay areas. In these regions the amount of accumulated crustal strain can be determined by survey, thus making long-range earthquake prediction possible.

At present it is feared that a Tokai earthquake centered in Suruga Bay is likely. The horizontal strain in Suruga Bay from the first survey in 1884 to the present has been characterized by a pronounced compression of 3 to $4 \cdot 10^{-5}$, in a direction perpendicular to the Sagami Trough during the last 90 years. Extrapolating from the Ansei Tokai earthquake of 1854 (the focal zone of which was said to have extended into Suruga Bay) to the present, the horizontal strain in Suruga Bay is thought to have reached almost $5 \cdot 10^{-5}$.*

The timing of great Pacific Coast earthquakes can be predicted by applying the concept used to predict earthquakes by critical strain to the amount of land subsidence before an earthquake. On the Boso Peninsula and the Muroto Cape, for instance, the pattern of land subsidence before an earthquake and uplift during an earthquake has been repeated. The amount of pre-earthquake subsidence, however, is not equal to the amount of uplift during the earthquake. The subsidence is reportedly 40% of the uplift on the Boso Peninsula and 80% on the Muroto Cape. Estimates (based on surveys, etc.) of the post-earthquake period required to reach 40 and 80% subsidence, respectively, can aid in predicting earthquakes. Since the steady subsidence velocity of Muroto Cape is 6 to 7 mm/y, for example, in the 90 years between the Ansei Nankai earthquake (1854) and the Nankai earthquake (1946), annual subsidence can be estimated at 60 cm or thereabouts. The amount of uplift on Muroto Cape at the time of the Ansei Nankai earthquake was estimated at 1.2 m. Thus the critical amount of

* Tsuboi's figure of $1 \cdot 10^{-4}$ as the critical point of crustal strain is limited to destructive inland earthquakes such as the Northern Tango and Northern Izu earthquakes with recurrence intervals of 1,000 years. Consequently it does not necessarily apply to great earthquakes on the Pacific plate boundary, which have recurrence intervals almost 1/10th those of inland earthquakes. Some think that the critical strain of great earthquakes on the plate boundaries is smaller than that of inland earthquakes and is about $5 \cdot 10^{-5}$.

122 LONG-RANGE PRECURSORS

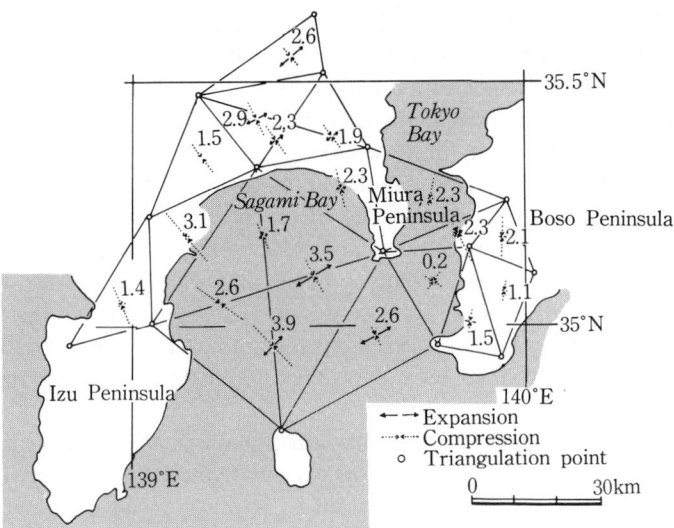

Fig. 6.11 Horizontal strain in Sagami Bay during the 50 years since the great Kanto earthquake.

The figures represent the maximum shearing strain in units of 10^{-5}. Triangles with relatively great strain, such as those with values of 3.9 or 3.5, include triangulation points on the Miura Peninsula where aftereffect movement is thought to be relatively great.

subsidence is 90 to 100cm. From calculations such as these it was possible to surmise before the Tonankai earthquake of 1944 that a Nankai earthquake would not take place for a long time since it seems to take at least 40 years for Muroto Cape to reach the critical subsidence point. Since Tokai earthquakes have always taken place at about the same time as Nankai earthquakes, it could also have been assumed that there wouldn't be any great earthquakes in the Tokai area either for a long time to come. These suppositions were proven wrong.

Thirty to 40 years may seem like a long time from the standpoint of a human life, but it is just a flash in terms of the time scale of such geological phenomena as earthquakes. Thus long-range earthquake prediction based on the critical strain of crustal movement is bound to be vague in human terms, no matter how precise the theory it comes from. Long-range prediction of great earthquakes is necessary, however, both for disaster prevention and for the strategic positioning of an effective observation network. By combining observations of seismic activity (seismic gaps, for instance) with observations of crustal movement, it is becoming possible to make long-range predictions of great earthquakes on the Pacific coast.

6.8 Crustal Movement Anomalies as Earthquake Precursors

It is well known that prior to the Niigata earthquake of 1964 there was an anomalous uplift of the ground in the Niigata district. Similar phenomena were reported at the time of the San Fernando earthquake (U.S.A.) of 1971 and the Haicheng earthquake (China) of 1975. A good many earthquakes in the past also seem to have been preceded by anomalous uplifting of the ground in and around the focal zones. The causal relationship between earthquakes and anomalous uplift is not exactly clear, but this uplift is one of the most commonly noted long-range earthquake precursors.

The dilatancy earthquake model explains ground uplift and variations in seismic wave velocity as being due to the minute cracks that form and grow in underground rocks before an earthquake actually strikes. Another model (Fig. 6.12) postulates a slow aseismic slide that causes the upper crust to uplift. Eventually this uplift causes an earthquake. In either case, anomalous uplift seems to occur as a surface displacement when crustal strain approaches the critical point and when the crust reaches a high-stress condition.

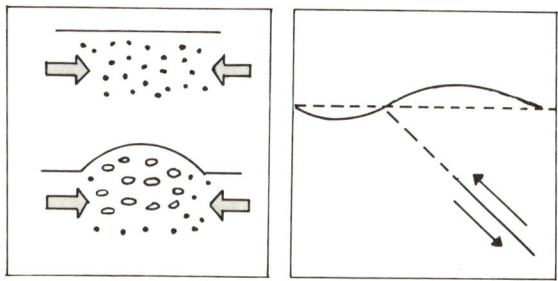

Fig. 6.12 Two models that explain precursory uplift.
The left-hand diagram shows the dilatancy model; the right-hand diagram, the press/slip model.

Precursory ground uplift has received attention in Japan since the repeated leveling surveys that were begun before World War II. The Sekihara earthquake (M 5.3) of 1927 is a well-known example. Two months before the earthquake the area was re-surveyed, and the bench marks near the hypocenter indicated an uplift of 2 cm. The uplift, however, was found by comparison with a survey undertaken in 1894; thus it is unclear whether or not this anomalous uplifting was actually precursory movement.

Before the Earthquake Prediction Project was established, leveling surveys were carried out every 20 to 30 years. When ground uplift was discovered during these surveys, it was difficult to tell, in many cases, just how long ago the anomalous condition had begun. Figure 6.13 shows the earthquakes since 1900 with a magnitude of more than 6 that were preceded by ground uplift. The hypocenters of these earthquakes had all been surveyed more than once, and the last re-survey had taken place within 10 years of the earthquake. The solid black circles represent those earthquakes that were accompanied by what is thought to be precursory phenomena, judging from the time of the last survey, the amount of movement shown, and the positions of the bench marks. The half-solid circles represent earthquakes that were accompanied by movement, but it is unclear whether or not this movement was precursory. The open circles

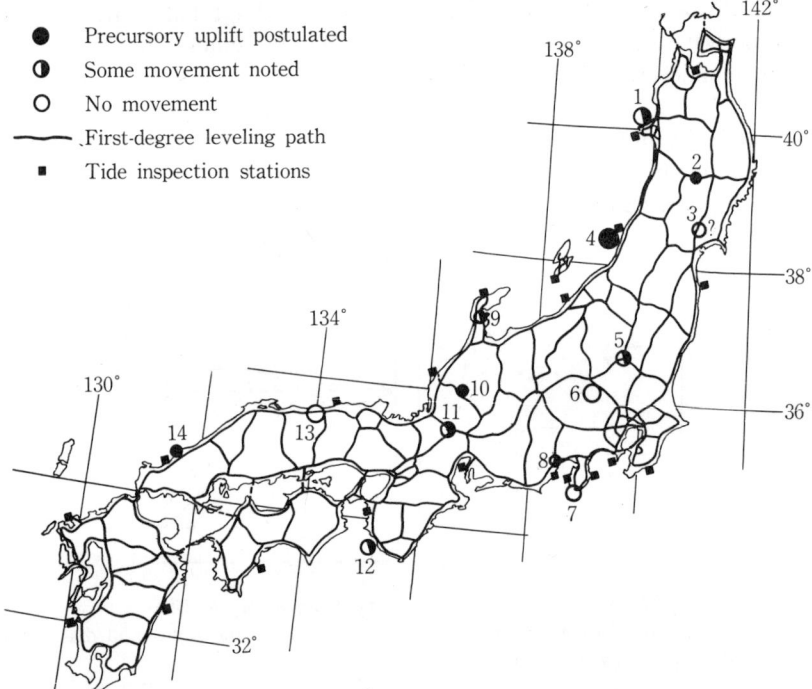

- ● Precursory uplift postulated
- ◐ Some movement noted
- ○ No movement
- ——— First-degree leveling path
- ■ Tide inspection stations

Fig. 6.13 Distribution of earthquakes accompanied by pre-earthquake ground uplift as of 1975 (Hokkaido is not included).
1. Oga earthquake (1939); 2. Southeastern Akita earthquake (1970); 3. Northern Miyagi earthquake (1962); 4. Niigata earthquake (1964); 5. Imaichi earthquake (1949); 6. Western Saitama earthquake (1931); 7. Izu Peninsula offshore earthquake (1974); 8. Shizuoka earthquake (1935); 9. Noto Peninsula earthquake (1933); 10. Northern Mino earthquake (1961); 11. Anegawa earthquake (1909); 12. Tanabe Bay offshore earthquake (1938); 13. Tottori earthquake (1943); 14. Northern Yamaguchi earthquake (1941). In addition to these cases, anomalous ground uplift accompanied by swarm earthquakes was noted in Wakayama (1920), Ito (1930), and Matsushiro (1965). Data on earthquake 10 are based on Tsubokawa's (1969) findings.

represent earthquakes that were not accompanied by pre-earthquake movement. From this figure it becomes apparent that even when such pre-earthquake movements are detected, it is difficult to judge whether or not they are precursory phenomena. In the cases of the Western Saitama earthquake (1931, M 7), the Tottori earthquake (1943, M 7.5), and the Izu Peninsula offshore earthquake (1974, M 6.9), no movement was noted. These earthquakes are attributed to transcurrent faults that run parallel to the leveling path of the survey.

For purposes of earthquake prediction, it is important to investigate those cases in which earthquakes did not occur in spite of the discovery of ground uplift. In my experience there are far more cases in which no link is found between ground uplift and earthquakes than there are cases in which there is a link. The prewar leveling re-surveys were not quite systematic enough, however, to come up with an accurate count of such cases. It has also been found that when strict care is applied to the evaluation of anomalous uplift, the percentage of the cases in which uplift is related to earthquakes increases, and so does the number of cases in which precursory phenomena were overlooked.

The gravest problem facing leveling surveys of ground uplift is apparent uplift that is really due to an accumulation of errors. When accidental errors have accumulated and anomalous movement is recorded along a 10- to 20-km wavelength, it becomes very difficult to distinguish real movement from movement resulting from errors, and many of those cases in which apparent uplift was not followed by earthquakes may be the result of such an accumulation of errors. Repeated surveys that will reliably detect anomalous uplift can be expected to improve the linking of anomalous uplift with earthquakes.

In addition to uplift, anomalous horizontal crustal movement might also be expected to be an earthquake precursor. However, there is not one example in Japan of anomalous horizontal strain being observed in the crust prior to an earthquake. An anomalous uplift accompanied by horizontal strain of about $1 \cdot 10^{-5}$ was discovered in the central part of the Izu Peninsula prior to the Izu Oshima offshore earthquake of 1978. The relationship between anomalous movement in central Izu and the Oshima offshore earthquake has not yet been fully explained, however.

The major reason for the absence of such examples is that there are very few cases in which there has been a pre-earthquake re-survey of triangulation points near the earthquake's focal zone—with the exception of the nationwide re-survey of the first-degree triangulation network after World War II. Another reason may be that anomalous horizontal movement is too slight for present survey techniques to record.

The present velocity of steady horizontal crustal strain is known to be 2 to $3 \cdot 10^{-7}$/y or, at most, $5 \cdot 10^{-7}$/y. Even if the velocity increased two-fold as a long-range earthquake precursor, the amount of anomalous movement would only be $5 \cdot 10^{-6}$ in 5 years or so, which is not all that different from the accuracy of the usual triangulation survey using a transit. Thus, even if such a survey were to be repeated every 5 years, it is doubtful that a horizontal anomaly could be detected and identified as a precursory phenomenon.

The transit used for triangulation surveys has recently been replaced by the light wave range finder. This measurement technique surveys distance by measuring, with great precision, the time light takes to travel between two points. Since light travels at an almost constant velocity within the atmosphere, a light wave range finder can survey a distance of 10 km with an error of only 2 cm—in other words, with an accuracy of $2 \cdot 10^{-6}$. Thus it is now possible to detect horizontal strain anomalies if they are more than $3 \cdot 10^{-6}$.

The accuracy of light wave range finders does not surpass that of first-degree leveling, however, which is $\pm$ 5 mm for distances of 10 km. Unless the anomalous movement is fairly great, therefore, the ability of the range finder to detect anomalous movement does not equal that of leveling.

6.9 Future Problems

By observing long- and short-term precursory phenomena, earthquake prediction is possible. A knowledge of crustal conditions, in addition to knowledge of earthquake phenomena *per se*, is crucial to such prediction. Information on the progress of crustal strain (displacement) is fundamental since an earthquake is a phenomenon in which strain accumulated in the earth's crust is relieved. For this reason the Earthquake Prediction Project considers the observation of crustal movement to be primary to its observation regimen.

The survey network established during the Meiji Era is still the most useful tool for this observation. The basic purpose of the network at the time it was established, however, was map-making, not earthquake prediction. Obviously the periodic re-surveying of this network alone cannot serve the purpose of detecting precursory anomalous movement. The density of the leveling network in particular warrants re-examination.

The first-degree leveling route system covers Japan in a grid. Since the length of each grid may be as great as 100 km, anomalous ground uplift cannot always be detected. At present the density of the leveling network is being increased in such areas as Southern Kanto and Tokai. Immediate attention should be paid to increasing the density of the network in other areas as well. The finer the network's grid, the more reliable leveling surveys will become at detecting anomalous movement.

In a country with a topography like Japan's, there are bound to be areas that are too mountainous for the establishment of bench-mark grids. In these areas, surveys must be supplemented by precise gravity measurements. By combining the leveling and gravity measurements, the vertical movement detected by leveling surveys can be double-checked at the same time that information on underground structure is being obtained.

Surveying horizontal movement with a light wave range finder presents one basic problem, and that is the improvement of survey accuracy. Errors in light wave surveys stem not so much from the instrument itself but rather from the corrections made due to the nature of light. The velocity of light varies according to the atmospheric temperature and according to its color. Using light sources of different colors—blue and red, for instance—to measure the same distance will result in a difference in the measurement according to the colors. Since this difference is a function of atmospheric temperature, the temperature can be accurately obtained from it and the measurement adjusted accordingly. A light wave range finder based on this logic is now being developed, and a great improvement in the accuracy of optical wave surveys is expected in the near future.

From the standpoint of detecting anomalous movement, the frequency with which the survey cycle is repeated bears further examination. In an orogenic belt like Japan, the earth is in constant motion. If anomalous movement is to be distinguished from normal movement, frequent surveys are essential. In order to recognize anomalies in ground movement prior to an earthquake, surveys must be repeated at least twice. This would mean that a survey with a five-year cycle could detect only anomalous movement that began at least ten years before an earthquake.

Several recent cases in Japan have borne this out. Before the Izu Oshima offshore earthquake in early 1978, an anomalous ground uplift was observed in the eastern part of the Izu Peninsula. At the time of the regular re-survey of bench marks in this area in 1968, no anomalous movement had been detected. From this evidence it can be concluded that a survey cycle of five years would not have detected this anomalous uplift. Uplift was observed before the Niigata earthquake because of the frequent surveys that were being conducted to investigate ground subsidence in the area.

In Japan and other areas where crustal movement is constant even in normal times, a change in the velocity of this movement can help to distinguish anomalous from normal movement. Unless the survey intervals are regular, however, survey errors can appear to be velocity changes. Therefore it is vital to repeat surveys at regular intervals if anomalous movement is to be detected.

An immediate countermeasure system is also needed in case anomalies in seismic activity or in crustal movement are discovered. This presents an unexpectedly serious problem given Japan's present budget and administrative system. After the 1925 Northern Tajima earthquake, A. Imamura was convinced that another great earthquake would strike in the areas marginal to the quake, and he requested that the Land Survey Division re-survey the leveling network in the area. Administrative procedures delayed the re-survey, however, and, to Imamura's dismay, the 1927 Northern Tango earthquake occurred before the resurvey could begin. Since the resurvey would have included the benchmark paths that traversed the focal zone of the Northern Tango earthquake, priceless information would have been gained for earthquake prediction research had the releveling been carried out as scheduled.

Although repeated surveys to detect anomalous crustal movement are fundamental to earthquake prediction, all anomalous movements are not necessarily earthquake precursors. There will also be regional and local differences in such movement even if it does portend an earthquake. If anomalous crustal movement can be detected, a diagram can be plotted in which the time of a future earthquake can be predicted from the relationship between the magnitude of the earthquake and the length of the precursory period. The size of the earthquake can be estimated from the extent of the crustal movement. It must be recognized, however, that such a diagram would contain many uncertain elements. The difficulty in earthquake prediction is the lack of information based on actual examples—information vital to the prediction of natural phenomena. The path to certain progress in earthquake prediction is to improve survey accuracy and reinforce the observation network in order to increase the number of cases in which precursory observations are available.

References

Ando, M., 1971: A fault-origin model of the great Kanto earthquake of 1923 as deduced from geodetic data, *Bull. Earthq. Res. Inst.*, **49**, 19–32.

Harada, T. and N. Isawa, 1969: Horizontal deformation of the crust in Japan: Result obtained by multiple fixed stations (in Japanese). *J. Geod. Soc. Japan*, **14**, 101–105.

Kanamori, H., 1972: Determination of effective tectonic stress associated with earthquake faulting; the Tottori earthquake of 1943, *Phys. Earth Planet. Interiors*, **5**, 426–434.

Matsuda, T., 1969: Active faults and earthquakes (in Japanese), *Kagaku*, **39**, 398–407.

Rikitake, T., 1974: Probability of earthquake occurrence as estimated from crustal strains, *Tectonophysics*, **23**, 299–312.

Sato, H., 1973: A study of horizontal movement of the earth crust associated with destructive earthquakes in Japan, *Bull. Geogr. Surv. Inst. Japan*, **19**, 89–130.

Tsuboi, C., 1932: Investigation on the deformation of the earth crust connected with the Tango earthquake of 1927, *Bull. Earthq. Res. Inst.*, **10**, 411–432.

PART III SHORT-TERM PRECURSORY PHENOMENA

Crustal strain increases with time: when it goes beyond a critical point, certain changes begin to take place—changes in which dilatancy probably plays a central role. The changes continue for several years or decades depending on the size of the earthquake. In the case of smaller earthquakes, they may take place only for a matter of days.

What happens when the final stage approaches? The crust of the area in question is being subjected to intense strain. Since the crust is intrinsically heterogeneous, the strain distribution must be heterogeneous too.

Several hours or several days prior to an earthquake, a fault begins to form in the focal area. This is a more dynamic, transitional stage than the rather static crustal alteration that functions as a long-range earthquake precursor. The trigger has already been pulled, so to speak.

When the long-range precursory period has been completed and the transitional stage, in which the fault is being formed, begins, what actually happens? The crust, as a whole, has been exposed to a great deal of strain. The strain has not been evenly distributed, however, throughout the crust's various regions. When the force in an area where the strain is extreme exceeds a critical point, another area, one with less strain, quickly begins to be filled with strain. As the strain distribution becomes more homogeneous, the maximum value of strain in the entire region increases and finally reaches the critical point. The displacement begins to form. In M 8 earthquakes the fault is completely formed in several tens of seconds—in short, the earthquake phenomenon is complete.

The above description may sound like an established fact, but it is really only a hypothetical description based on various theories. The work of Sacks, who developed the original version of the bore-hole strainmeter, was mainly taken into consideration in developing this model. Transitional phenomena can be observed by various methods. The detection of these phenomena constitutes the observation of short-term earthquake precursors.

Chapter 7 Continuous Observation of Crustal Movement

Shigeji Suyehiro

7.1 The Meaning of the Continuous Observation of Crustal Movement

7.1.1 The Earthquake Mechanism

Force works on a part of the earth's crust, causing deformation or strain. When this strain reaches approximately 10^{-4} (which corresponds to the strain that will bend a pole 100 m long 1 cm), the strength of the rocks that constitute the crust exceeds an ultimate limit, and rupture starts. This rupture takes the form of a fault, and an earthquake occurs at that moment. In short, the formation of a fault is an earthquake.

There is no doubt about the earthquake mechanism described above—at least insofar as earthquakes with magnitudes greater than 7 are concerned. It has been proven through the analysis of seismic waves generated by earthquakes and through surveys of crustal movement both before and after earthquakes. The release of strain has been confirmed by survey.

7.1.2 An Accurate Understanding of Earthquakes

The discussion thus far points to the fact that merely looking at earthquakes as vibrations of the earth does not provide a comprehensive understanding, but rather looks only at the aftereffect of crustal rupture. Earthquake phenomena must be seen in their totality as a process in which seismic energy accumulates in the form of progressive strain that finally reaches the stage of rupture. From the standpoint of earthquake prediction, the stages prior to the point of rupture are far more important than the phenomena that occur after the rupture has taken place.

What kinds of processes, then, take place during this prerupture stage? Consolidating our present knowledge with our past experience, they can be classified as follows:

(a) Progressive strain that is almost constant,

(b) Arrival of the critical point,
(c) Beginning of the principal rupture,
(d) Occurrence of principal rupture.

Stage (a) is the accumulation of seismic energy, and it progresses rather slowly. It takes great Pacific coast earthquakes of the M 8 category 100 to 200 years to reach stage (b). Inland earthquakes of up to M 7 take even longer.

At stage (b), conditions—including the progress of the strain—are no longer constant as the critical point approaches. And yet there is still time before the principal rupture stage is actually reached. The length of this "grace period," according to the dilatancy theory, seems to depend on the scale of the principal rupture—or, in short, on the magnitude of the earthquake.

Stage (c) is the immediate precursory period before the principal rupture. The principal rupture does not occur suddenly, but seems to be preceded for several days (or, in some cases, several hours) by precursory phenomena. At this stage the principal rupture has begun in earnest, and there is no turning back.

The observations and examples of actual surveys that seem to support this model have been compiled in Table 7.3 at the end of this chapter. (See also Table 7.2.) Regrettably, the mechanisms of (b) and (c), which are the most crucial to earthquake prediction, are not yet quite clear. Compared to the clearly defined mechanism of the origin and propagation of seismic waves, research on this stage is very underdeveloped. The future development of earthquake prediction techniques hinges on the research that will elucidate this part of the earthquake mechanism.

7.1.3 Earthquake Surveillance

Stages (a), (b), and (c) do not involve the vibration phenomenon. Although they are accompanied by minor and micro earthquakes, the phenomena that take place in these stages are principally changes in crustal strain.

Since stage (a) is thought to be in constant progress over a long period of time, repeated geodetic surveys are the most effective way of observing it. As this method covers an extensive area, it is possible to detect the extent of energy accumulation, thus enabling one to speculate on the magnitude of future earthquakes.

Once the process reaches stages (b) and (c), however, the phenomena, according to recent research, cease to progress constantly in terms of either time or space. If this is the case, then the periodic repetition of surveys may miss some significant information. Therefore it is vital to continuously observe changes in strain. In fact, some of the examples at the end of this

CONTINUOUS OBSERVATION OF CRUSTAL MOVEMENT   135

chapter demonstrate this point. In these examples, some of the (b) and (c) processes that are especially crucial to short-range prediction were detected by the continuous observation of crustal movement or by other, similar, methods. This fact alone points to the indispensability of continuous observation.

7.1.4   Continuous Crustal Movement Observation

As the process advances from (a) to (b) to (c), the velocity of the strain change accelerates, and uniformity in time and space is lost. Particularly when the process reaches (c), extremely sudden movement can occur, like the movement that preceded the Hamada earthquake of 1872 and the Tonankai earthquake of 1944 (see Table 7.3).

Judging from the observations to date, monitoring of stage (a) depends entirely upon repeated surveys; monitoring of stage (b) requires that repeated surveys be supplemented with continuous observation. When the process has reached stage (c), continuous observation is again necessary for definite short-range prediction.

When the "Blueprint," *Prediction of Earthquakes: Progress to Date and Plans for the Future,* was published in 1962, the aims of the continuous observation of crustal movements were described as follows: (1) To fill in the gaps in geodetic surveys with continuous observation, (2) To study the earthquake mechanism through observation of the release of strain that accompanies the principal rupture.

The successful results of this project are discussed later in this chapter, but these two objectives are not enough for the future. The elucidation of stage (c), mentioned earlier, needs to be included also. It was believed a generation ago that earthquakes are sudden events. But in fact the principal rupture is preceded by several hours or days of precursory phenomena. The detection and elucidation of this process holds the key to knowledge of the earthquake mechanism and to effective short-term earthquake prediction. In this sense, further research into stage (c) is a new and significant objective for continuous crustal movement observation.

7.2   Continual Observation of Crustal Movement in the Past

7.2.1   Method of Observation

The continuous observation of crustal movement has a long history. Its subjects, from the very beginning, were the expansion and tilting of the ground. The amount of movement in both cases was usually very small—10^{-7}, which corresponds to a $1/100$ mm extension of a pole 100 m long in the case of expansion, and 0.1 sec, which corresponds to a 0.05 mm movement of a pole 100 m long in the case of tilting. The speed of the movement is so slow that days or months are used as the unit. Considerable effort and patience were necessary in order to develop instruments stable and sensitive enough to measure such movement effectively. After 50 years of research and observation, the method now recognized as standard uses the quartz tube extensometer and the water tube tiltmeter in conjunction with horizontal vault observation. The quartz tube extensometer measures the expansion of the ground directly, using a quartz bar as a measure stick. The water tube tiltmeter detects changes in the tilting of the ground by comparing the levels on either end of a long water-filled pipe, based on the principle of the communicating vessel (see Fig. 7.1). The longer the quartz bar or the water tube is, the better the sensitivity of the instruments. A minimum of 20

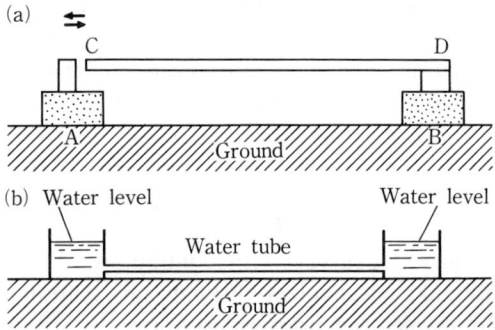

Fig. 7.1 Diagrams showing the principles used in the quartz tube extensometer (a) and the water tube tiltmeter (b).

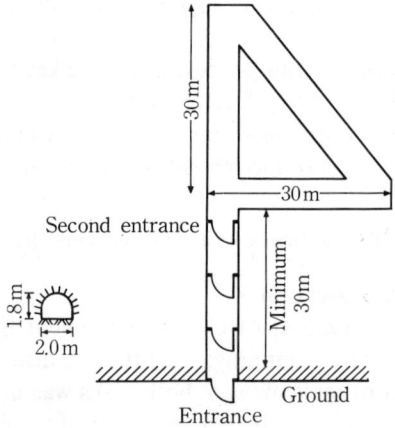

Fig. 7.2 An example of a horizontal vault observation station.

to 30 m is necessary. In order to minimize the effect of temperature change, the horizontal vault must be dug quite deep, and many partitions installed to shut out the outside air. The instruments require a constant temperature. An example of such a horizontal vault is shown in Fig. 7.2.

Site selection and construction of horizontal vault observation stations is never a simple task. Japan, however, has approximately 30 such stations, as is shown in Fig. 7.3.

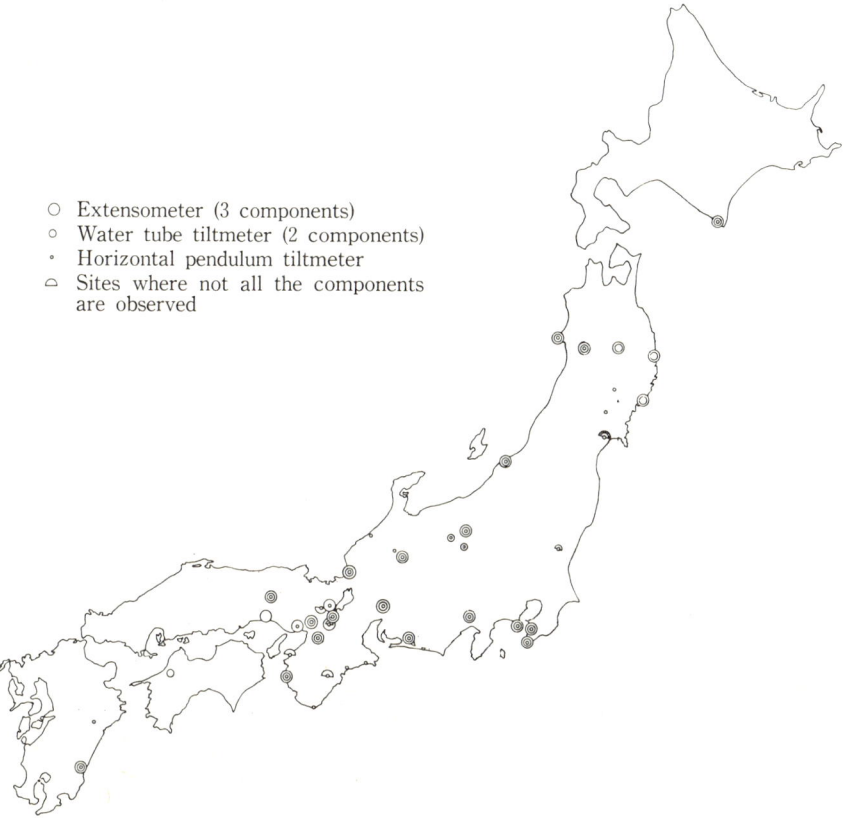

Fig. 7.3 Distribution of horizontal vault observation sites in Japan.

7.2.2 The Results Thus Far

As a result of many years of observation from these stations, our knowledge of this subject has increased considerably.

Complementary observations. In Matsushiro, Southern Kanto, Inuyama, Matsuyama, and Niigata, the results of geodetic surveys were

138 SHORT-TERM PRECURSORY PHENOMENA

compared with the results from the continuous observation stations. The qualitative agreement was excellent. These results demonstrate the reliability of continuous crustal movement observation. They also show that even single-location observation can be effective in watching stage (a), the constant progress of crustal strain.

Detection of anomalies shortly before an earthquake. Here the concern is with a part of the process that cannot be detected by the repeated surveys of the past—the part that concerns stage (b), the arrival of the critical point, and stage (c), the beginning of the principal rupture. Innumerable phenomena have been reported as suspected earthquake precursors in the past. As is also shown in Table 7.3 at the end of this chapter, these are the examples of earthquake precursors detected by crustal movement and other observations that took place anywhere from several hours before to immediately before the earthquake itself:

Kanto earthquake (1923)	Tilting
Tottori earthquake (1943)	Tilting
Tonankai earthquake (1944)	Tilting
Central Gifu earthquake (1969)	Strain and tilting change

The following anomalous changes were observed several days to several years prior to an earthquake:

Daishoji offshore earthquake (1952)	Tilting
Yoshino earthquake (1952)	Strain and tilting change
Odaigahara earthquake (1960)	Strain and tilting change
Hyuga Sea earthquake (1961)	Tilting
Niigata earthquake (1964)	Tilting
Eastern Akita earthquake (1970)	Tilting
Atsumi Peninsula offshore earthquake (1971)	Strain and tilting change
Nemuro Peninsula offshore earthquake (1973)	Strain and tilting change
Central Wakayama earthquake (1973)	Strain change
Izu Peninsula offshore earthquake (1974)	Strain change
Aichi Southern Coast earthquake (1975)	Strain change
Eastern Yamanashi earthquake (1976)	Strain change
Hamamatsu Vicinity earthquake (1976)	Strain change

At the time of the Matsushiro swarm earthquakes (1965–), violent tilting was observed just prior to the increase in seismic activity. A strain change was also reported preceding a small earthquake in the vicinity of the Yamasaki Fault. For details, please refer to the Table 7.3.

Reviewing the observation results during the last five years, when there were a greater number of observation stations, it is clear that certain kinds of precursory phenomena are almost always observed by some of the observation stations prior to earthquakes above a certain magnitude.

Strain step. Although this phenomenon does not function as a short-term earthquake predictor, it does accompany the principal rupture and is useful for studying the earthquake mechanism. At the time of the Central Gifu earthquake of 1969, for example, strain steps were observed at some 16 observation sites, offering hope that the source mechanism can be understood through these observations.

Discovery of migratory crustal movement. The discovery of migratory crustal movement is another of the fruits of continuous observation. This phenomenon was first discovered in the 1960s based on observations in Aburatsubo and Nokogiriyama using water tube tiltmeters. Later reports of this phenomenon came from a group of observation stations in the Tohoku district, and even from Peru. The migration is as slow as 20 to 40 km per year, and it is extremely difficult to explain at this time in geophysical terms. It is thought to have something to do with the fact that earthquake phenomena are no longer constant once stage (b) is reached.

Detecting anomalous movements. Rain and upwelling water are the major sources of noise in horizontal vault observations. Their influence depends strongly on the location. Expansion can sometimes reach an order of 10^{-6}; there can be a delay of several days, however, before it finally shows up. Some consider this discrepancy to be a kind of precursory signal in areas that are prone to numerous small earthquakes. There is a problem, therefore, in quantifying the detection threshold. A workable detection threshold according to Kasahara's definition is shown in Table 7.1 below. The detection threshold of anomalies that occur several hours to several days before an earthquake can be defined as 10^{-7} or so, in terms of expansion or compression.

Table 7.1 Practical Detection Threshold

Type of change	Tilting	Expansion
Instantaneous change (strain-step, etc.)	$0.1 \sim 0.01''$	$10^{-7} \sim 10^{-8}$
Periodical change (hour $\sim$ day)	$0.01 \sim 0.001''$	$10^{-8} \sim 10^{-9}$
Secular variation	$1 \sim 0.1''/y$	$10^{-6} \sim 10^{-7}/y$

7.2.3 Problems in the Horizontal Vault Observation Method and Prospects for the Future

Spatial distribution. Crustal movement is, of course, never uniform, but it does have continuity over space and time. Geodetic surveys can cover an extensive area spatially, but they are sporadic. One of the objectives of continuous observation is to fill in these gaps in time. This objective has not yet been completely achieved since, at present, the observation sites are not spaced densely enough. Since the start of the Earthquake Prediction Project, however, the number of observation sites has increased, and the instances of anomalous precursory phenomena for earthquakes above a certain magnitude have multiplied.

The few observation sites where these phenomena were noticed are not those closest to the origin of the phenomena, however, and the phenomena appeared in various ways. Thus, one may be able to make a "post-earthquake prediction," tracing these phenomena back after the earthquake occurs. Determining which of the various phenomena are related to an earthquake "before" it occurs is the objective; this is a difficult but crucial problem. Meaningful earthquake prediction requires either that more information be obtained from at least several observation sites in order to improve the dependability of the observation of anomalies, or that observation incidents be increased by test field-type research—even if it is done by only one observation site—so as to establish, to some extent, an empirical rule. Increasing the distribution density of the observation sites seems to be the only way to improve the situation, since destructive earthquakes do not necessarily occur frequently in one location.

In the narrow sense of the term, earthquake observation is about seismic waves and their propagation. If the sensitivity of an observation site is good, then that one site can perform this function over a considerable area. Where crustal movement observation is concerned, however, only the areas adjacent to the site can be observed. Thus, the capacity to represent activity over a large area is rather limited. On the other hand, Japan's crustal structure is so complex that areas with homogeneous crustal movement are, at the largest, blocks with sides of 10 to 20 km.

Now that a dense geodetic survey network, with a grid of 8 to 10 km, is being established in areas thought to be significant to earthquake prediction, it is time to consider establishing continuous observation sites of a similar density. The future of the research on strain steps, as well as on migratory crustal movement, hinges on the density of such an observation network.

Time resolving power. In the past, there was a fixed idea that crustal movement is always very slow. The most rapid changes relating to crustal movement were thought to be the earth's tides. Consequently, the feeding speed of the recording paper used for continuous observation was, at one point, as slow as one rotation per week, using 30 cm of recording paper. Changes, at that rate, could be detected in terms of hours at the most.

Once the crustal movement begins to lead toward a principal rupture, however, it seems to progress with considerable speed. During the Hamada earthquake of 1872, for instance, a sudden uplift was observed in the coastal areas 20 minutes before the earthquake. In order to record such phenomena, the observation records must be in terms of minutes rather than hours.

Although observation techniques have recently been modernized to include digital systems, etc., their time resolving power is still not high enough to meet the objectives of all the observations needed.

Sensitivity. It goes without saying that the elucidation of the underlying principles of earthquake phenomena, such as the accumulation of strain, is the right direction for earthquake prediction to take. In order to achieve this goal, continuous observation, as well as an increased number of earthquakes actually observed, is necessary. On the other hand, great earthquakes occur only rarely. Thus, if enough actual earthquake examples are to be observed, magnitude 4 earthquakes must be included in the observations, and the number of observation sites in the grid must be increased.

Once smaller earthquakes are included, however, the size of anomalous phenomena decrease logarithmically, making it necessary to increase the sensitivity of the observations. Today, such new instruments have been developed as the well-type strain gauge that can decrease instrumental noise to 10^{-13} within the band of a one-hour cycle.

In the future, the instrumental noises of continuous strain observation and the distribution of background noise should be examined according to frequency, as is being done with earthquake measurements. By these means it may be possible to find a band that is geophysically significant and superior in S/N ratio.

Recording method. Continuous observation originally emphasized the static side of the investigation of crustal movement, taking a long, careful look at the process in order to clarify the nature of the movement itself.

This aspect should continue to be investigated in the future. And yet, if we are to utilize the results for earthquake prediction, and in particular for short-term earthquake prediction, it is important to discover and analyze the anomalous phenomena that occur just before earthquakes take place. As is indicated in Table 7.3, many earthquakes take place within hours after a short-term anomaly is discovered.

It is necessary, then, to revise the recording method so that a constant watch can be kept over the observation results. The universities concerned have already started to use telemeters for the observation of microearthquakes as well as for the continuous observation of crustal movement. This is done by concentrated recording via telemeter, using more than one site as the nucleus. The Japan Meteorological Agency had already chosen this method when it set up the strain observation network in Tokai and the

Southern Kanto district recently. The method needs further automation, and its efficiency could be improved.

7.3 The Embedded Volume Strainmeter System

7.3.1 Objectives

In order to gather new information on crustal movement that was difficult to obtain using the old horizontal vault-type method of extensometers and tiltmeters, an "embedded volume strainmeter" has been newly developed, based on a completely different principle. Its characteristics are as follows:

(1) The instrument, being embedded, takes less space than the horizontal vault-type instrument, and is much simpler to set up. Therefore it is economical and better suited to the establishment of a dense observation network.

(2) Data will be centrally controlled, as they are telemeter-transmitted by stable telephone lines. The central facility will allow for continuous observation of the raw data, and the terminals can be controlled with various commands.

(3) In addition to the dot strip chart recorder at the central facility and the analog recorder at each terminal, there is a magnetic tape recorder set up at the central facility. The recorder makes it possible to trace any sudden changes in crustal movement by improving the time resolving power of the data recording method.

(4) The stability of the instrument is good and there is no initial drifting.

(5) The instrument withstands large acceleration well, and its normal operation is not affected by strong earthquakes.

(6) The instrument is highly sensitive and can detect very small crustal movements.

These characteristics not only will contribute greatly to short-term earthquake prediction by improving the detection of precursory phenomena, but also will help to elucidate the strain step phenomenon that takes place at the time of an earthquake. The instrument will also be helpful in long-range seismic wave research, functioning as a long-period seismograph. Although the observation period has not yet been long enough to verify its effectiveness where secular variation is concerned, it is expected to be quite useful.

7.3.2 The Principle of the Instrument

First, a vertical observation well 15 cm in diameter and 50 to 300 m deep is drilled into bedrock. Since a core-boring technique developed for oil drilling and geological exploration is used for this operation, it is simple as well as economical. The land surface needed for this observation facility is minimal. At the bottom of the well, a stainless steel pipe is fixed to the well wall with cement. The pipe is 114 mm in outer diameter, 3 mm thick, and 4 m long (see Fig. 7.4). The cement is an expansive type that keeps the pipe under constant pressure so that the strainmeter can follow accurately the contraction and expansion of the bedrock around it.

The sensor (S) at the lower part of the pipe is filled with vacuum-distilled silicon oil. In the upper part of the pipe is a divider with a hole (R) 20 mm long with a sectional area of 0.063 mm². Above this divider is an elastic bellows (B). Although silicon oil fills the section above the divider, the silicon oil inside B and S is completely independent of the silicon oil inside section A. In the upper part of A there is a space filled

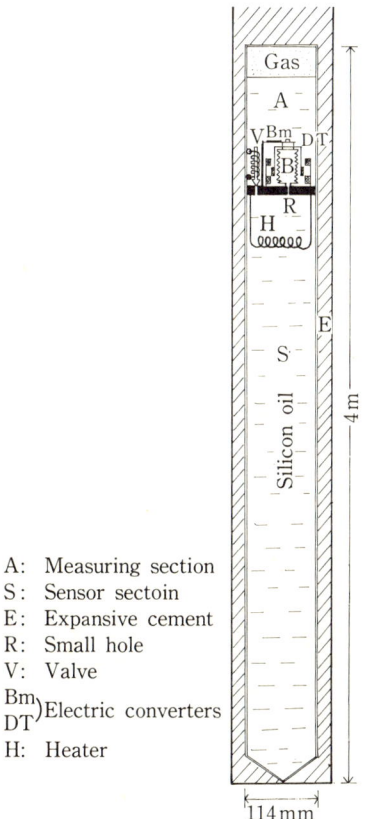

A: Measuring section
S: Sensor sectoin
E: Expansive cement
R: Small hole
V: Valve
Bm) Electric converters
DT
H: Heater

Fig. 7.4 Diagram of embedded strain gauge.

with inactive gas (argon) of one atmospheric pressure (volume approximately 550 cc). Therefore, the oil in section A has a free surface.

When the bedrock around the pipe contracts, S contracts as well, squeezing the oil out through the hole (R) into the bellows (B). With the volume of silicon oil that flows into B as ΔV and the effective sectional area of B as B_a, the upper section of B is lifted up by as much as $\Delta V/B_a$. When the surrounding bedrock expands, on the other hand, the reverse process takes place and the upper part of B sinks as much as $\Delta V/B_a$.

With the combination of the sensor oil and the bellows, the change in volume strain in the surrounding bedrock can be converted to the vertical movement in the upper section of B. The specifications of the present instrument are as follows:

Volume of sensor section (S): 3.07×10^4 cc;
Sectional area of bellows (B): 1.95 cm².

Calculating the compression ratio of silicon oil and the elasticity constant of the piezo electric bimorph element Bm, which functions as a plate spring, the relationship between the volume strain change and the vertical movement of the upper B section $\Delta \chi$ (cm) can be represented by the following equation:

$$\text{Strain change} = 7.33 \times 10^{-5} \cdot \Delta \chi.$$

Consequently, a strain change of 10^{-8} corresponds to a vertical movement of B of 1.4μ.

7.3.3 Converters

Two kinds of converters are provided to detect the vertical movements of B. The first is a differential transformer DT with a working frequency of 4kc and with its electronic circuitry above the ground. In order to prevent the amplifier from being affected by changes in the direct current resistance of the cable, which is caused by variations in temperature or in inter-cable capacity, a low impedance primary coil is used, and a secondary coil is used for current detection. After rectification, the direct output is 1V per 1mm of vertical movement of B (7.3×10^{-6} in strain change). The frequency characteristic here is flat to infinity.

The second converter consists of two affixed piezo electric elements in the form of lead zirconate plate called bimorphs (Bm), which have a great electromotive force against bending. By bending a 45 mm-long plate 1 mm, 24V of electricity are generated. Mechanically speaking, Bm is a plate spring and therefore also restricts the movement of B to the strictly vertical. Its frequency characteristic is flat up to about a 25-minute period. If the period is longer, the output becomes proportional to the velocity of the strain change—in other words, it becomes a velocity characteristic.

The subject of the DT converter is the long period changes ranging from hourly changes to secular variations. The Bm converter, which is 100 times more sensitive than the DT converter, concentrates on the comparatively short period variations in the zone between seismic wave recordings and the earth's tide.

7.3.4 Fixing the Instruments to the Bottom of the Observation Well

In order to faithfully transmit the strain change within the surrounding bedrock to the strainmeter, the gap between the meter and the bedrock wall needs to be completely filled with cement, and the continuity of the whole section should be maintained. This is essential if the strainmeter is to follow the strain changes in the bedrock. A constant tension must be present between the meter and the bedrock. For this reason, a chemically expansive cement is used that expands as it hardens in order to create a primary tension of approximately 0.5 kg/cm^2.

The bottom of the observation well where the strainmeter is installed is bare bedrock. If there is a weak area near the surface, casing pipes are driven in to prevent collapse. The bare vault is washed with clean water. The cement is mixed well with a suitable amount of inflating agent, sand, and water. This mixture—at this point, a mortar—is quite fluid, with a specific gravity of approximately 2.5. It is poured into a cylindrical container and then lowered gently into the observation well which is filled with clean water. The cylinder has a valve on the bottom that opens automatically, due to its own weight, when it reaches the bottom of the well. The cylinder is then lifted gradually and the mortar flows out, forming a mortar column at the bottom of the well. The water is pushed into the upper part of the well and does not mix with the mortar, nor does the sand and water in the mortar separate during this process.

After the cylindrical receptacle is lifted out of the well, the strainmeter is lowered. Its specific gravity is about 5.0, which adds enough weight to the meter that it sinks to the bottom of the mortar column (see Fig. 7.5). When this process is complete, there is about 1m of mortar over the top of the meter. When the mortar is hardened, its strength is 400 to 500kg/cm^2.

This process is pre-checked above the ground using a transparent cylinder with the same inner diameter as the actual observation well, and with apparatus identical to the actual ones being used.

The strainmeter starts recording the violent contraction strain change caused by the expanding mortar 24 hours after it has been installed. It also records a strain change on the order of 10^{-10} which is attributable to microseisms. These recordings confirm that the continuity of the meter and the bedrock has been established.

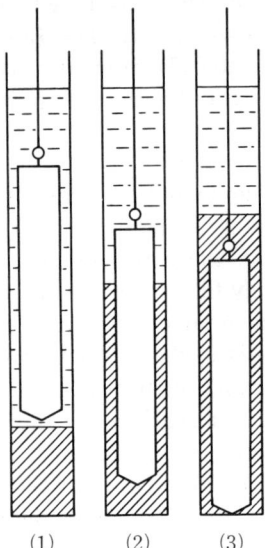

Fig. 7.5 Installing the instrument at the bottom of the well.

7.3.5 Confirmation of Integrated Value and Insurance of Large Dynamic Range

Since the two converters, DT and Bm, should not be deformed beyond a certain point to ensure their linearity, stoppers are provided at the upper and lower ends of the bellows 1.5 mm off the center. This corresponds to 1×10^{-5} in strain change. When the integrated value of the strain reaches this point, no further strain change in the same direction can be recorded.

In order to measure further strain change, an electromagnetic value, V, in the dividing wall (which is closed with a pressure of 8.7 kg/cm²) is opened by remote control when the integrated strain change reaches near this limit, so that further strain change can be recorded. This provides a short circuit between the sensor oil S and the oil A (with a free surface above the upper part of the divider). The hydrostatic pressure that causes the displacement in the converters is removed, and the converter assembly, which constitutes an oscillatory system, immediately returns to its mechanical zero position. The valve then can be closed and everything can start again from zero. As the volume of the casing for the sensor S is 3.07×10^4 cc, the amount of silicon oil shift corresponding to a critical strain of 10^{-4} (which is thought to trigger rupture) is about 3 cc. This is less than 1/100 of the gas at the top of the instrument. By opening and closing the valve, strain change even beyond the critical strain can be measured.

The maneuverability of the valve makes it useful for more than expanding the dynamic range of the converters. If the integrated value of the strain change, which is measured electrically from a mechanically-set zero position, is correct, then it should agree with the value that is set back to zero by opening the valve. The operation of the valve, then, can serve to detect any electric drift in direct current amplifiers. In other words, it allows secular variations to be confirmed, regardless of the amount of strain change.

At present the valve maneuvering operation is performed when the integrated strain change value reaches 5×10^{-6}, or after 6 months. The integrated value, up to that point, is in a precise agreement of 10^{-8}/month or less with the amount retrogressed by opening the valve. This agreement demonstrates that the system, including the electric circuits, is working well. The amount of drift found at the time of the valve maneuver, no matter how minute, is used to correct the secular variation.

7.3.6 The Stability of the Instruments

Stability of the converter assembly. Although the spring constant of the bellows is very small, the constant of the bimorph, Bm, as a plate spring, is quite large. The spring constant of the converter as a whole, therefore, is 100 g/mm. Since the mass of the movable parts is approximately 20 g, the period of the vibration system is as short as 0.029 seconds. The converter assembly, moreover, is immersed in silicon oil with a viscosity of 10 centistockes, with the two coils of DT acting as a dash-pot type damper ($h \doteq 1$). Hence the mechanical stability of the converters is excellent. During the instrument's actual operation, the compressive nature of the silicon oil in the S section works as a force of restitution, which further shortens the comprehensive natural period. Drift that might be attributable to the instrument itself is, therefore, out of the question—a fact confirmed by observation.

Stability against strong vibration. When the area in which a strainmeter is embedded is subject to strong seismic movements, the strainmeter invariably records a great strain change in a short period of time. As

$$\text{Strain change} = \frac{\text{the velocity amplitude of seismic movement}}{\text{seismic wave velocity}}$$

a short period strain change of 10^{-5} is imposed in the case of intensity IV (JMA scale). If the strain meter is unstable under such conditions, then the observation of the strain step that accompanies earthquakes is greatly affected.

In order to protect the converter assembly from such a great strain change over a short period, a small path, R, is provided. The hydrostatic resistance and the compressive nature of the silicon oil in the sensing section makes it work like a CR filter—*i.e.,* it acts as a "high cut" with a time constant of about 6 sec. Sensitivity decreases to 66%. A vibration test of intensity V or thereabouts was performed at the Matsushiro Earthquake Observatory using explosives, but no instability on the part of the strainmeter was found.

Short-range stability. In order to test the short-range stability of the instrument, the valve was left open for one month at the Choshi Observation Point. As was mentioned before, this corresponds to reducing the input to zero by short-circuiting the signal input side. No DT drifting was found at the graduation of 5×10^{-8} (strain). As for Bm, a slight fluctuation of 1×10^{-9} (strain) was found in a one-hour period by the flashing of a thermostatic oven heater contained in the above-ground electric circuits. The line at the zero mark did not move, however.

The noise characteristic of the strainmeter itself, in the shorter band that covers from several seconds to one hour, was 5×10^{-11} (strain) for DT, and 1×10^{-12} for Bm.

Long-range stability. Test observations of the embedded strainmeter began in December 1970 at the Matsushiro Earthquake Observatory. In January 1971, immediately after the tests were completed, a comparison analysis was made of data recorded by strainmeter and pendulum seismograph at the time of an earthquake in West New Guinea. The same analysis was repeated six years later, in January 1977, at the time of another earthquake in New Guinea. The results showed no alteration in sensitivity. This shows that the inner structure and the combination of the strainmeter and the surrounding bedrock did not alter during this period.

7.3.7 The Difference in Actual Measurements between Embedded Strainmeter and Quartz Tube Extensometer

In the case of the horizontal vault method, the bedrock is invaded, as it were, when the tunnel is dug in the mountainside. As a result, a certain transition time seems to be required before the stress distribution becomes stable, regardless of the instrument. Observation by the quartz tube extensometer, on the other hand, does not affect the subject of the observation at all since the observation is made by simply holding the quartz measuring stick against the rock surface.

In the case of the embedded strainmeter, the adjustment of the meter and the bedrock around it becomes a crucial point, since they contract and expand together as one body. The apparent rigidity of the embedded strainmeter is designed so that the "order" will match that of the rocks. In some cases, however, they may not be the same, depending upon the nature of the surrounding rocks. If the rigidity of the rocks is greater than that of the strain meter, the observed amount of strain change becomes greater than the actual value—similar to the situation with the bare rock vault. If the opposite is true, then the observed amount will be smaller than the actual amount of strain change.

The strainmeter has equipment to detect and check the converter's movement and sensitivity by heating the sensor oil for a short period of time with a built-in heater, thus making it expand. However, there is no comprehensive method of testing the whole set-up, including the surrounding rocks. Quantitative discussion of the observation results may have to wait until this problem is resolved.

Given the records of the earth tide, distant earthquakes, and the influence by oceanic tides, and also given the comparison of the embedded strainmeter and the quartz tube extensometer at Matsushiro, it can be speculated that the strainmeter is recording strain change that is 0.5 to 2.0 times the actual crustal volume strain change. They are, however, about the same order of magnitude.

Since the primary purpose of the strainmeter system is to catch the comparatively violent precursory phenomena, small discrepancies are not a problem. So far, the discussions of the amount of strain change, including the secular variation, caught by continuous observation have not been strictly quantitative. Quantitative examination will, of course, be necessary in the future. By that time, comprehensive sensitivity can probably be defined by the analysis of the various records obtained by embedded strainmeters.

7.4 The Development of an Observation Network in the Tokai and Southern Kanto Districts

An observation network using embedded strainmeters was set up in the area shown in Fig. 7.6 and Table 7.2. The outputs of DT and Bm are transmitted to the Japan Meteorological Agency by telemeters. The value of Bm

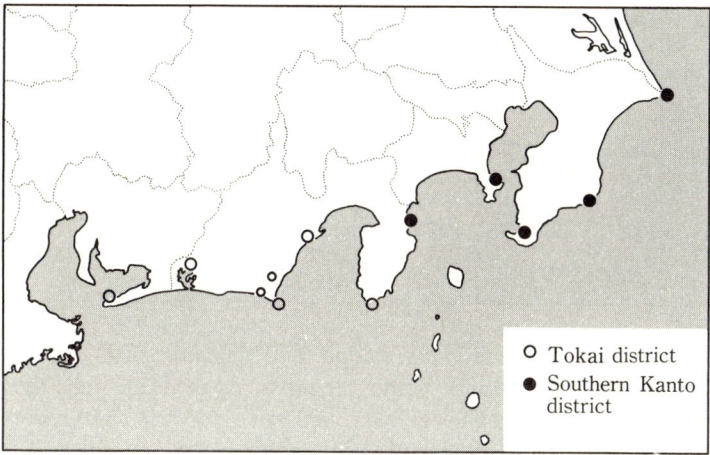

Fig. 7.6 Development of an embedded strainmeter network in the Tokai and Kanto districts.

SHORT-TERM PRECURSORY PHENOMENA

Table 7.2 Strainmeter Observation Network

Location	Depth	Date established	Date observation began	Recording method (local)	Recording method (Tokyo Central)	Distance from neighboring site	Rock type
Irako	141 m	9/ 5/'75	4/1/'76	Continuous analog 2 cm/hr	Visible dot strip & digital tape		Black schist (Palaeozoic)
Mikkabi	51	10/11/'75	4/1/'76	Continuous analog 2cm/hr	Visible dot strip & digital tape	54 km 56	Clayslate (Palaeozoic)
Hamaoka	250	2/21/'77	4/1/'77	Continuous analog 2cm/hr	Visible dot strip	10	Mudstone (Tertiary)
Omaezaki	208	11/ 1/'75	4/1/'76	Continuous analog 2cm/hr	Visible dot strip & digital tape	15	Mudstone (Tertiary)
Haibara	250	2/23/'77	4/1/'77	Continuous analog 2cm/hr	Visible dot strip	30	Mudstone (Tertiary)
Shizuoka	60	11/19/'75	4/1/'76	Continuous analog 2cm/hr	Visible dot strip & digital tape	56	Sandstone (Palaeogene)
Irozaki	133	2/ 2/'76	4/1/'76	Continuous analog 2cm/hr	Visible dot strip & digital tape	56 55	(Andesite) (Brecchia)
Ajiro	120	9/ 2/'76	4/1/'77	Continuous analog 2cm/hr	Visible dot strip & digital tape		Basalt (Lava)
Yokosuka	146	9/ 8/'76	4/1/'77	Continuous analog 2cm/hr	Visible dot strip & digital tape	60	Mudstone (Neogene)
Tateyama	190	8/ 4/'76	4/1/'77	Continuous analog 2cm/hr	Visible dot strip & digital tape	30	Mudstone
Katsuura	180	9/21/'76	4/1/'77	Continuous analog 2cm/hr	Visible dot Strip & digital tape	45	Mudstone (Neogene)
Choshi	100	12/24/'76	4/1/'77	Continuous analog 2cm/hr	Visible dot strip & digital tape	78	Sandstone (Palaeozoic)

from each site is recorded every second, and the value of DT is recorded every 5 seconds on magnetic computer tapes. At the same time, the value of DT and Bm is recorded every 30 seconds on a dot strip chart recorder as a visual record, with the observation sites color-coded.

The sensitivity of these instruments is $5 \times 10^{-9}/V$ for Bm and $5 \times 10^{-7}/V$

for DT. The sensitivity of the records on magnetic tape is 0.1mV unit and, on the dot strip chart recorder, 1V/8mm.

Five observation sites in Tokai and five in Southern Kanto are independent systems. Access to each site, the read-out of the observed value, the AD conversion, transmission, and recording of tapes are all controlled by a central computer. This computer also handles the surveillance of operational conditions, an alarm if a disturbance occurs, printout of the operation records, and an alarm when the observation value exceeds the norm and becomes anomalous. Under this system, each point in the network is under constant surveillance.

7.5 Observation Results

7.5.1 The Earth's Tide and Secular Variation

Figure 7.7 shows the record of the earth tide by DT at Choshi. The strain change and oceanic tide at Irozaki is represented in Fig. 7.8. In this figure the amplitude of the strain change is on the order of 10^{-7}. Bm shows the velocity characteristic during this period. There is a phase lag of $\pi/2$ between DT and Bm. Because Irozaki is located on the tip of a cape that protrudes into the ocean, the record at Irozaki is greatly affected by the ocean loading. There is no time delay between the tide and the strain change, as is apparent in the figure.

The strain changes since observations began at those 10 sites are shown in Fig. 7.9. There seems to have been almost no movement at the Irako and Choshi sites. The movement at Mikkabi and Shizuoka is great because the vaults are shallow. All the others show continuous contraction. Examination of the rocks in which the strainmeters are embedded reveals that where little movement is shown the rocks are old and hard, and there is a great deal of contraction in the new mudstone. Taking into consideration the fact that the instruments themselves are free of drift, and that there has been no movement recorded at Choshi where the rocks are hard, the contractions observed in the mudstone area probably result from creep or the collapse of pores in the rocks caused by the pressure surrounding the meters.

In order to analyze this creep phenomenon, a core sample was taken at Katsuura where the rate of contraction is great. The sample was then dipped into water, which it absorbed quickly. This experiment explains the contraction phenomenon as follows: in homogeneous areas of new mudstone such as Katsuura, the section that was dry before the boring operation became moist when water seeped down from near the surface via the vault. The dry condition of the section was confirmed by examining the core sample taken at the time of the operation. As long as water continues to seep, the contraction attributable to the expansion of the surrounding rocks—an expansion

152 SHORT-TERM PRECURSORY PHENOMENA

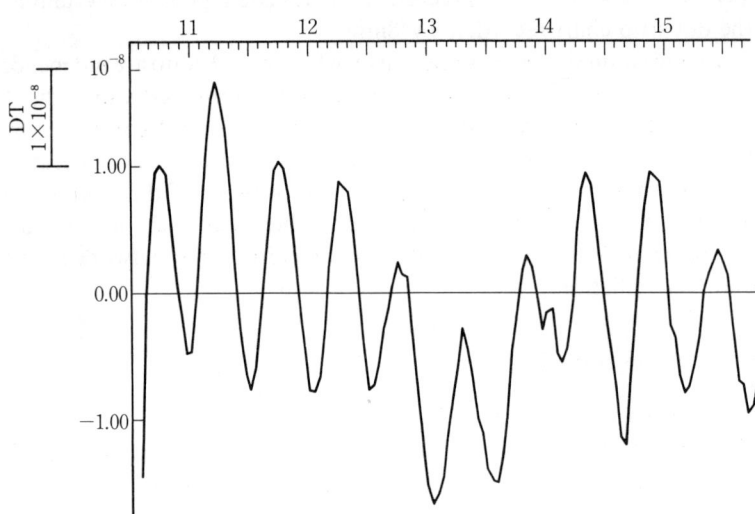

Fig. 7.7 Earth's tide (DT) at Choshi, Chiba Prefecture (at the northeasternmost tip of the Boso Peninsula), November 11–15, 1977.

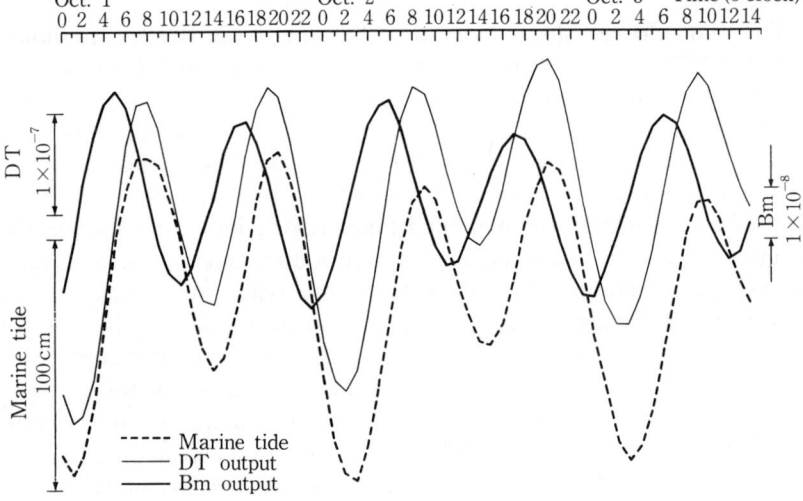

Fig. 7.8 Earth's tide (DT and Bm of embedded strainmeter) and marine tide at Irozaki (at the tip of the Izu Peninsula), Oct. 1–3, 1977.

Atmospheric pressure coefficient: 0.8×10^{-8}/mb. Strain change coefficient by marine tide: 0.24×10^{-8}/mb. The strain gauge is set up 80 m below sea level.

CONTINUOUS OBSERVATION OF CRUSTAL MOVEMENT 153

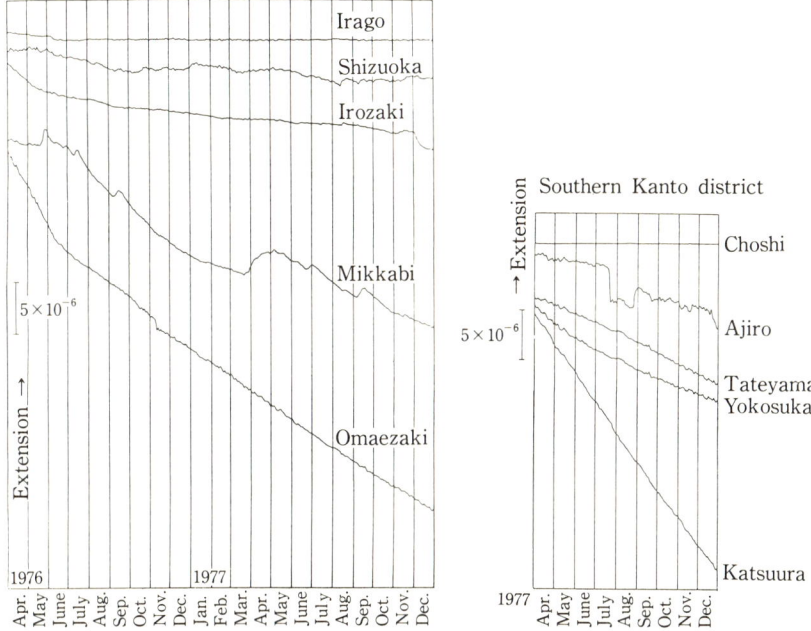

Fig. 7.9 Secular strain variation in the Tokai district, 1976–77.

brought on by the absorption of water—will continue indefinitely. In order to test this assumption, the space between the hard rock wall and the casing pipe in the Katsuura observation well was filled with cement in mid-March 1979. As a result, the rate of contraction from that time on has become much slower, due to a decrease in the amount of water seeping to the surrounding rocks, although the supply was not entirely cut off.

Given the results of the test at Katsuura, the full-hole cementing method, in which the rocks surrounding the instruments are kept completely free of well water, has been employed since 1979 in the construction of additional observation wells positioned in weak ground.

7.5.2 The Effect of Meteorological Changes

Rain. The effect of rain and spring water is great in the case of the horizontal vault extensometer and water tube tiltmeter, for it can amount to from 10^{-7} to 10^{-6} in linear strain. The embedded strainmeter, however, is not affected, even after continuous rainfall of about 100mm, as is shown in Fig. 7.10. The only exceptions are the shallow wells in Mikkabi and Shizuoka. As a result of this experience, the wells in Southern Kanto are all more than 100 m deep.

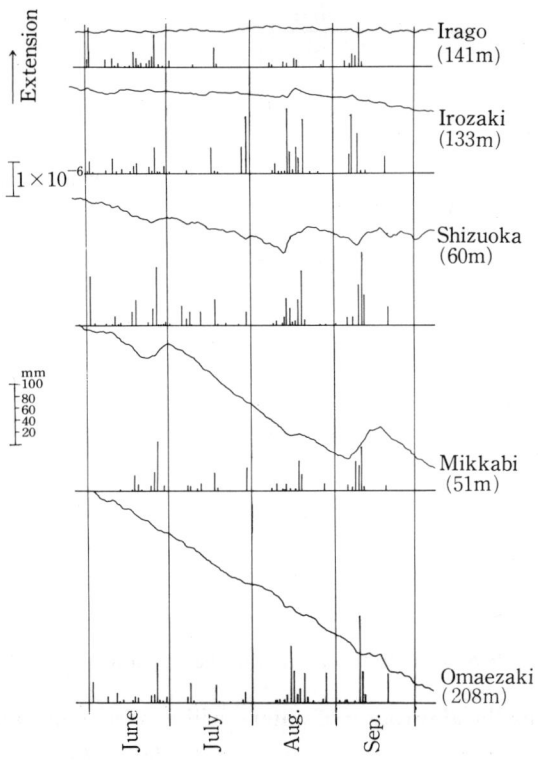

Fig. 7.10 Correlation between strain and rainfall at five points in the Tokai district, June~September 1977.
Distances in parentheses are the depths of the embedded instruments.

Atmospheric pressure. Horizontal vault-type observation wells are seldom affected by atmospheric pressure. In the case of the embedded instruments, however, the influence of atmospheric pressure changes must be considered, since these instruments are highly sensitive. The distinction between earthquake precursors and strain change caused by incidents such as the passage of a front with a great barometric gradient should be clearly made. The correlation between strain and atmospheric pressure is demonstrated in Fig. 7.11. The comparison was made during a period with little rain.

The atmospheric pressure coefficient in soft areas such as Omaezaki was double the coefficient recorded in areas with hard rocks such as Irako. An air pressure change of 20mb or thereabouts can take place in a comparatively short period of time; corrections can easily be made in this case, however, because there is no time lag different from the case of rain.

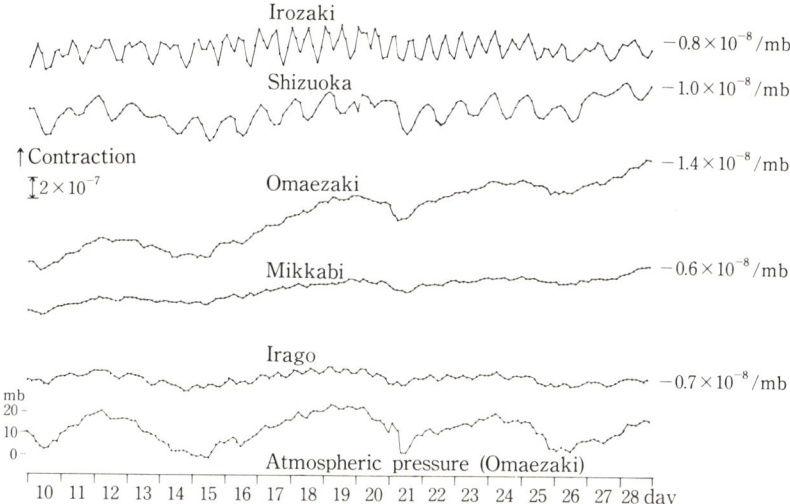

Fig. 7.11 Correlation between strain and atmospheric pressure at five points in the Tokai district, Feb. 9–28, 1977.

7.5.3 Spatial Representation of One-point Observation

At the time of an earthquake, extensive information can be obtained about seismic activity, even from a one-point observation. This is because seismic waves propagate and one has only to wait for them to arrive.

Crustal movements, on the other hand, do not propagate like seismic waves. Even with a quartz tube extensometer in a horizontal vault 100 m long, strain can be measured only at one particular spot. How representative strain change is, when it is measured at any one spot, is an important problem since this determines the density of the observation network.

Figure 7.12 is a record of a distant earthquake. It shows that the correlation between surface waves over a 30-second period at each observation site is very good. Thus, if seismic surface waves were to affect the Tokai district for a 30-second period, the response at each site would be uniform. Therefore we can make one point, for instance Omaezaki, represent the entire Tokai district.

On the other hand, each site records a strain change that corresponds to ground noise over a period ranging from several minutes to one hour. The amplitude of the noise differs depending on the location, but it is between 10^{-9} and 10^{-8}. The cause of this phenomenon has not yet been explained. An example is shown in Fig. 7.13. As is apparent in the figure, an extremely good correlation is seen only between Hamaoka, Omaezaki, and Haibara. This seems to be quite different from the results caused by distant earthquakes.

Furthermore, the sudden change in strain mentioned in the following sec-

156 SHORT-TERM PRECURSORY PHENOMENA

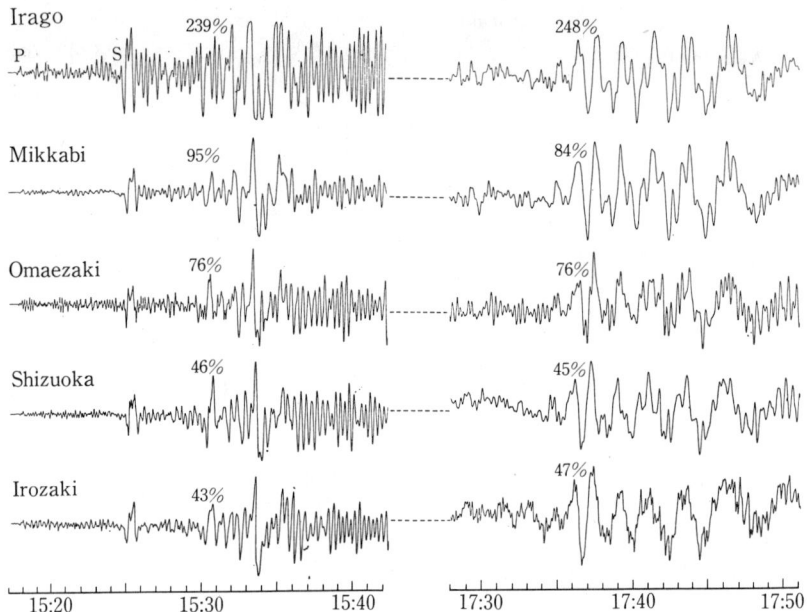

Fig. 7.12 The Sumbawa Island earthquake of Aug. 19, 1977, as recorded by strain gauge meters in the area.
Percentage figures represent comparisons with the average amplitude.

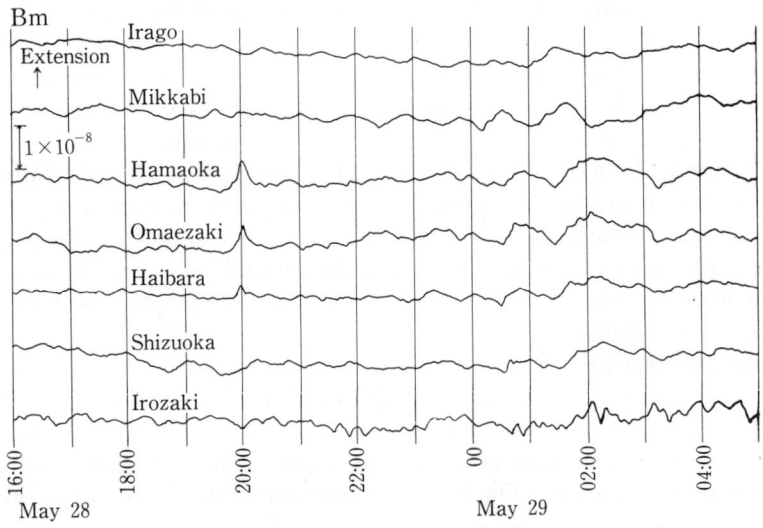

Fig. 7.13 Strain noise recorded at stations in the Tokai district, May 28–29, 1977.

tion never appears at more than two points simultaneously. At the time of a sudden change that lasted several hours in Omaezaki in November 1976, the neighboring observation sites at Hamaoka and Haibara had not yet been added to the network.

According to past observation results, the instruments react uniformly, to a considerable degree, to short-period vibrant strain change. When the period is prolonged beyond 40 to 50 minutes, however, or becomes non-vibrant, the area that reacts uniformly as a unit is limited to a square of 20 km or so. Detailed information on this problem will require further observation in the future. For this reason the two additional observation sites near Omaezaki will be extremely useful.

7.5.4 Sudden Strain Change

There has been a preconception that crustal change, although it is not uniform, is always very slow. Sudden change at the time of an earthquake, caused by the release of strain, has recently been drawing attention. Sudden but aseismic movements, however, are rarely reported.

It was found, using an embedded strainmeter to make highly sensitive observations with high time resolution, that crustal strain sometimes changes in a very short time in some areas. These changes include both step-type changes that are complete in several minutes and drift-type changes that take from several hours to two days. There are others also, such as the small step-type changes that are recorded often, but only by the highly sensitive Bm (bimorph).

Since the instruments that recorded these changes continued to function smoothly, recording tides and earthquakes, and since the integrated values, including the sudden changes, were tested by opening and closing the valve, the sudden changes must be regarded as real phenomena.

Examples of strain changes of the slower type at Omaezaki and Ajiro are shown in Figs. 7.14 and 7.15 respectively. Figure 7.16 shows examples of the step-type changes from the records at Ajiro made just before a nearby earthquake. Figure 7.17 shows the small structure of the step-type changes recorded at Shizuoka. It is intriguing to note that slow and rapid small changes are intermingled with both the drift-type and step-type changes.

7.5.5 Correlation between Sudden Changes and Earthquakes

There have been several instances in which earthquakes took place anywhere from several hours to several days after aseismic sudden strain changes occurred in the same area. The major examples are the following:
(1) The Eastern Yamanashi earthquake (June 16, 1976) occurred after frequent step-type strain changes took place in Shizuoka (see Fig. 7.18).

158 SHORT-TERM PRECURSORY PHENOMENA

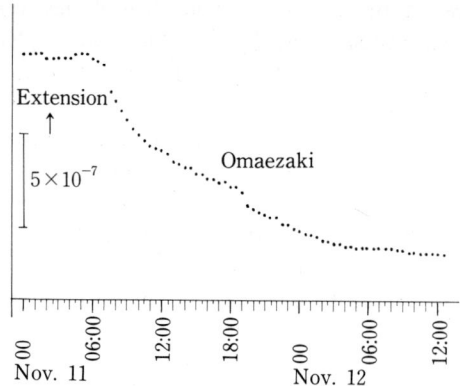

Fig. 7.14 Strain change of the slower type recorded at Omaezaki (at the southern tip of Shizuoka Prefecture), Nov. 11–12, 1976.

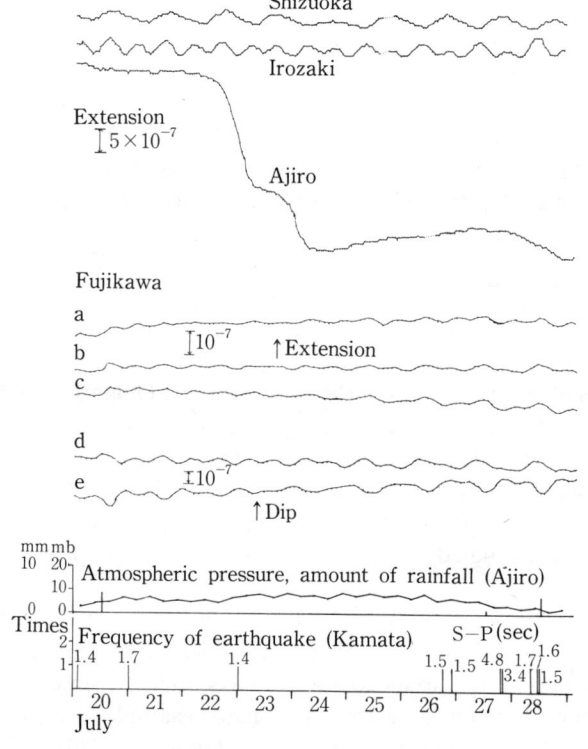

Fig. 7.15 Slower type strain change at Ajiro on the Izu Peninsula, recorded at stations in the area July 20–28, 1977.

    a: Extensometer #1   d: Water tube tiltmeter #1
    b: Extensometer #2   e: Water tube tiltmeter #2
    c: Extensometer #3

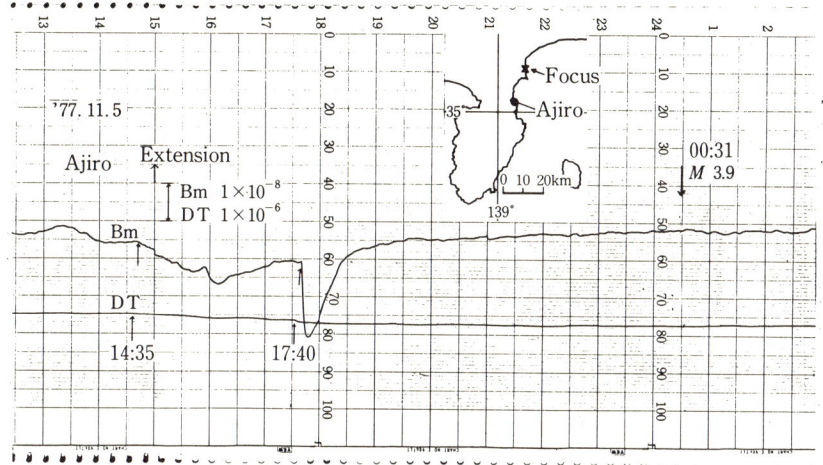

Fig. 7.16 Data recorded by an embedded strainmeter at Ajiro prior to the Nebukawa earthquake of Nov. 6, 1977 (00: 31 hours, M 3.9).

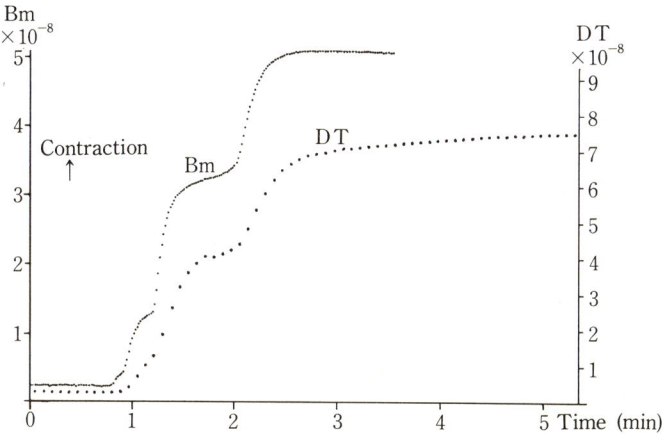

Fig. 7.17 Sudden strain change recorded at Shizuoka in March, 1976.

(2) The Ise Bay earthquake (July 26, 1976) occurred after step-type strain changes took place in Irago (see Fig. 7.19).

(3) The Nebukawa earthquake (Nov. 6, 1977) occurred after step-type and subsequent rapid drift-type strain changes took place in Ajiro (see Fig. 7.16).

(4) The Izu Oshima offshore earthquake (Jan. 14, 1978) occurred one

160   SHORT-TERM PRECURSORY PHENOMENA

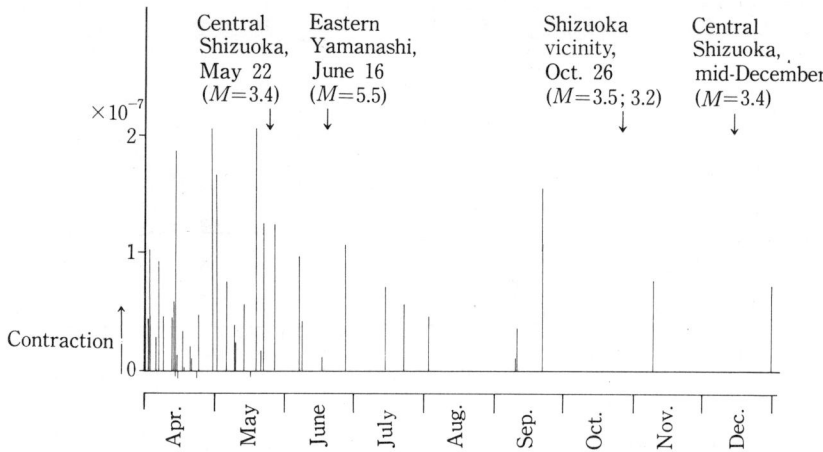

Fig. 7.18 Strain step changes in Shizuoka before a series of earthquakes in 1976.

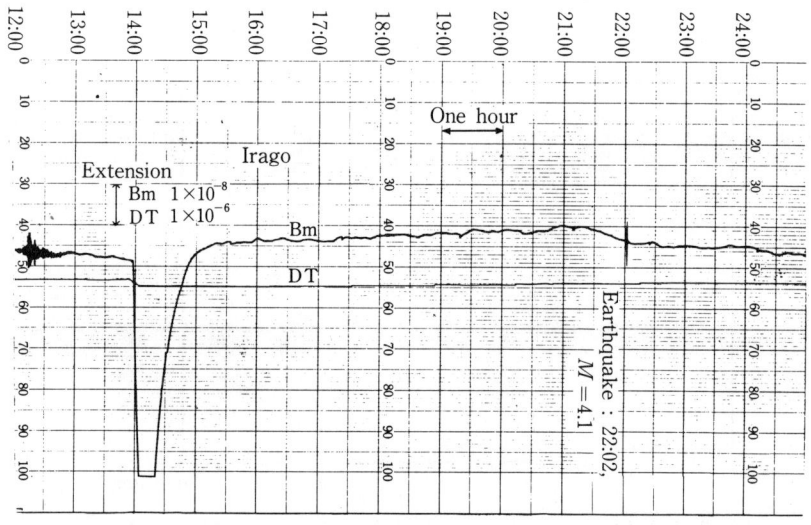

Fig. 7.19 Strain step changes at Irago (Aichi Prefecture) preceding an earthquake in Ise Bay, July 26, 1976.

month after previously unknown contractual changes and five days after a rapid stretch change was noted.

On the other hand, there are instances where considerable strain changes that were not followed by earthquakes—Omaezaki (Nov. 11, 1976; see Fig. 7.14), for instance, or Ajiro (July 22, 1977; see Fig. 7.15).

CONTINUOUS OBSERVATION OF CRUSTAL MOVEMENT 161

The physical mechanism of the initial motion that leads to rupture is not yet clearly understood, making the physical interruption of these sudden strain changes difficult. One hypothesis is the "pre-slip" theory; another is that stress redistribution occurs immediately prior to the earthquake. The latter is based on the idea that the strain energy caused by tectonic forces do not accumulate uniformly, but rather in localized clusters, rather like the onset of freeway congestion. This energy is thought to regroup around the principal rupture point via redistribution just before the earthquake.

In any case, it is known that strain distribution undergoes a sudden and local change, according to the data gathered by the embedded strainmeter observation network.

7.6 Toward High-density Continuous Observation

Continuous observation, based on the results from observations over a

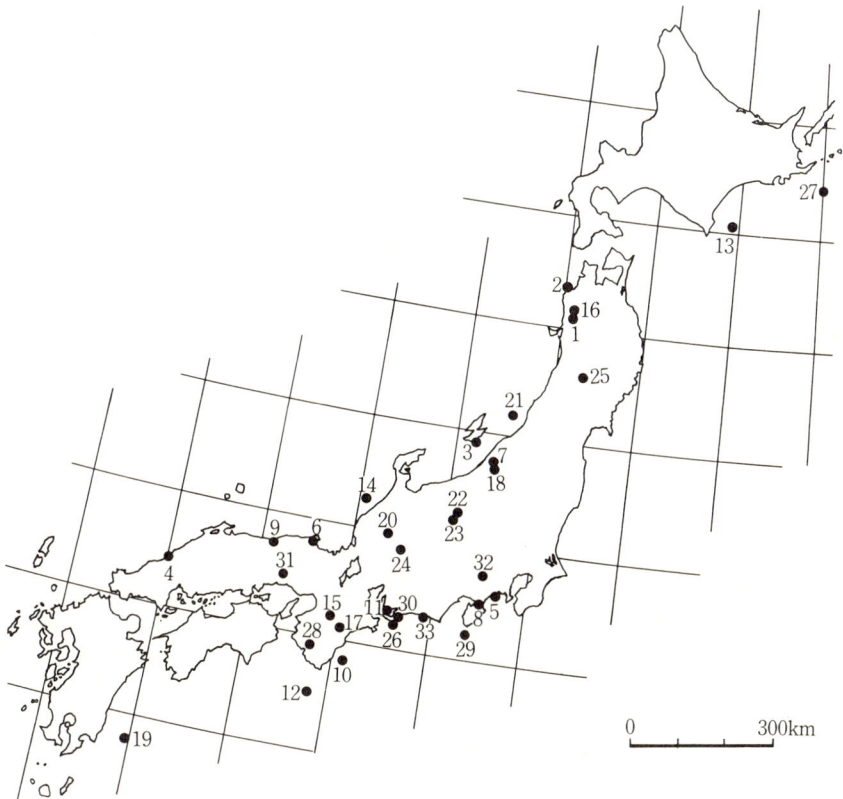

Fig. 7.20 Earthquakes preceded by anomalous crustal movement. Numbers refer to the list in Table 7.3.

long time, is entering a new dimension. Namely, static and statistical investigation, is becoming dynamic and capable of observing current conditions in real time.

The purpose of observation is earthquake prediction. Thus it is crucial to be able to make a judgment before the earthquake actually takes place. When a great earthquake is imminent, past experience clearly indicates that there will be some degree of anomalous activity. Therefore, even if the empirical laws, or the laws governing these particular phenomena, are not sufficiently well-understood, a judgment can still be made.

This alone is not sufficient, of course. Ultimately the pre-earthquake mechanism must be elucidated, and for this purpose a high-density continuous observation network to observe crustal movements will play a significant role.

Table 7.3 Examples of Observed Anomalous Crustal Movement Preceding Earthquakes [based on studies by Y. Suzuki of the JMA]

No.	Year	Name of earthquake	M	Observation method	Observation site	Time when anomalous crustal movement appeared before earthquake	Description [Source or reporter]
1	1694	Noshiro earthquake	7.0	Observation	Omori-shimohama (Noshiro City)		A large tree which had been completely buried in the sand at Kudamajiri, Omori-shimohama, surfaced overnight. Also, in early May, a stone lantern in front of the Sanno Gongen Sanctuary toppled though no wind was blowing at the time. [C. Kishino]
2	1793	Western Tsugaru earthquake	6.9	Observation	Ajigasawa	A few hours	The sea level at Ajigasawa dropped, and people were expecting tidal waves. Instead of the waves, however, an earthquake first hit the area. People fled to the beach, but big waves were coming in. As a result, many children were drowned. [Tsugaru

Table 7.3. *Continued*

No.	Year	Name of earthquake	M	Observation method	Observation site	Time before earthquake	Description [Source or reporter]
3	1802	Sado earthquake	6.6	Observation	Ogi Harbor	Approx. 4 hours	Almanac] A strong temblor hit Ogi Harbor and the bay dried up. People were worried about floods and tidal waves. At around 2:00 p.m. an earthquake hit. Many houses were reported destroyed. [Sado Chronicle]
4	1872	Hamada earthquake	7.1	Observation	Hamadaura	Approx. 20 min.	There was a low tide prior to the earthquake. A report came in that the base of Tsurushima Island was exposed, and the fishermen could pick abalone by hand. When they returned home, an earthquake hit accompanied by tsunami. [Hamada Observatory]
5	1923	Great Kanto earthquake	7.9	Tide level	Aburatsubo	10 years	Scientists had been viewing the vertical movement of the crust obtained from the average monthly tide level at Aburatsubo. They noted that while the level had been subsiding steadily since the Meiji Period, it stopped subsiding about 10 years prior to the earthquake and remained at the same level when the event took place. [Geographic Research Institute]
				Tilt change	Mitaka	Approx. 4 years	Examination of the Mitaka Rhomboid line revealed that the dimension increased about 1/10000 4 years before the earthquake. When the earthquake hit, the increase jumped to 3/100000. [Geo-

Table 7.3 *Continued*

No.	Year	Name of earthquake	M	Observation method	Observation site	Time before earthquake	Description [Source or reporter]
				Tilting	Tokyo	1 month	graphic Research Institute] At the University of Tokyo, the Omori seismograph showed an anomalous tilt to the west from July 31 to Aug. 1. A violent tilt took place again between 4:00 a.m. and noon on Sept. 1. [A. Imamura]
						Approx. 8 hours	
6	1927	Northern Tango earthquake	7.5	Observation	Sanzu	Approx. 2–3 hours	Some fishermen were out fishing and returned about 4:00 p.m. On the way back they noticed that a rock they had never seen before was exposed above water. They wondered about it, but did not dwell on it. [A. Imamura]
7	1927	Sekihara earthquake	5.3	Leveling	Leveling Point 3755–3760		After discounting the extensive movement here, what was left seemed to represent a truly radical change. There was a bulge that reached as high as 2.7 cm. [A. Imamura]
8	1930	Northern Izu earthquake	7.0	Leveling	Leveling Point 9333–9332		There was a bulge of approx. 3 cm with its center 2 to 4 km north of Itocho. A bulge that was determined to be an earthquake precursor had the sam shape, except that its volume was about 50% greater [C. Tsuboi]
9	1943	Tottori earthquake	7.4	Tilting	Ikuno Mine	Approx. 6 hours	The horizontal pendulum titlmeter in the Ikuno mine 60 km southeast of the epicenter recorded an angle of 0.1 sec 6 hours before the earthquake. [K. Sassa *et al.*]
10	1944	Tonankai earthquake	8.0	Leveling	Kakegawa	Approx. 10 years	The tilting of Kakegawa slowed down nearly 10 years prior to the earth-

Table 7.3 *Continued*

No.	Year	Name of earth-quake	M	Observation method	Observation site	Time before earth-quake	Description [Source or reporter]
				Leveling	Kakegawa	Several hours	quake. This tendency is thought to be a precursor or a small-scale beginning of the radical uplift that occurred at the time of the earthquake. [Geographic Research Institute] A "closing error" that surpassed the measurement error by more than 6mm/km was discovered at the time of observation in the northern part of Kakegawa on the day of the event from 10:10 a.m. to 12:00 p.m. [Geographic Research Institute]
11	1945	Mikawa earth-quake	7.1	Tide level	Senma, Nishiura, Maeshiba	Approx. 4–5 years	The results of the tide examination indicated that the ground had been subsiding from 1931 through 1940. It then began to uplift and reached a maximum of 20 cm until recent years. [N. Miyabe]
12	1946	Nankai earth-quake	8.1	Leveling	Kii Peninsula and Shikoku		The slow uplift and subsidence that continued for 30 years before the earthquake shows a negative correlation with that at the time of the event. This seems to indicate a close relationship between pre-earthquake crustal movements and the event itself. [Geographic Research Institute]
				Leveling	Muroto Cape	11 years	There was no indication that the subsidence of Muroto Cape had accelerated since 1935. [I. Tsubokawa]
				Tide level	Uwajima	Approx.	Rather, the movement

166 SHORT-TERM PRECURSORY PHENOMENA

Table 7.3 *Continued*

No.	Year	Name of earthquake	M	Observation method	Observation site	Time before earthquake	Description [Source or reporter]
				Tide level	Tosashimizu	5–6 years	seemed to have slowed down. Examination of the vertical movement of Uwajima from the average annual tide range (Hososhima–Uwajima) clearly reveals a radical change of about 20 cm from 1941 to the time of the event. [N. Fujita]
						4–5 years	Examination of vertical movement in Tosashimizu obtained from the average annual tide range (Hososhima–Tosashimizu) seemed to indicate an uplift of several centimeters 4 to 5 years before the earthquake, compared to the average prior to that. [Geographic Research Institute]
13	1952	Tokachi offshore earthquake	8.1	Tide level	Hachinohe	Approx. 5 years	Examination of the vertical movement at Hachinohe obtained from the annual average tide range (Miyako–Hachinohe) revealed a continuous uplift and subsidence of Hachinohe against Miyako since 1947. This movement seems to have been earthquake-related. [N. Fujita]
14	1952	Daishoji offshore earthquake	6.8	Tilting	Ogoya	3 months 8 days	The tiltmeter showed a remarkable change 3 months and especially 8 days before the event. Immediately after the event, the direction of tilt turned almost perpendicular to the epicenter. [E. Nishimura]
15	1952	Yoshino earth-	7.0	Strain	Mt. Osaka	Approx. 10	The strain gauge showed a strain change of 2.5 ×

Table 7.3 *Continued*

No.	Year	Name of earthquake	M	Observation method	Observation site	Time before earthquake	Description [Source or reporter]
		quake				months	10^{-6} in amplitude about 10 months prior to the event. It returned to the original position within a year after the event. [K. Sassa *et al.*]
				Tilting	Yura	About 10 days	3 days before the earthquake the direction of tilt reversed and turned toward the epicenter. [E. Nishimura]
16	1955	Futatsui earthquake	5.7	Leveling	Leveling Point 5856–5857		Examination of the change between 1942 and 1949 revealed that the tilt, which had been slightly upward toward the east prior to this period, drastically changed to a downward tilt toward the east. This tendency returned to normal completely during the period from 1949 to 1955. [S. Miyamura, *et al.*]
17	1960	Odaigahara earthquake	6.0	Strain, tilt	Various observation sites in Kii Peninsula	Several months	Several months prior to the earthquake, anomalies such as expansion and tilt changes were observed at observation sites at Kishu, Ushio Cape, Yura, and Oura. Anomalous changes were seen in late Oct. and again around Dec. 10 to 20. [E. Nishimura]
						Several days	
18	1961	Nagaoka earthquake	5.2	Leveling	Leveling Point 3759	Approx. 3 years	Between 1958 and 1961 the hitherto subsiding portion of BM 3759 was elevated approximately 5 cm. Around the epicenter, the uplift was believed to be either the same or greater. [J. Okada]
19	1961	Hyuga Sea earth-	7.0	Tide level	Hososhima	Approx. 4 years	Examination of the vertical movement at Hososhima obtained from the

Table 7.3 Continued

No.	Year	Name of earthquake	M	Observation method	Observation site	Time before earthquake	Description [source or reporter]
							average difference of sea level (Uwajima—Hososhima), showed that an anomalous uplift started in 1957. In the 4 years before the earthquake took place in 1961, the velocity of the uplift was + 14.5 mm/y. [T. Dambara]
				Tilting	Makinomine	Approx. 12 days	The horizontal pendulum tiltmeter registered anomalous movements starting on Jan. 10, and obvious anomalous tilting began on Feb. 15 in reverse direction of the epicenter. The tilting then reversed again on Feb. 23, 4 days before the earthquake took place. [E. Nishimura]
20	1961	Northern Mino earthquake	7.0	Leveling	Katsuyama–Ono		Reviewing the changes between 1949 and 1961, the amount of change recorded during the period 1929–1949 almost completely disappeared. Judging from the uplift that occurred from 1949 to immediately before the earthquake between Katsuyama and Ono, an uplift of 30 to 40 cm must have occurred around the epicenter. [I. Tsubokawa]
21	1964	Niigata earthquake	7.5	Leveling	Niigata	Approx. 10 years; 4–5 years	In 1954 or thereabouts— approximately 10 years before the earthquake— an anomalous uplift was noted. This uplift reached its maximum in 1959 and maintained this condition for 2 to 3 years, followed by the earthquake itself. [T. Dambara; Geogra-

CONTINUOUS OBSERVATION OF CRUSTAL MOVEMENT 169

Table 7.3 *continued*

No.	Year	Name of earthquake	M	Observation method	Observation site	Time before earthquake	Description [Source or reporter]
				Tilting	Mase	6–7 years; approx. 4 years	phical Research Institute] The change registered by the water tube tiltmeter showed that there was active crustal movement from 1957 to 1958. The north–south component especially dropped sharply from 1960 on reaching the lowest point in early 1964. It then began to rise again shortly thereafter. [Yahiko Crustal Movement Observatory]
				Tide level	Nezugaseki	6–7 years; approx. 4 years	The general trend based on the average monthly tide range (Kashiwazaki–Nezugaseki), is similar to the results reported by Tsubokawa *et al.* Compared to the slight subsidence 1 to 2 years prior to the earthquake reported by Tsubokawa and others, the subsidence in this case seems to have leveled off. [Geographic Research Institute]
22	1965	Matsushiro swarm earthquakes		Tilting	Matsushiro		The water tube tiltmeter registered "tilt steps" frequently. At that time, changes in the minute structure of the steps that are said to begin changing before an earthquake were detected. [Earthquake Research Institute, University of Tokyo]
23	1967	Omi earthquake	5.1	Leveling	Leveling Point 5090–5128		Surveys were performed 3 times—in May, Aug., and Nov. of 1967. At Omi Village, 2/3 of the uplift that occurred during the May-to-Aug. period seemed to have settled

Table 7.3 *continued*

No.	Year	Name of earthquake	M	Observation method	Observation site	Time before earthquake	Description [Source or reporter]
24	1969	Central Gifu earthquake	6.6	Strain, tilting	Inuyama	Approx. 290–300 days; 50–60 days; several hours	between Aug. and Nov. This corresponds approximately to the general concept of crustal movement that results in earthquakes. [I. Tsubokawa] No change was registered on quartz tube extensometer and water tube tiltmeter for 250 days after the observation began. Then a rapid change appeared during the 240-day period that followed. The direction of change reversed shortly before the event. Several hours before the earthquake, the extensometer registered a great extension. [K. Iida et al.]
				Strain, tilting	Kamidakara	Approx. 8–9 months	The quartz tube extensometer and the water tube tiltmeter registered a remarkable change during the period from the end of 1968 to early 1969. [Kamidakara Crustal Movement Observatory]
				Optical wave	Fuchi		The results of the optical wave survey demonstrated a remarkable contraction throughout the base line in 1968, which seems to be related to the earthquake. [Earthquake Research Institute, University of Tokyo]
25	1970	Southeastern Akita earthquake	6.2	Leveling	Leveling Point 5540–5556		There was a trend toward subsidence from Kitakami to Yokote during the period between 1897 and 1934, but this ceased during the 1934–1956 period.

Table 7.3 continued

No.	Year	Name of earthquake	M	Observation method	Observation site	Time before earthquake	Description [Source or reporter]
							In the following period from 1956–1966, an area 15 km wide on the prefectural border uplifted 2 cm. [Geographic Research Institute]
				Tilting	Nibetsu	Approx. 60–70 days	Comparison of the predicted value—calculated from a year's data based on the observed value—with the results obtained by water tube tiltmeter indicated a difference greater than the normal predicted error in Aug. 1970 or thereabouts. [Ishii]
26	1971	Atsumi Peninsula offshore earthquake	6.1	Strain, tilting	Inuyama	Approx. 250 days	From the changes that appeared on the quartz tube extensometer and the water tube tiltmeter, it is apparent that approximately 250 days elapsed from the first appearance of the rapid change to the actual event. [K. Iida et al.]
27	1973	Nemuro Peninsula offshore earthquake	7.4	Strain, tilting	Erimo	Approx. 1 year; approx. 1 month	About a year before the earthquake, the quartz extensometer and the water tube tiltmeter indicated a change different from the changes during the preceding period. Furthermore, the nature of the change altered the extensometer over the several months before the earthquake. [Erimo Crustal Movement Observatory]
28	1973	Central Wakayama	5.9	Strain	Kishu		The strain velocity shown on the extensometer indicated extension within the

Table 7.3 *continued*

No.	Year	Name of earthquake	M	Observation method	Observation site	Time before earthquake	Description [Source or reporter]
		earthquake					north-south component (attributable to rain during mid-Nov.) and some anomalous contraction in the east–west component. Some additional anomalies may have occurred shortly before the earthquake. [I. Ozawa]
29	1974	Izu Peninsula offshore earthquake	6.9	Strain	Fujikawa	Approx. 7 months; approx. 2 months	The changes of the quartz tube extensometer remained the same for 3 years, and then, in Nov. 1973, it began to register a radical change. In March 1974, the movement changed from a rapid extension to a contraction. The earthquake struck not too long after that. [Fujikawa Crustal Movement Observatory]
30	1975	Aichi Southern Coast earthquake	4.3	Strain	Toyohashi		The results of examination of crustal response to rain based on extensometer records indicated that there was an anomalous response before the earthquake. This can be considered one of the earthquake precursors. [T. Yamauchi et al.]
31	1975	Microearthquakes (Yamasaki Fault)		Strain	Yasutomi		Examination of the expansion changes observed at the observation vault that traverses the fractured zone within the Yamasaki Fault revealed a frequent phenomenon in which the seismic activities around the fault became active when a pronounced short-range change appeared after a rainfall. [Disaster Pre-

Table 7.3 continued

No.	Year	Name of earthquake	M	Observation method	Observation site	Time before earthquake	Description [Source or reporter]
32	1976	Eastern Yamanashi earthquake.	5.5	Strain	Shizuoka		vention Research Institute, Kyoto University] Since the beginning of the observations in this area in April 1976, steps were frequently recorded by the embedded strain gauge. This phenomenon subsided about 10 days before the event. After the event, the frequency of the steps' appearance was greatly reduced. [JMA]
				Strain, tilting	Fujikawa		From Jan. through May 1976, a great many peculiar wave patterns were recorded by the extensometer and the well-type tiltmeter. This phenomenon, however, cannot be considered to be an earthquake precursor. Rather, it may be related to movement between the Itoi River and the Shizuoka Tectonic Line. [Fujikawa Crustal Movement Observatory]
33	1976	Hamamatsu vicinity earthquake	3.5	Strain	Toyohashi		Anomalies in this case are almost identical to those accompanying the earthquake on the southern coast of Aichi Prefecture (#30). No anomaly was found during and after 30 mm of rainfall, but anomalous change was found during and after 26 mm of rainfall. The earthquake occurred soon after this finding. [T. Yamauchi et al.]

References

Sacks, I. S., S. Suyehiro, A. T. Linde and J. A. Snoke, 1978: Slow earthquakes and stress redistribution, *Nature*, **275**, 599–602.
Sacks, I. S., S. Suyehiro, D. W. Evertson and Y. Yamaguchi, 1971: Sacks–Evertson strainmeter, its installation in Japan and some preliminary results concerning strain steps, *Pap. Meteor. Geophys,* **XXII**, Nos. 3–4.

Chapter 8   Changes in Groundwater Level and Chemical Composition

Hiroshi Wakita

Earthquakes are among the most vigorous of the earth's activities that occur near the ground surface. During the process that leads to each earthquake, various geochemical changes take place as a result of the interaction between rocks and groundwater. The interaction is due to the accumulation of strain within the bedrock as the rupture point approaches.

Although research on these changes is relatively new and still unsystematized, it is clearly becoming one of the most significant fields of earthquake research.

8.1   Earthquakes and Changes in Groundwater

The fact that the movement or migration of groundwater is an important earthquake-related phenomena is well known. Thus, for the purposes of earthquake prediction, it is essential to know what changes take place in groundwater near the earth's surface at the time of earthquakes.

The following is a summary of the changes observed so far in groundwater (which includes hot spring water, well water, and spring water) both before and after earthquakes:

(1) Increase or decrease in the amount of flow; rise or drop in groundwater level;
(2) Change in water temperature;
(3) Changes in chemical composition, color, taste, smell, or clarity of water;
(4) Gushing or seeping of mineral oil, gas, water, etc.;
(5) Spouting of water or sand;
(6) Formation of eddies and bubbles;
(7) Changes in the isotopic abundance of elements.

Most of these changes are seen immediately after earthquakes, but some of them can be clearly observed before earthquakes, which makes them important for earthquake prediction. The gushing water and spouting sand that frequently follow an earthquake occasionally show a fixed direction or quadrant sector, making them an intriguing research subject, especially in terms of their correlation to the earthquake source mechanism. During the 1965 Matsushiro swarm earthquakes, for example, there was a great gush of groundwater in regions near the earthquake zone. Cases such as this provide strong direct evidence that can lead to a clarification of the cause of earthquakes. Thus research in this field cannot be dismissed as unrelated to earthquake prediction.

The levels of well, lake, and river water show coseismic step changes followed by slower-type changes, as do the levels of gas and oil wells. The changes that took place during the great Alaska earthquake of 1964 (M 8.4) are typical. During this earthquake, changes were reported in the recorded water levels of a great many wells on the North American Continent as well as in England, Belgium, Denmark, Egypt, Libya, Israel, South Africa, the Philippines, Australia, etc. [Vorhis, 1968].

Figure 8.1 demonstrates the correlation between the earthquake source mechanism and the regional distribution of the coseismic rise and fall of the well water level as observed at the time of the Izu Peninsula offshore earthquake of 1974 (M 6.9) [Wakita, 1975]. In the quadrant that contracted due to the earthquake, the water level rose; in the quadrant that expanded (dilatation), the water level dropped.

Changes in the groundwater level and the amount of flow before and

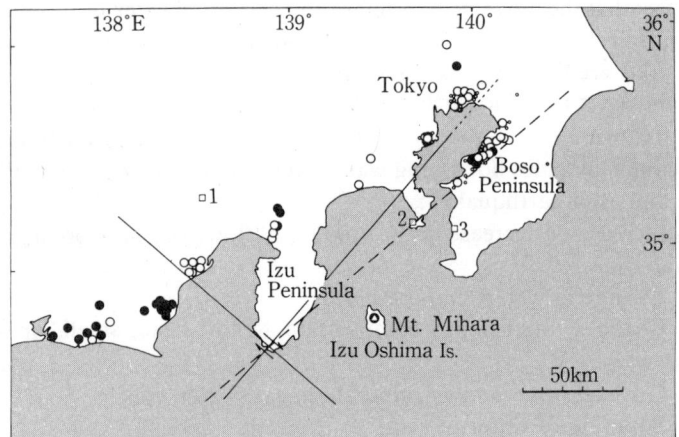

Fig. 8.1 Coseismic water level change at the time of the Izu Peninsula offshore earthquake on May 9, 1974 [Wakita, 1975].
1, 2, and 3 indicate crustal movement observatories at Fujigawa, Aburatsubo and Nokogiriyama, respectively.

after earthquakes are thought to be due to: the deformation of the groundwater system caused by crustal movement resulting in the tilting, expansion, or contraction of the ground; changes in the pore pressure of underground rocks; and the plastic deformation of the aquifer. There is a positive correlation between phenomena (1) and (2) in that the water temperature generally rises when the amount of water flow increases and falls when it decreases.

According to historical descriptions, when a great earthquake occurs off the coast of Nankaido or Saikaido—and this happens repeatedly—hot springs in Dogo, Iyo (now Ehime Prefecture), in Yunomine, Kii (now Wakayama Prefecture), and in Kumano (Mie Prefecture) in most cases cease to flow for a while. The changes in water level and temperature at Dogo Hot Springs that occurred during the Nankai earthquake of 1946 were interpreted as due to the changes in the path of groundwater within the shallow layer; there was no essential change in the hot spring's source itself [Rikitake, 1949].

In the case of the Fushimi earthquake of 1596, a change in the water temperature of the Arima Hot Springs, Hyogo Prefecture, was noted from the time of the earthquake to 1899, when there was a tremor accompanied by a rumbling noise. According to Omori [1917], the flow path of the spring tends to become blocked with insoluble precipitation. Every time an earthquake or a rumbling tremor occurs, the path opens up, causing a change in the ratio of spring water that is mixed with groundwater in the shallow layer, thus increasing the amount and temperature of the spring water.

Cases in which a change in the chemical composition of groundwater, phenomenon (3), has been quantitatively recorded are scarce. But there are numerous reports on the resulting changes in the taste and color of groundwater. When sediments are formed, carbon dioxide is released, causing the water to turn white and muddy. The release of hydrogen sulfide, sulphur dioxide, etc., causes a discoloration of the water and an unpleasant odor. These changes are related to phenomenon (4). Animal and plant life, especially creatures that live in the water, are greatly affected by oil or various gases seeping through the water.

Spouting sand and water, phenomenon (5), are almost always observed at the time of a great earthquake. Generally they occur in alluvial plains immediately after the earthquake. There have been a few cases, however, in which these phenomena occurred on platforms where the ground is solid. They rarely occur in mountainous areas. These phenomena take various forms. In some cases, clear water spouts forth from underground, and in others sand erupts in various colors. The variety of colors is amazingly rich: red, black, gray, white, brown, yellow, and blue have been reported. These small eruptions have been known to bury a well or form a tall mudcone.

Sometimes sand and water spout in a line along the fissure zone; at other times the small eruptions do not have any connection with the fissure. Most sand eruptions come from several meters underground.

Phenomenon (6) may overlap with phenomenon (5). Although the formation of eddies and bubbles in well and spring water has not often been reported in Japan, it is considered to be a precursory phenomenon and is therefore specified as one of the items that the public should recognize.

There was an example of phenomenon (7) during the Matsushiro swarm earthquakes, where the changes in the isotopic composition of erupting groundwater changed over time in a highly anomalous way.

Table 8.1 is a compilation of the groundwater anomalies observed at the time of major earthquakes in Japan. The data source for earthquakes prior to the Meiji Era was *Data—Comprehensive List of Destructive Earthquakes in Japan* [Usami, 1975] and a few other sources. Very few studies have been performed that emphasize anomalies in groundwater related to the occurrence of an earthquake. There are a few documents summarizing various precursors [Imamura, 1977, and Miyabe, 1933]. Table 8.1 does not cover all the earthquakes that have been reported; it should be supplemented continuously in the future.

As the table shows, there are an unexpected number of precursors reported among earthquake-related phenomena. It is as yet difficult to obtain a general rule for what kind of earthquake will be associated with which precursor in any given well or hot spring.

8.2 New Precursory Phenomena

In recent years, changes in the concentration of various trace elements in the groundwater have been recognized as a precursory phenomenon of an earthquake. The best-known of these changes is the anomalous changes in radon concentration. In the USSR and China, geochemical research and observations related to earthquakes are progressing to the point where a considerable number of destructive earthquakes can be successfully predicted. In view of this development it is important to describe what radon is.

8.2.1 Radon

Radon is the only radioactive rare gas that exists in nature. Generally radon is referred to ^{222}Rn, which has a half life $(T_{1/2})$ of 3.825 days. There are two other radon isotopes with shorter half lives: ^{220}Rn, $T_{1/2}$

= 54.5sec., and ^{219}Rn, $T_{1/2} = 3.92$sec. ^{222}Rn is a nuclide belonging to the uranium (^{238}U) decay series, produced by the radioactive decay of the direct parent nuclide radium (^{226}Ra), and is an inactive gaseous element that emits α particles.

As is shown in the radioactive decay series in Fig. 8.2, radon produces four nuclides, ^{218}Po, ^{214}Pb, ^{214}Bi, and ^{214}Po, before it becomes ^{210}Pb with its relatively long half life ($T_{1/2} = 22$ years). The radioactivities of these daughter nuclides, with their shorter half lives, reach equilibrium within 3 to 4 hours in a closed system and become equal to the radioactivity of the parent nuclide, radon. As a consequence, by the time equilibrium is reached, 3 α rays, 2 β rays, and numerous γ rays are emitted per one radioactive disintegration of radon. Radon measurements are made by measuring one of these (α, β, γ) rays.

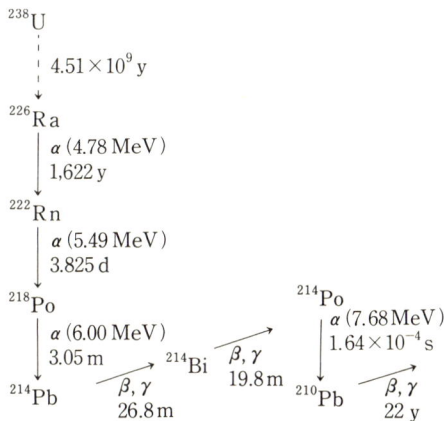

Fig. 8.2 Radioactive decay scheme of ^{222}Rn, α ray energy, and half-lives.

As rocks in the crust contain relatively large amounts of uranium (10^{-13} to 10^{-12} Ci/g), approximately the same amount of radon as of uranium is present in the earth's rocks and soil. Radon produced in rocks in the crust either reaches the ground surface directly as a gas, or else dissolves in the groundwater. It is then released into the atmosphere. Groundwater, river water, sea water, and the atmosphere will therefore each contain a radon concentration that is proportional to the strength and distance of the source, the solubility of radon in water, the degassing rate, residence time, and others.

180 SHORT-TERM PRECURSORY PHENOMENA

Table 8.1 Geochemical Anomalies Accompanying Major Earthquakes in Japan (Precursory changes are shown as ◎)

No.	Earthquakes			Accompanying phenomena							Remarks
	Date	Epicentral location(s) (names)	M H	Groundwater					Gas Release	Spouting sand, water	Others	
				Flow rate		Temp.		Quality	Clouding			
				Increase	Decrease	Rise	Fall					
1	Nov. 29, 684	Kyushu, Shikoku, Tokai districts	8.4		○							Hot springs in Iyo (Ehime): Water stopped flowing.
2	Jul. 10, 863	Toyama and Niigata Prefs.	7.0	○								Toyama, Niigata: Groundwater erupted in the area.
3	Oct. 9, 1257	Southern Kanto district	7.0	○							○	Kamakura (Kanagawa): Fissures formed and water gushed out. Near-Nakashimomabashi (Kanagawa): Fissures formed and blue flames seemed to spew forth from the fissures.
4	Aug. 3, 1361	Kinki district, Shikoku	8.4		○							Yunomine Hot Springs in Kii (Wakayama): Water stopped flowing.
5	Jan. 21, 1408	Kii Peninsula, Wakayama Pref.	7.0		○							Kumano-Hongu Hot Springs (Wakayama): Water stopped flowing.
6	Sep. 20, 1498	Throughout Tokaido	8.6		○							Yunomine Hot Springs (Wakayama): Water stopped flowing until Oct. 8.
7	Sep. 4, 1596	Oita Pref.	6.9		○						◎	Uryu Island (Oita): Village wells dried up. Foreshocks began on July 3.

CHANGES IN GROUNDWATER LEVEL AND CHEMICAL COMPOSITION 181

#	Date	Location	Mag.						Remarks
8	Sep. 5, 1596	Kyoto and its vicinity	7.0	○	○				Arima Hot Springs (Hyogo): Flow of water increased and the water became very hot.
9	Nov. 26, 1614	Takada, Niigata Pref.	7.7		○				Dogo Hot Springs (Ehime): Water stopped flowing.
10	Oct. 18, 1644	Honjo, Akita Pref.	6.9	○					Innai Village (Akita): Fissures were formed. Groundwater gushed out.
11	Dec. 29, 1685	Ehime Pref.	5.9		○				Dogo Hot Springs (Ehime): Water stopped flowing.
12	Jan. 4, 1686	Hiroshima and Ehime Prefs.	7.0			○	○		Dogo Hot Springs (Ehime): Water turned muddy and yellow.
13	Jun. 19, 1694	Noshiro, Akita Pref.	7.0					○	In the vicinity of Hirosaki (Aomori): Fissures formed and gravels and sand blew out.
14	Dec. 12, 1694	Western Kyoto Pref.	6.1		○			○	Miyazu (Kyoto): Fissures were formed and mud spouted.
15	Oct. 28, 1707	Throughout Honshu, Shikoku, Kyushu (Hoei earthquake)	8.4		○		○	○	Dogo Hot Springs (Ehime): Water stopped flowing. Yunomine, Yamaji, Ryujin, Seto-Kanayama (Wakayama): Hot springs stopped flowing. Mt. Fuji erupted, forming Hoei Crater.
16	May 13, 1717	Hanamaki, Iwate Pref.	7.6					○	Hanamaki (Iwate): Fissures appeared and mud erupted.
17	Dec. 18, 1723	Western Fukuoka Pref.	6.2		○			○	Yanagawa (Saga): Mud erupted.
18	Oct. 7, 1731	Iwashiro, Western Fukushima Pref.	6.6						Obara Hot Springs (Miyagi): A landslide occurred and the spring source disappeared.
19	May 20,	Toyama and	6.6			◎		○	Gochi (Niigata): Clear water was muddied one

SHORT-TERM PRECURSORY PHENOMENA

No.	Earthquakes Date	Epicentral location(s) (names)	M	H	Flow rate Increase	Flow rate Decrease	Temp. Rise	Temp. Fall	Quality	Clouding	Gas release	Spouting sand, Water	Others	Remarks
19	1751	Niigata Prefs.												day before the earthquake. Several wells in this area also became muddied. Shibata (Niigata): Fissures appeared in a field and blew sand.
20	Oct. 31, 1762	Sado Is., Niigata Pref.	6.6									○		Niigata: Fissures appeared, and sand and water gushed from fissures.
21	Mar. 8, 1766	Tsugaru, Western Aomori Pref.	6.9									○		Tsugaru (Aomori): Fissures appeared here and there, and blue sand erupted.
22	May 21, 1792	Mt. Unzen, Nagasaki Pref.	6.4								◎		◎	Foreshocks and rumbling began October of the preceding year, followed by an earthquake on Jan. 18. Mt. Fugen (Nagasaki) erupted.
23	Jun. 29, 1799	Kanazawa, Ishikawa Prefs.	6.4									○		Kurigasaki (Ishikawa): Sandy soil cracked in an octagonal pattern and water spouted.
24	Jul. 10, 1804	Yamagata and Akita Prefs.	7.1			◎				◎	◎	○	◎	Nagaoka, Odaki (Niigata): Prior to the earthquake, well water decreased and got muddy. Rumbling began late in May. Many fissures formed in Sakata and its vicinity (Niigata). Well water then gushed out 3 m high in the air.
25	Sep. 25, 1810	Eastern haf of Oga	6.6									○	◎	Hachirogata (Akita): The water had been changing to red, black, or clear (from the fact

CHANGES IN GROUNDWATER LEVEL AND CHEMICAL COMPOSITION 183

		Peninsula, Akita Pref.			that a great many gray mullets were found dead, some speculate the oil had been seeping from the bottom of the lake) since early August Rumbling started in May. Mud spewed from fissures in Kanpu Mountain.
26	Dec. 18, 1828	Niigata Pref.	6.9	○	In the basin of Shinano River (Niigata): Water and blue sand spewed from fissures.
27	Aug. 19, 1830	Kyoto and its vicinity	6.4	○	Mud gushed out from fissures.
28	Dec. 7, 1833	Yamagata, Akita and Niigata Prefs., Sado Is.	7.4	○	Shibata (Niigata): Fissures formed, spewing water and sand.
29	Feb. 9, 1834	Ishikari, Hokkaido	6.4	○	Ishikari (Hokkaido): Mud gushed out from fissures.
30	Apr. 22, 1841	Suruga, Shizuoka Pref.	6.4	○	Shimizu (Shizuoka): Water gushed out from fissures.
31	May 8, 1847	Northern Nagano and Western Niigata Prefs. (Zenkoji earthquake)	7.4	○ ○	Bessho Hot Springs (Nagano): Water stopped flowing. Kagai, Matsushiro (Nagano): Hot water gushed 1.8 m high on the day of the earthquake, 0.9 m next day and 15 to 21 cm the following day from its water source.
32	May 13, 1847	Kubiki, Niigata Pref.	6.5	○ ○	Kubiki (Niigata): Fissures formed and mud spewed forth.
33	Dec. 24, 1854	Throughout Honshu, Kyushu	8.4	○	Yunomine, Dogo, Kii-Kanayama (Wakayama): Hot springs stopped flowing, and then resumed flowing gradually beginning in Feburary and

184 SHORT-TERM PRECURSORY PHENOMENA

No.	Date	Earthquakes Epicentral location(s) (names)	M H	Flow rate Increase	Flow rate Decrease	Temp. Rise	Temp. Fall	Groundwater Quality	Groundwater Clouding	Gas release	Spouting sand, water	Others	Remarks
		(Ansei Nankai earthquake)											March. Two years later, they returned to normal.
34	Nov. 11, 1855	Tokyo and its vicinity (Ansei Edo earthquake)	6.9	◎	◎			◎	◎			◎	Asakusa Kuramae (Tokyo): Water began gushing out from the floor of a tea-house 4 to 5 days before the earthquake. Near Honjo (Tokyo): The well water got muddied in the morning of the day of the earthquake, and the water tasted salty. There were remarkable changes in water levels in wells all over the place. Eihei-cho (Tokyo): Water was seen flowing out of alleys before the earthquake. Workers digging a well in Fukagawa (Tokyo) had to stop working around noon on the day of the earthquake due to violent rumbling underground.
35	Apr. 9, 1858	Western Hokuriku district	6.9								○		Toyama: Fissures formed, spouting water.
36	Jul. 23, 1886	Border of Nagano and Niigata Pref.	6.1		○								Nozawa Hot Springs (Nagano): Water stopped flowing.
37	Jul. 28, 1889	Kumamoto Pref.	6.3							○	○	○	Akita-gun (Kumamoto): Sand spouted 40 to

CHANGES IN GROUNDWATER LEVEL AND CHEMICAL COMPOSITION

No.	Date	Location	M							Description
38	Jan. 7, 1890	Saikawa Basin, Nagano Pref.	6.3			○				50 days after the earthquake, traces of eruption of high temperature steam were discovered in 60 areas within some 10 square kilometers in Kimpozan, Ninotake and Sannotake Mountains (Kumamoto). Grass and trees were burnt in a 100 to 400 square meter area. Kakeyu (Nagano) and Kawarayu (Gunma): The amount of spring water increased and the water temperature rose.
39	Oct. 28, 1891	Aichi and Gifu Prefs. (Nobi earthquake)	8.4			○	◎	○	◎	Nagoya and its vicinity: The well water drastically decreased or muddied several days before the earthquake. Early in the morning of the day the earthquake hit, farmers pulling bean plants out of the ground, found steam rising from the holes of the plants they just pulled. Three days before the earthquake, there were two strong earthquakes, and abnormal behavior was observed in animals such as mudfish, phesants, crows, foxes, winged ants, and doves. After the earthquake, well water either increased, dried, or muddied and some wells spurted mud.
40	Sep. 7, 1893	Near Chiran Village, Kagoshima Pref.	6.4			○	◎○	○	◎	Near Ibusuki Hot Springs (Kagoshima): From July 13 or thereabouts, the amount and temperature of the spring water decreased. It returned to normal before the earthquake. After the event, the well water either increased or decreased and changed colors (muddy gray, white).
41	Jun. 20, 1894	Tokyo Bay Area	7.5			○		○	○	Minami-Hirayanagi Village (Saitama): Fissures formed in the paddy fields, and mud gushed out

SHORT-TERM PRECURSORY PHENOMENA

No.	Date	Earthquakes Epicentral location(s) (names)	M	H	Flow rate Increase	Flow rate Decrease	Temp. Rise	Temp. Fall	Quality	Clouding	Gas release	Spouting sand, water	Others	Remarks
42	Oct. 22, 1894	Shonai Plain, Yamagata Pref. (Shonai earthquake)	7.3				◎						◎	from the fissures. Konosu and Shobu (Saitama): Fissures were numerous and spouted mud. Well water either decreased or increased and muddied in many instances. Sakata and its vicinity (Yamagata): River water decreased about 20 days before the earthquake and the wells dried up. Fukuura (Yamagata): The sea level dropped about 45 cm, 14 to 15 days before the earthquake. After the event on the Shonai Plain, sand spouted in many areas. Spouting water was observed in many coastal areas where violent tremors were experienced.
43	Aug. 31, 1896	Border of Akita and Iwate Prefs. (Rikuu earthquake)	7.5		○	○	○	○	○			○		Sotokotomo Village, Senpoku-gun (Akita): Water in Umenoyu Hot Springs decreased to 1/4 the normal amount and the water turned almost cold. Three years after the earthquake, the temperature began to rise, and 9 years after the event, it returned to normal. Oshuku, Tsunagi, Osawa, etc. (Iwate): Hot springs stopped flowing. Namari and Yuda (Iwate): The amount of water in the hot springs de-

CHANGES IN GROUNDWATER LEVEL AND CHEMICAL COMPOSITION

	Date	Location	Magnitude				Description
							creased slightly. In Yunosawa (Akita): The amount of water and the temperature in the hot springs increased (similar changes were noted at the earthquake occurred in 1856). Basin of the Kawaguchi River (Akita): Several cold and hot springs appeared accompanied by white sediment locally called "Yunohana" (or flower of the water).
44	Jan. 17, 1897	Northern Nagano Pref.	6.3			○	Fissures formed in numerous fields, accompanied by gushing sand and mud. In many areas, colored sand, mud and groundwater spouted. Some wells were buried by sand.
45	Feb. 20, 1897	Sendai offshore, Miyagi Pref.	7.8			○	Hanamaki (Iwate): Fissures formed, spouting mud water.
46	Apr. 3, 1898	Mishima, Is., Yamaguchi Pref.	6.8		○		For several days, the well water remained unfit for human consumption.
47	Apr. 23, 1898	Iwate offshore	7.8			○	Hanamaki (Iwate): Fissures formed, accompanied by spouting sand.
48	May 26, 1898	Near Muika-machi, Niigata Pref.	6.7		○	○	Muikamachi (Niigata): Fissures formed in fields accompanied by gushing sand.
49	Aug. 10, 1898	Near Fukuoka City, Fukuoka Pref.	6.5	◎		◎	Imajuku and Kafuri Village (Fukuoka): Anomalies were found in the ocean—the ocean water temperature increased and there were tsunamis. On the coast of Hakata Bay and Karatsu Bay in Itoshima Peninsula, the beaches became soft, and anomalous fish behavior was observed.

SHORT-TERM PRECURSORY PHENOMENA

No.	Date	Epicentral location(s) (names)	M	H	Flow rate Increase	Flow rate Decrease	Temp. Rise	Temp. Fall	Quality	Clouding	Gas release	Spouting sand, water	Others	Remarks
50	Nov. 13, 1898	Kiso River Basin, Aichi Pref.	6.5						○	○				Yellowish sand and water spouting from the ground smelled like calcium sulfate. Kuroda-machi, Fukuzawa-machi in Nakajima-gun, Ogaki-cho (Aichi): The river water turned grayish white.
51	May 28, 1902	Kushiro offshore, Hokkaido	7.4							○				Shibecha (Hokkaido): The river water turned muddy and was temporarily unfit for human consumption.
52	Aug. 10, 1903	Western part of Mt. Norikura, Gifu Pref.	5.7			○								Hatahoko Village (Gifu): The well water dried up.
53	May 8, 1904	Near Muika-machi, Niigata Pref.	6.9			○						○		Ikazawa Village, Minami Uonuma-gun (Niigata): Blue sand erupted from the fissures in the roadway.
54	Aug. 14, 1909	Near Anegawa, Shiga Pref. (Gono earthquake)	6.9		○							○	○	Anegawajiri (Shiga): Mud water spouted 2.5 m high from 6 holes. Centred in the Higashi-Asai-gun (Shiga): Anomalies such as an increase and decrease in well water were observed. The water level of the Anegawa River increased and there

CHANGES IN GROUNDWATER LEVEL AND CHEMICAL COMPOSITION

No.	Date	Location	M					Description
55	Jul. 24, 1910	Mt. Usu, Hokkaido	6.5		O		O	seemed to be a correlation with the water level of Lake Biwa. Luminescence was observed in Ibuki Mountain.
								Lake Toya (Hokkaido): Hot springs were formed.
56	Sep. 8, 1910	Onishika, Hokkaido	5.6				O	Onishika Village, Rumoe (Hokkaido): A fissure formed in the ocean floor 9 m deep, and sea water gushed from the ocean surface.
57	Jan. 12, 1914	Sakurajima, Kagoshima Pref.	6.1		◎		◎	On the southeastern side of the Sakurajima volcano (Kagoshima): Hot springs spouted 1m high in the morning of the 12th. On the beach on the north side of the volcano, water gushed out on the 12th at 8:30; and again on the 22nd. Well water rose 1m on the morning of the 12th.
58	Mar. 15, 1914	Senpoku-gun, Akita Pref. (Akita Senpoku earthquake)	6.4	O			O	Umenoyu Hot Springs, Sotootomo Village, Senpoku-gun (Akita): Hot springs completely dried up after the earthquake. Spouting sand was observed.
59	May 23, 1914	Izumo, Shimane Pref.	6.3		O		O	Tamatsukuri Hot Springs (Shimane): The amount of flowing water increased as much as three times the nomral amount and the temperature also increased.
60	Jul. 14, 1915	Kurino-Yoshimatsu, Kagoshima Pref. (West of Mt. Kirishima)					O	Yunono (Kagoshima): Boilding mud erupted 3 m high.
61	Nov. 26,	Kobe,	6.3				O	Arima Hot Springs (Hyogo): The water temper-

SHORT-TERM PRECURSORY PHENOMENA

No.	Date	Epicentral location(s) (names)	M	H	Flow rate Increase	Flow rate Decrease	Temp. Rise	Temp. Fall	Quality	Clouding	Gas release	Spouting sand, water	Others	Remarks
	1916	Hyogo Pref.												ature rose 1°C to register 53.4°C. (At the time of the rumbling tremor in July, 1899, the temperature of the hot springs rose from 37°C to 47.9°C by October of the following year.)
62	Nov. 11, 1918	Omachi and its vicinity, Nagano Pref.	6.1		○	○	○						○	Asama Hot Springs (Nagano): The amount of water increased, and the temperature of the water rose. Jokoji, Kibune, Matsuzaki, Enshoji (Nagano): The well water level dropped. Nagahata, Yashita, Ohira (Nagano): The water flow increased. Luminescence was observed in Shinpi Mountain Range.
63	Nov. 1, 1919	Miyoshi and its vicinity, Hiroshima Pref.			○	○								In and around this area the well water either decreased or increased.
64	Dec. 8, 1922	Chijiwa Bay, Nagasaki Pref.			○	○			○	◎	○	○	○	Water, sand and mud spouted. Steam and sulfurous acid gushed out and formed strongly acid puddles containing hydrogen sulfide. Spring water either dried up or erupted.
65	Sep. 1, 1923	Southern Kanto district	7.9		◎	◎			◎		◎		◎	Oyu, Atami Hot Springs (Shizuoka): The geysers erupted with increased force the day before the

CHANGES IN GROUNDWATER LEVEL AND CHEMICAL COMPOSITION 191

(Great Kanto earthquake)	earthquake. The force of the eruptions further increased after the event and continued for one week. Aoki-yu and Seizaemon-yu increased their flow. The water in Kona Hot Springs (Izu Peninsula) became whitish. Wells dried up in Shinagawa and Ryoshi-machi (Tokyo) in July. (Similar incidents were reported at the time of the 1855 Edo earthquakes. They returned normal after the events.) The well in a temple called Harusame-an began smelling of metal or some kind of herb in June or July. Well water was reported muddy near Takinogawa (Tokyo) before the earthquake. Well water near Imaizumi in Minami Hatano Village (Kanagawa) decreased. The same was reported in Shitaya, Ota Village (Kanagawa), two months prior to the earthquake. Lake Yamanaka (Yamanashi) became muddy from May and June on. Water within the City of Shizuoka got muddy a few days before the event. Kadokawa, Yoshihama Village (Kanagawa): Well water turned bluish and muddy on Sept. 1. About 180 m off the southern coast of Misaki, Jogashima (Kanagawa), bubbles about the size of beans or thumb tips were seen rising. One of the tributaries on the Kanogawa River in Kami-Kano Village (Shizuoka) turned to indigo blue. This phenomenon had been observed prior to every foreshock. A hot spring in the same area increased its flow as well as its temperature. A week before the earthquake, well water levels shot up or dropped and the well

No.	Date	Epicentral location(s) (names)	M	H	Flow rate Increase	Flow rate Decrease	Temp. Rise	Temp. Fall	Quality	Clouding	Gas release	Spouting sand, water	Others	Remarks
66	May 23, 1925	Northern Hyogo Pref. (Northern Tajima earthquake)	7.0		○	◎				○				Toyooka (Hyogo): The amount of water in the Toyooka Channel with its source in the south of the Genbu Caves drastically decreased several days prior to the earthquake. After the event, spouting sand, water and muddy water were reported in the areas between Kinosaki and Tsuiyama (Hyogo).
67	Jul. 4, 1925	Miho Bay, Shimane Pref.	6.3										○		Water and fine sand spouted from fissures and wells were buried.
68	Aug. 10, 1925	Hita, Northern Oita Pref.				○								○		Fissures in the ground and anomalies in ground water were reported.
69	Mar. 7, 1927	Northwestern Kyoto Pref. (Northern Tango earthquake)			○	○	○				◎		○	○	Kamitokumitsu, Tokumitsu Village; Mitsu, Shimazu Village (Kyoto): 3 days before the earthquake, well water was found to be muddy. The water returned to normal 3 weeks after the event. Near Kizu Village (Kyoto): A string of

Note: Unlabeled top-of-page note reads: "water in some wells muddied in Kinugasa Village, Miura-gun (Kanagawa): Low rumblings were heard off the coast of the Miura Peninsula."

CHANGES IN GROUNDWATER LEVEL AND CHEMICAL COMPOSITION

No.	Date	Location	M	Depth								Remarks
70	Aug. 6, 1927	Miyagi Pref. offshore	6.9		O		◎		◎			new hot springs formed after the earthquake. They were almost in a straight line going in an eastward direction 30°N. The well water showed various anomalous conditions such as decrease, increase, and muddiness. Spouting sand and water were also reported. Tottori City: The temperature of ten hot springs increased by 1 to 2°C, and the amount of water erupting increased. The water got muddy and bluish white in color during the 3 hours after the event. Formation of the Gomura and Yamada faults.
71	Oct. 27, 1927	Central Niigata Pref. (Sekihara earthquake)	5.3							O	◎	Aone Hot Springs (Miyagi): The temperature of the water increased before the earthquake and the water turned white. The amount of water flowing from the Sakunami Hot Springs increased after the event and its temperature went up 4°C. Mud water spouted.
72	Nov. 26, 1930	Northern Izu Peninsula (Northern Izu earthquake)	7.0	0–5						O	O	Nishida, Miyamoto Village (Niigata): In a paddy field jets of petroleum gas gushed from holes (in a straight line N60°W 300 m long), spouting blue sand and oil. Foreshocks and crustal movement (uplift) were reported.
73	Dec. 20, 1930	Miyoshi and its vicinity, Hiroshima Pref.	6.0	20	O					O		Komatsu-cho and other towns (Shizuoka): Water spouted. Miyoshi (Hiroshima): Well water dried up or became muddied. No increase of well water was reported.

SHORT-TERM PRECURSORY PHENOMENA

No.	Date	Epicentral location(s) (names)	M	H	Flow rate Increase	Flow rate Decrease	Temp. Rise	Temp. Fall	Quality	Clouding	Gas release	Spouting sand, water	Others	Remarks
74	Feb. 17, 1931	Urakawa and its vicinity, Hokkaido	6.8	40									○	Near the mouth of the Niikapu River (Hokkaido): Numerous fish were found dead on the beach on the 19th.
75	Mar. 9, 1931	Southeastern coast of Aomori Pref.	7.6	0	○					○				Yunokawa Hot Springs (Aomori): The water got muddy temporarily and the amount of water increased.
76	Sep. 21, 1931	Central Saitama Pref. (Western Saitama earthquake)	7.0	10–20	○					○		○		Extensive fissures formed, and spouting sand and water (1.2 m high in some places) was observed. The color of the sand was blue, brown and black. This was attributable to a combination of quartz and feldspar in most cases. Well water was muddied and the amount of water increased over extensive areas.
77	Mar. 3, 1933	Off the coast of Eastern Tohoku district (Sanriku offshore earthquake)	8.3			◎				◎			◎	Coastal villages such as Oshima, Karakuwa in Motoyoshi-gun, (Miyagi), and Otsukirai in Kesen-gun, and Funakoshi in Shimohei-gun (Iwate): Water levels dropped and water got muddy up to 20 days before the earthquake. The tide level also dropped 2 days before the event.

#	Date	Location	M	Depth						Description
78	Sep. 21, 1933	Noto Peninsula, Ishikawa Pref.	6.0	15				○		Akauragata Beach (Ishikawa): Fissures formed and spouted sand and mud.
79	Mar. 21, 1934	Amagi Mountain, Izu Peninsula	5.5	0–10	○	○	○			According to research reports, changes in temperature, water clarity, and water flow were noted in hot springs throughout Izu Peninsula. The possibility of an interrelationship between these changes and the earthquake source mechanism is a point of speculation.
80	Jul. 11, 1935	Shizuoka City and its vicinity (Shizuoka earthquake)	6.3	10	◎					Rendaiji Hot Springs (Izu Peninsula): The water level rose 70 cm about 5 months prior to the earthquake. After the earthquake the water level dropped drastically—as much as 262 cm by Aug. 5. Anomalies in well water were noted around Shimizu City (Shizuoka). Water spouted near the banks of the Abe River (Shizuoka) before the event.
81	Feb. 21, 1936	Nara Pref. and Osaka (Kawachi Yamato earthquake)	6.4	20	◎			◎○		Domyoji Village (Osaka): There was a sudden increase in well water shortly before the earthquake. People were startled and ran outside to find the fields covered with water. Spouting mud was also noted.
82	mid-Nov. 1936	Osarizawa-Hanawa swarm earthquakes, Akita Pref.			○	○				Yuze Hot Springs (Akita): The amount and temperature of the spring water dropped.
83	Nov. 11, 1936	Wakamatsu City and its			○					Wakamatsu City (Fukuoka): The well water dropped 1 to 1.3 m.

No.	Date	Earthquakes Epicentral location(s) (names)	M	H	Accompanying phenomena Groundwater Flow rate Increase	Flow rate Decrease	Temp. Rise	Temp. Fall	Quality	Clouding	Gas release	Spouting sand, water	Others	Remarks
		vicinity, Fukuoka Pref.												Kii coast (Wakayama): The well water level changed.
84	Jan. 12, 1938	Tanabe Bay offshore, Wakayama Pref.	6.7	20	○									
85	May 23, 1938	Shioyazaki offshore, Fukushima Pref.	7.1	10										Iwashiro-Atami, Yumoto and Iizaka Hot Springs (Fukushima): Anomalies were noted in hot springs.
86	May 29, 1938	Lake Kussharo and its vicinity, Hokkaido	6.0	20	○			○			○			Wakoto Hot Springs (Hokkaido): The water flow increased and the water temperature dropped. The fumaroles on the northern coast of Wakoto Peninsula increased in number and force after the earthquake.
87	May 1, 1939	Oga Peninsula, Akita Pref.	7.0	0	○				◎		◎			Hachimori Beach (Akita): Innumerable octopi (*octopus ocellatus*) beached in a sort of "drunken" condition from the afternoon of the day prior to the earthquake through the morning of the day of the event. On Nakamura Beach (Akita): Octopi, which are seldom seen in normal times, beached one after another on the morning of the day of the earthquake. Accord-

CHANGES IN GROUNDWATER LEVEL AND CHEMICAL COMPOSITION 197

88	Jul. 15, 1941	Nagano City and its vicinity, Nagano Pref.	6.2	5–20	◎		◎	○	○	ing to A. Imamura, this phenomenon was due to seeping gas or groundwater. Iwadate (Akita): The bottom of the sea became extremely muddy on the day before the earthquake. In four areas, including the town of Kitaura and Yumoto Hot Springs (Akita): New hot springs emerged after the event. Kagai Hot Springs in Terao Village, Hanishina-gun (Nagano): The water got muddy and the water flow increased before the earthquake. It returned to normal after the earthquake. Yamada Hot Springs in Takai, Kamitakai-gun (Nagano): The amount and temperature of the spring water increased before the event. Kami-Suwa Hot Springs (Nagano): A case of decreased temperature and a case of decrease in the amount of water flow were reported before the event (225 cases were studied). In the areas along the Chikuma River (Nagano): Wells were buried due to spouting sand and mud. Tochio-mata Hot Springs (Niigata): Radon concentration changed.
89	Mar. 4 1943 Mar. 5 1943	Tottori offshore	6.1 6.1	20 20	◎	○	◎	◎	◎	Yoshikata (Tottori): 30 minutes before the earthquake, CO_2 springs became white and muddy. The amount of water increased by 1.5 times. Yoshioka Hot Springs (Tottori): The temperature of the water decreased slightly and the water seemed muddy the day before the the earthquake. Changes in well water levels and clarity were noted extensively throughout

No.	Earthquakes				Accompanying phenomena								Remarks
	Date	Epicentral location(s) (names)	M	H	Groundwater								
					Flow rate		Temp.		Quality	Clouding	Gas release	Spouting sand, water	Others	
					Increase	Decrease	Rise	Fall						
90	Sep. 10, 1943	Tottori City and its vicinity (Tottori earthquake)	7.4	10					◎	◎		○		the area. Iwai Hot Springs (Tottori): The water temperature increased. The temperature did not change in hot springs within Tottori City but the clarity decreased. The amount of hot spring water in Hamamura (Tottori) dropped after the earthquake but soon returned to normal. Misasa and Matsuzaki Hot Springs (Tottori): A slight increase of water temperature was reported. Yoshioka Hot Springs (Tottori): The water became muddy and white the day before the earthquake. Spouting sand, water, and amomalies in hot springs were noted.
91	Dec. 7, 1944	Off the coast of Tokaido (Tonankai earthquake)	8.0	0-30	○	○	○			○				Anomalies in groundwater were noted after the earthquake.
92	Jan. 13, 1945	Southern Aichi Pref. (Mikawa earthquake)	7.1	0									◎	Foreshocks were noted. Water increases and dried up water sources were reported. In some places, hot water spouted from the ground.

CHANGES IN GROUNDWATER LEVEL AND CHEMICAL COMPOSITION *199*

93	Dec. 21, 1946	Off the coast of Shikoku and Southern Kii Peninsula (Nankai earthquake)	8.1	30		◎	◎	In areas along the coast of the Kii Peninsula (Wakayama) and the coast of Shikoku: Well water either got muddy or dried up before the earthquake. Owase (Mie): Well water got muddy and its level dropped the day before the event. Inami (Wakayama): Well water dried up. Tomoura (Tokushima): The level of well water dropped considerably 5 days before the earthquake, and the day before the event the water became muddy. Urado (Kochi): Well water dried up 2 days before the event. This happened also on Aosuna Island near Cape Ashizuri (Kochi) one week before the earthquake.
94	Jun. 28, 1948	Fukui Plains (Fukui earthquake)	7.3	20	○	○	○	Fukui Plains: Anomalies were noted in wells over an extensive area. Water decreased or increased in amount and clarity, and turned reddish or whitish in color. Spouting sand and mud was noted. Changes in the water level were recorded in Wajima (Ishikawa) where water level rose after the earthquake.
95	Dec. 26, 1949	Imaichi, Tochigi Pref. (Imaichi earthquake)	6.4 6.7	Very shallow Very shallow		○	◎ ◎	Imaichi (Tochigi): The water level of a well in a wood factory dropped so drastically in mid-November of that year that the pump could not be used. Sulfur compounds were found in Kinugawa Hot Springs (Tochigi) beginning in mid-August A similar phenomenon occurred during the earthquake. In extensive areas, the well water level changed after the earthquake. The west half of Itaka Village experienced an increase in the water level while the water level dropped in

200 SHORT-TERM PRECURSORY PHENOMENA

No.	Earthquakes			Accompanying phenomena								Remarks
	Date	Epicentral location(s) (names)	M	H	Groundwater							
					Flow rate		Temp.		Quality	Clouding	Gas release	Spouting sand, water	Others	
					Increase	Decrease	Rise	Fall						
96	Aug. 22, 1950	Vicinity of Mt. Sambe, Shimane Pref.	5.3	30						○				The well water near the epicenter became white and muddy.
97	Mar. 4, 1952	Off the coast of Tokachi, Hokkaido (Tokachi offshore earthquake)	8.1	45	○	◎	○	◎		○	○		◎	Meakandake Mountain (Hokkaido): Rumblings had been heard since July of the preceding year. (This is a rare incident for this mountain. It occurred also at the time of the Sanriku earthquake of 1933.) The amount and temperature of spring water on the shores of Lake Akan (Hokkaido) decreased beginning in December of the preceding year. The water in hot springs and around ponds near the Bokke Mud Volcano stopped erupting (observed Jan. 20). After the earthquake both the amount and temperature of water increased. The smoke from Meakandake Mountain changed from its normal white and gray to blackish gray, and the amount of smoke

Remarks (row above no. 96): the east half. Izumi Village (Tochigi): A jet of gas spouted from an outcropping of andesite on a steep incline of 35°, 100 m high on the west bank of the Miyagawa River.

CHANGES IN GROUNDWATER LEVEL AND CHEMICAL COMPOSITION 201

#	Date	Location	M	Depth						Description
98	Mar. 7, 1952	Daishoji offshore, Fukui Pref.	6.8	20	○					increased. Well water and hot springs throughout Hokkaido showed anomalies such as muddiness, decrease or increase of amount and increased temperature. Kitakata Village and the areas along the lake in Hamasaka (Fukui): Water dried up or increased. Mud water and sand spouted from the ground. The amount and temperature of the water at Katayamazu Hot Springs (Ishikawa) showed little change either before or after the earthquake.
99	Oct. 19, 1955	Downstream R. Yoneshiro, Akita Pref. (Futatsui earthquake)	5.7	0–10	○	○	○			Yoneshiro (Akita): Spouting, clouding of groundwater and an increase of cold water springs were noted. The level of the Yoneshiro River dropped approximately 15 cm at the time of the earthquake, then went back to normal after 15 minutes (recorded by automatic water gauges).
100	Sep. 30, 1956	Southern Miyagi Pref.	6.1	20	○					Ohara Hot Springs (Miyagi): At the moment the earthquake hit, the water flow ceased. It then resumed with an increased amount of water.
101	Jan. 31, 1959	Teshikaga and its vicinity, Hokkaido	6.2	20 6.1	○	○	○			Teshikaga (Hokkaido): Within the area where there were strong shocks, most of the wells became muddy and the water level decreased, except for Wakoto where the well water level increased. River water changed color. Spring water in hot springs near Teshikaga either decreased or dried up. The hot springs on the line connecting Wakoto and Nibuse on the

202 SHORT-TERM PRECURSORY PHENOMENA

No.	Date	Earthquakes Epicentral location(s) (names)	M	H	Accompanying phenomena Flow rate Increase	Decrease	Groundwater Temp. Rise	Fall	Quality	Clouding	Gas release	Spouting sand, water	Others	Remarks
														shore of Lake Akan, increased, but in other areas little change was noted in water flow. The temperature of the water rose in the areas where the amount of water increased.
102	Feb. 2, 1961	Nagaoka and its vicinity, Niigata Pref.	5.2	20	○		○				○	○		Sangoya Hot Springs (Niigata): Sandy hot water (970m layer, shale) with an increased amount, temperature (42.0°C → 42.5°C) and gas content, erupted immediately after the earthquake. Fine sand and mud water erupted from the 310 m layer. Cone-shaped fumaroles in the shape of mud volcanoes were seen in various places.
103	Apr. 30, 1962	(Northern Miyagi earthquake)	6.5	0	○					○				Extensive groundwater anomalies were noted. Well water increased and clouded. Very few wells decreased in water level. In the west of Furukawa City (Miyagi), a small-scale uplift 30 cm in diameter was formed several days after the earthquake. This phenomenon is interpreted as related to the changes in groundwater.
104	May. 7, 1964	Oga Peninsula offshore, Akita Pref.	6.9	0	○					○		○		Gokomyo (Akita): Clouded water and spouting mud were noted.

105	Jun. 16, 1964	Near Awa-shima Island, Niigata Pref. offshore (Niigata earthquake)	7.5			○	In the low, damp land in Niigata City and Sakata City (Yamagata): Erupting sand and water were observed several minutes after the earthquake. In some areas the water spouted as high as 2 m and the erupted sand made mounds with an average height of 20 to 30 cm; in extreme cases the mounds were as high as 70 to 80 cm.
106	Aug. 3, 1965 –1967	Vicinity of Matsushiro, Nagano Pref. (Matsushiro swarm earthquakes)		○	○	○	Matsushiro (Nagano): Erupting water and fissures were noted from the second active period on (Mar. ~ Jul. 1966). The erupting water contained a great deal of Cl^-, Ca^{2+} and CO_2—a chemical anomaly—and the amount of water was unusually great, $i.e.$ $10^7 m^3$. The changes in chemical composition and isotopic abundance of the water that accompanied the change in seismic activity was reported. Continuous release of a large amount of helium (350 ppm—measured in 1976) in a "spot" pattern along the fault was observed. The $^3He/^4He$ ratio of this helium is as high as that of upper mantle material.
107	Feb. 21, 1968	Ebino-cho, Miyazaki Pref. (Ebino earthquake)	5.7 0				
108	May. 16, 1968	Tokachi offshore, Hokkaido (Tokachi offshore earthquake)	7.9		○		Noboribetsu Hot Springs (Hokkaido): The amount of water increased during the one day period after the event.

No.	Date	Earthquakes - Epicentral location(s) (names)	M	H	Flow rate - Increase	Flow rate - Decrease	Groundwater Temp. - Rise	Groundwater Temp. - Fall	Quality	Clouding	Gas release	Spouting sand, water	Others	Remarks
109	Sep. 9, 1969	Central Gifu Pref.	6.6		○					○		○		Wara Village (Gifu): Immediately after the earthquake, the well water became muddy, then 2 to 3 hours later it became cloudy. Groundwater increased and became brown and sandy.
110	Jun. 17, 1973	Nemuro Peninsula offshore	7.4						○					
111	May. 9, 1974	Izu Peninsula, Shizuoka Pref. (Izu Peninsula offshore earthquake)	6.9		○		○							Coseismic water level changes were recorded at water level observation wells in the Kanto and Tokai districts. A correlation was found between the distribution of the wells that showed a water level change and the earthquake source mechanism. At hot springs south of Kamo Village and Kawazu (Izu Peninsula), an increase in the amount and temperature of spring water were generally noted. The spring water became cloudy during a two week period at Daisenzan, Hatage Hot Springs (Izu Peninsula).

112	Apr. 21, 1975	Near Mt. Kuju, Western Oita Pref.	6.4								
113	Jan. 14, 1978	(Izu Oshima offshore earthquake)	7.0	4.3		◎			◎		◎	Hiekawa, Nakaizu-cho (Izu Peninsula): A change in the continuously-monitored radon concentration was recorded. Omaezaki (Shizuoka): The recorded water level of a well (500 m) had dropped approximately 20 cm since Dec. 28 of the preceding year. Then it returned to normal and the earthquake occurred. At that time, a coseismic water level drop was recorded. Foreshocks occurred on Jan. 13.

8.2.2 Unit of Radon Concentration

Radon concentration is represented by the radioactive unit Ci (Curies). One Ci is defined as a nuclide that has a disintegration number of 3.7×10^{10} per second. A unit called "eman" is sometimes used for groundwater and soil gas. One eman is equivalent to 10^{-10} CiRn/l.

8.2.3 Emanation of Radon

If one assumes that radium, the parent nuclide of radon, is homogeneously distributed in rocks, then radon will be produced homogeneously in those rocks as well. Taking into consideration radon's half life (3.825 days) and the fact that both the recoil distance of ^{222}Rn within rocks ($\sim 3 \times 10^{-6}$cm for the average mineral), and the molecular diffusion coefficient ($\sim 10^{-20}$cm^2/sec [Tanner, 1964]) are extremely small, it is clear that most radon decay by itself in rocks, and that only the radon that is near the surface escapes outside rocks.

For the release rate of radon, the stability of the mineral structure and crystal structure play a more important role than the uranium content in rocks. In other words, radon emanation depends largely on the degree of fragmentation of rocks. The ratio between the total amount of radon produced in rocks and the amount of radon that is released into the atmosphere is called the emanation power.

Radon in groundwater is largely supplied by rocks in aquifer since the amount of radium in groundwater is extremely low. The radon originates from the portion in the surface of rocks, as mentioned above. As the release rate of radon from rocks with certain fixed surface dimensions is constant, the groundwater that comes in contact with those rocks will also have a certain fixed concentration of radon, ignoring abnormal circumstances such as preseismic conditions.

Taking into consideration the diffusion coefficient of radon in water, 1.14×10^{-5}cm^2/sec (18°C), radon can move a distance of only 4 cm or thereabouts in completely still water within the same amount of time as its half life. In actuality, however, groundwater does move, and there are substances such as carbon dioxide and nitrogen migrating upward. Thus radon that is formed in one region can in fact move to other regions with considerable speed. The cracks and fissures in fault strands will act as passages to the ground surface for the gaseous substances including radon.

8.2.4 Changes in Radon Concentration Prior to Earthquakes

The concentration of radon in groundwater does not change very much under normal conditions, except for daily and annual changes, as far as the supply is constant. How, then, does change occur in radon concentration prior to an earthquake?

Unfortunately, there is no complete answer to this question at this time. According to the dilatancy model, minute cracks and pores form in rocks in the hypocentral region before an earthquake, and these, in effect, result in an increase in the surface area of the rocks. A successive large-scale migration of groundwater is thought to be quite effective in extracting radon from the rocks.

Besides the above effects, a fluctuation in the radon concentration near the ground surface will be expected on the basis of assuming the possible pre-earthquake changes that may be taking place in the groundwater system due to crustal movement such as contraction, expansion, tilting, the deformation of the aquifer, and changes in permeability due to crustal stress variation.

In addition to these factors, one interpretation suggests that the convection of fluid formed by regional geothermal gradients can cause variations in the radon concentration in soil gas [Mogro-Campero and Fleischer, 1977].

Basic research on the mechanism of radon emanation is being conducted in China using explosion seismology in field experiments and rock deformation experiments in the laboratory. Such research is badly needed in Japan as well.

8.2.5 Radon Measurement in Groundwater

Discrete measurement. The most suitable method of measuring the radon concentration in groundwater is the toluene extraction–liquid scintillation counting method [Noguchi, 1964]. This newly developed method is extremely convenient, highly sensitive, and reliable. For actual field measurements, a modified version of this method [Wakita et al., 1976] is being used, replacing the cumbersome electroscope-type measurements. In the field, definite amounts of groundwater are introduced to a specially designed separatory funnel which contains 30 ml of a scintillator with a toluene base. The funnel is shaken vigorously until the mixture turns milky. The solubility of radon in toluene is about 50 times higher than that in water at room temperature. Radon in water is extracted into the toluene phase according to its distribution coefficients. For example, about 80% of the total radon in 300 ml of water is extracted into 30 ml of the liquid scintillator. After the mixture settles into two phases again, a known volume of the upper organic phase is transferred to a counting vial. During the sampling and extraction procedure, care must be taken as much as possible to keep the groundwater from coming in contact with air and to keep radon in the system.

After radioactive equilibrium is reached, the radioactivity in the counting vial is counted by a commercial type liquid scintillation counter. Three alpha particle emissions and two beta transitions from radon and its daughter nuclides are counted. For the absolute counting of these α and β rays, the zero extrapolation method is applied. The integral counting rate at the threshold value of zero is obtained from the integral counting rates at three different bias settings in the β ray region. One fifth of the total counting rate (N_{total}: counts per minutes) at the zero threshold is equivalent to the intensity of the radon activity.

In this method the radon concentration in groundwater (C_0) is calculated by the following equation:

$$C_0 = \left(\frac{1}{D_t} \times \frac{V_{\text{air}}}{V_w} + \frac{V_t}{V_w} + \frac{D_w}{D_t}\right) \times C_t \quad (1)$$

$$C_t = N_{\text{total}} \exp(\lambda t)/(5 \times 222 \times V_{vt})$$

where C_t is the radon concentration of scintillator in the counting vial. V_{air}, V_t and V_w are the volumes of the air, liquid scintillator and groundwater in the separatory funnel, respectively. V_{vt} is the volume of the liquid scintillator in the counting vial. λ is the decay constant of ^{222}Rn and t is the elapsed time between the extraction time of radon into toluene and the middle of the counting period. D_w and D_t are the distribution coefficient for water and toluene, respectively. They are approximated as follows:

$$D_w = 9.12 / (17.9 + T) \text{ and, } D_t = 18.3 \exp(-T/46.0)$$
$$5°C < T < 40°C.$$

Continuous measurement. A description of the measuring system for continuous measurements of radon concentration in groundwater has already been published by Noguchi and Wakita [1977]. Pumped groundwater is introduced into a ZnS scintillation chamber that is highly sensitive to α rays and continuously measures the radioactivity of the radon and its daughter nuclides. The continuous radon measurement system was developed, with the University of Tokyo playing a leading role, and it is now sold commercially and used by various organizations.

Measurement of radon in soil gas. To measure the radon content in soil gas, a hole 40 to 50 cm deep is dug on the ground and a thin film of cellulose nitrate is left for one to four weeks in the hole. (Usually Kodak's LR115-type II film is used for this purpose.) The tracks left on the film by the α ray from radon and its daughter nuclides in the soil gas are etched with a 10% solution of sodium hydroxide and enlarged for counting under the microscope.

In the U.S.A., the Terradex Corporation's film called "Radon Cup" is used for this purpose. The service of track counting after the recovery of the cup in the field is included in the price of the cup. The cost is considerably higher, but the user can simply leave the plastic cups, which are already equipped with film, for the prescribed length of time, and then recover them.

8.3 The Tashkent Earthquake

Changes in the radon concentration in groundwater became world-famous as an earthquake precursor as a result of the Tashkent earthquake that struck the Capital of Uzbekistan, USSR, in 1966. The concentration of radon in groundwater samples from deep wells had been increasing for several years prior to the M 5.5 earthquake. Immediately before the earthquake the concentration reached a maximum of three times the norm. It was reported that the anomalous concentration recovered to the normal level after the earthquake (Fig. 8.3) [Ulomov and Mavashev, 1971]. The hypocenter of the Tashkent earthquake was relatively shallow (approximately 8 km). Due to an observation well 2 km deep that was positioned directly above the hypocenter, similar changes in water temperature and head pressure were also observed. The radon concentration pattern seen in Fig. 8.3 resembles well the crustal movement pattern observed and recorded prior to the 1964 Niigata earthquake (M 7.4) in Japan.

At present, comprehensive research is being carried out by the Seismo-hydrogeochemistry Department of the Seismological Institute of Tashkent, including studies on the gases contained in groundwater (Rn*, He*, Ar, N_2, O_2, H_2*, CO_2*, CH_4, etc.), its chemical composition (F*, Cl*, Hg*, U, Mg, Ca, Fe, etc.), the abundance of isotopes ($^2H/^1H$, $^{13}C/^{12}C$, $^{18}O/^{16}O$, $^{40}Ar/^{36}Ar$, $^{234}U/^{238}U$, etc.), water temperature*, pH*, Eh*, water level*, and head

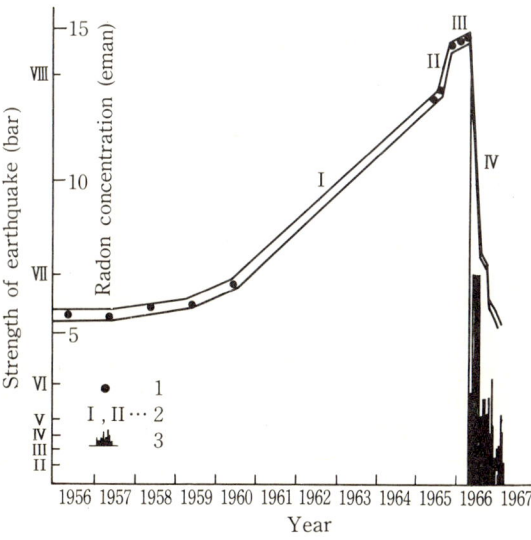

Fig. 8.3 Radon concentration changes in groundwater prior to and after the 1966 Tashkent earthquake [Ulomov and Mavashev, 1971].

pressure*. Combining geochemical, geophysical, and hydrological research has yielded successful results.

Once the changes in radon concentration were noticed in 1966, research was organized on changes in other parameters, such as gas components, chemical and isotopic compositions in groundwater, water temperature and head pressure, and hydrological parameters. As the research progressed, numerous precursory phenomena were reported. By 1972 the significance of research on groundwater had been recognized by seismologists and a Seismohydrogeochemistry Department was established in the Seismological Institute of Tashkent, Academy of Science, Uzbek SSR. The size of this department is indicated by the fact that 70 of the Institute's 600 members belong to it. There are said to be approximately 40 geochemists and hydrogeologists making observations and doing research at Tashkent alone. According to Prof. Sultankhodjaev, Director of the department, who visited Japan in the spring of 1978, they are now able to predict the time of earthquakes.

8.4 Earthquake Prediction in China

China has entered a seismically active period recently. Several destructive earthquakes have already occurred, not just in the earthquake-prone regions in the southwest, but also in the Hebei district which was thought to be relatively seismically stable. Table 8.2 shows the eleven earthquakes of $M \geq 6.8$ that occurred in China between 1966 and 1976, and their geochemical precursors. The earthquake prediction research that was supported so strongly in the Hebei district after the Yingtai earthquake of 1966 paid off a decade later, when various precursory phenomena that occurred in the stages leading up to five earthquakes were observed and, as a result, long-term, intermediate, and short-term predictions were announced. Four out of the five earthquakes which took place in 1975 and 1976 were predicted successfully. The Tangshan earthquake (July 28, 1976) was the exception. Scientists could not issue a warning immediately before this earthquake, we are told, partially due to the fact that no obvious precursors, such as foreshocks, were observed, and partially due to the political turmoil of the times. Despite the appearance of failure, prediction of the Tangshan earthquake can be regarded as almost 100% successful in terms of precursory phenomena, which were recorded.

The successful prediction of these earthquakes was the result of a system that gathers precursory information in one place and studies it compre-

---

* indicates that remarkable precursory changes were observed.

CHANGES IN GROUNDWATER LEVEL AND CHEMICAL COMPOSITION 211

Table 8.2 Geochemical Precursors and Earthquake Prediction in China from 1966 to 1976

Earthquake (location)	Date	M	Prediction	Geochemical Precursors
Yingtai (Hebei Province)	Mar. 8, 1966 Mar. 22, 1966	6.8 7.2		These are the earthquakes that triggered earthquake prediction in Hebei Province.
Luhuo (Sichuan Province)	Aug. 30, 1967	6.8		
Bohai (Bohai Bay)	Jul. 18, 1969	7.4		An increase was observed in the radon concentration in many wells in the Tientsin district.
Donghai (Yunnan Province)	Jan. 5, 1970	7.7		This earthquake triggered the earthquake prediction research in southwestern China.
Luhuo (Sichuan Province)	Feb. 6, 1973	7.9		A spike-like change (+120%) was observed at Guzan in Sichuan Province ($\Delta = 200$ km) seven days prior to the earthquake.
Yongshan-Daguan (Yunnan Province)	May 11, 1974	7.1		Radon concentration decreased (approx. 30%) at sichang, Sichuan Province ($\Delta = 140$ km).
Haicheng (Liaoning Province)	Feb. 4, 1975	7.3	○	Many wells along the tectonic strands showed anomalies. Positive anomalies in radon concentration (20 to 40%) were noted in wells within 200 km of the epicenter. The distribution of wells that showed water level changes and clouding prior to the earthquake demonstrate the regional regularities.
Longling (Yunnan Province)	May 29, 1976	7.5 7.6	○	Changes in radon concentration were seen in many wells, hot springs, and spring water. (Some cases were $\Delta = 460$ km). The amount of flowing rate of springs decreased ($\Delta = 60$ km). Increase in water temperature by 10°C observed for hot springs, etc.
Tangshan (Hebei Province)	Jul. 28, 1976	7.8 7.1	(○)	A 20% increase in radon concentration was noted in 20 wells within 200 km of the epicenter. The change was simultaneous. Eruption of groundwater, gases, and oil. Occurrences of 77 % of these anomalies concentrated in the period 3-5 days before the earthquakes.
Songpan-Pingwu (Sichuan	Aug. 16, 1976 Aug. 23, 1976	7.2 7.2	○	Anomalies in radon concentration were found in 10 out of the 24 wells

212 SHORT-TERM PRECURSORY PHENOMENA

Table 8.2 (continued)

Earthquake (location)	Date	M	Prediction	Geochemical Precursors
Province)				being observed. Among them, the farthest well from the epicenter was 550 km away. Various precursory phenomena were observed in groundwater.
Yanyuan-Ninglang (on the border of Sichuan and Yunnan Provinces)	Nov. 7, 1976 Dec. 13, 1976	6.9 6.8	○	Anomalies in radon concentration were observed in observation wells 160 and 270 km from the epicenter. Noted were water level changes, etc.

hensively. It is said that there was highly effective cooperation between the professionals and amateurs. Geochemical research and observation focusing on groundwater played a crucial role. Included among the measurements were radon concentration, the groundwater level, the amount of water flow, and the water temperature.

8.4.1 Radon

Simultaneous and extensive changes in the radon concentration of groundwater were observed in all the earthquakes successfully predicted by the Chinese. Long-term anomalies began two or three years before the earthquakes and continued until just before they occurred. The pattern of anomalous radon concentration varies from case to case: there were both positive and negative changes, violent vertical changes and spike-like changes. In short, anything that is not normal can be considered anomalous (Fig. 8.4).

In general, the anomalous pattern of radon concentration is complex. Obvious anomalies are extremely rare. Patterns vary from well to well, but it should be noted that the periods in which anomalies occur and return to normal seem to be roughly the same. Also, in many cases, there is a correlation between anomalies in radon concentration and anomalies in geomagnetism, ground resistivity, crustal movement, and small and intermediate earthquakes. The pattern of anomalous radon concentration is thought to reflect the processes of crustal stress accumulation, release, and recovery.

Radon concentration anomalies usually extend 200 to 300 km from the epicenter and, in some cases, as far as 600 km. The amplitude of change in the concentration ranges from 20 to 100%. In the case of the Longling earthquake, a correlation was noticed between the epicentral distance,

CHANGES IN GROUNDWATER LEVEL AND CHEMICAL COMPOSITION 213

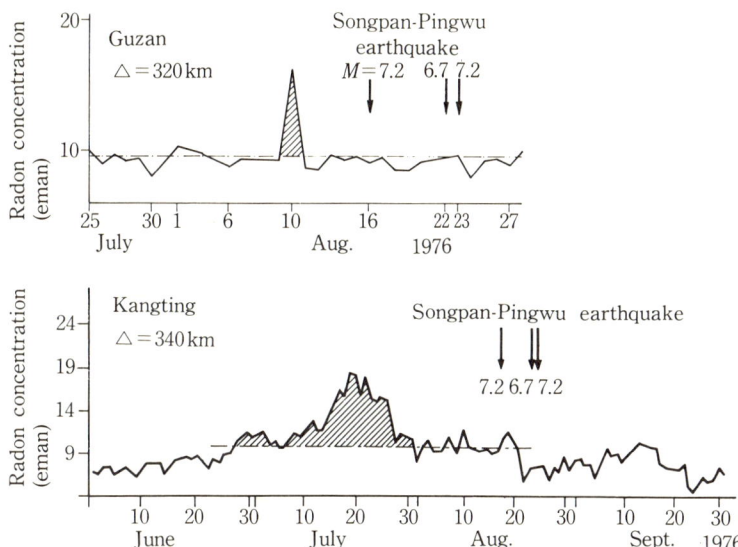

Fig. 8.4 Radon concentration anomalies prior to the Songpan-Pingwu earthquakes.

the magnitude of change, and the time the anomaly was observed. In other words, the closer the well to the epicenter, the greater the amplitude of the anomaly and the earlier it appeared.

8.4.2 Groundwater

Certain kinds of wells are remarkably sensitive to the crustal strain that precedes earthquakes. Some of the observed changes that were common to all of the five earthquakes mentioned above will be discussed here.

The spouting of well water accompanied by a thunderous boom, gushing of gas and oil, and changes in the taste, color, and temperature of the groundwater just before an earthquake are examples of such phenomena. These anomalies appeared, in many cases, in the vicinity of the epicenter, mostly during the one or two days preceding the earthquake.

Anomalies in groundwater do not, of course, appear in all wells. Wells where anomalies have been noted are scattered among numerous other wells that show no change at all. It is difficult to arrive at a general rule that would predict which wells are prone to precursory phenomena and which are not. It can be said, however, that such phenomena are more pronounced in wells situated in bedrock, at the ends of faults, or at the intersection of two faults.

8.5 Present Geochemical Studies in Japan

Geochemical study focusing on earthquake prediction in Japan was begun in late 1973 by the Faculty of Science, University of Tokyo. It is now being done by universities, government agencies, and local organizations as well.

At first the emphasis was on the establishment of the research methodology—the selection of observation wells and radon measurement methods, and especially the development of continuous measurement equipment. Now deep wells for observation have been drilled at Omaezaki in the Tokai district and other strategic points, and we are awaiting new results with high expectations.

At the time of the Izu Oshima offshore earthquake (M 7.0) that occurred on January 14, 1978, anomalous changes were observed in the radon concentration of groundwater before the earthquake. The change, the first observed in Japan, appeared in an artesian well at Hiekawa in Nakaizu-cho, about 25 km from the epicenter. Continuous radon measurements had been carried out in this particular well during the two years preceding the earthquake. The radon concentration, which had been decreasing at an abnormal rate since late October of the preceding year, dropped sharply six days before the earthquake. After recording a minimum value 13% lower than the normal value for eight hours, it returned to a normal level four days before the earthquake struck. The radon concentration then increased approximately 6% after the earthquake occurred. This period of anomalous radon concentration in general coincides with the period during which anomalies were recorded by the JMA's borehole strainmeters positioned at Ajiro and Irozaki, Izu Peninsula [Wakita et al., 1980].

8.6 Geochemical Studies for the Future

What direction should future geochemical studies take with regard to earthquake prediction? Some of the intriguing prospects will be discussed in the following sections.

8.6.1 Faults and the Gas-release Phenomenon

It is well-known that faults are directly related to earthquakes. In order to estimate the positions of future earthquakes, it is very important to know exactly where the active faults are. Faults are mainly examined by neotectonic study. Such methods are not suitable for faults that cannot be recognized from the topography, such as latent faults that are hidden beneath thick layers of sediment. Considering the fact that many of the major cities in Japan are situated on just such alluvial plains, a more effective method for detecting these faults is needed.

A large number of active faults have already been identified in Japan. The questions, then, are whether or not there is a way to tell how long it has been since these faults last moved, and which faults are likely to become active next. Geochemical methods may provide effective answers.

Since the earth was formed, various gases have continued to be driven from the earth's interior into the atmosphere in various degrees. This release of gas is more pronounced at volcanoes and oceanic ridges, but it is a common phenomenon; it is far from a uniform phenomenon, however, even if significant areas such as volcanoes are not included. Areas with greater degrees of fragmentation, including faults, would be likely to have a higher degree of gas release. Focusing on elements that are not particularly abundant in the atmosphere, such as helium, hydrogen, and radon, the measurement of gas released on the ground surface would seem to be an effective means of obtaining some clues to the questions posed earlier.

Recently the existence of areas with a higher (3×10^4 times) helium flux than that of the normal crust along the Matsushiro earthquake fault has been reported [Wakita et al., 1978].

8.6.2 $^3He/^4He$ Ratios

Helium has two isotopes. Their mass numbers are 3 and 4. These two isotopes were first formed with a fixed ratio at the time of the nuclear synthesis and incorporated into the earth's interior. The $^3He/^4He$ ratio has been steadily decreasing ever since because ^4He is supplied by α decay of natural radioactive nuclides. The $^3He/^4He$ ratio in the upper mantle is estimated at $\geq 10 \times 10^{-6}$. The crust lost a portion of ^3He and ^4He when the substance supplied by the upper mantle solidified. Since then ^4He has been added from uranium and thorium, thus lowering the $^3He/^4He$ ratio in the crust ($< 0.1 \times 10^{-6}$). There is a great difference between the $^3He/^4He$ ratio in the crust and in the upper mantle: the ratio is more than 100 times higher in the upper mantle.

The measurements of the $^3He/^4He$ ratio in the He obtained at the Matsushiro fault disclosed a ratio as high as that of the upper mantle. Given this fact, geochemical researchers hypothesize that the Matsushiro earthquake swarms were triggered by the water released into the upper crust at the time of a diapiric uprise of magma in the upper mantle [Wakita et al., 1978].

The results of recent $^3He/^4He$ measurements (which require highly refined techniques) throughout the world have been very interesting. Areas with high $^3He/^4He$ ratios are: the East Pacific Rise, the Mid Atlantic Ridge,

Iceland, the Aleutians, Kamchatka, the Kuriles, Japan, Hawaii, Mexico, Antarctica, Ethiopia, and the Red Sea. Thus there would seem to be a vital correlation between the higher $^3He/^4He$ ratio and plate movements. A great deal of information is likely to be gained from $^3He/^4He$ measurements—information different from that gained from geophysical studies, and which will illuminate the conditions of occurrences of earthquakes along the Japanese Island arc.

8.6.3 Miscellaneous Aspects of Research

Research on the isotope ratios of various elements, or on the abundance of trace elements in groundwater, is indeed central to earthquake prediction. The basis for such research has yet to be firmly established, however. When discussing the migration of groundwater, it is necessary to have identified specific elements or isotopic ratios that will indicate whether the water comes from a certain depth, whether it is magma-type water, etc. The fact that a volatile element like mercury exists in high concentrations in the seawater near oceanic ridges and in volcanic gas seems to indicate one direction for future research.

Furthermore, it is necessary to establish close ties between basic experiments and field observations. For instance, the comparative analysis of actual observation data and data from laboratory experiments that simulate conditions in the earth's crust, subjecting water and rocks to controlled pressure under controlled temperature, should prove highly useful.

References

Mogro-Campero, A. and R. L. Fleischer, 1977: Subterrestial fluid convection—A hypothesis for long-distance migration of radon within the earth, *Earth Planet. Sci. Lett.*, **34**, 321–325.

Noguchi, M. and H. Wakita, 1977: A method for continuous measurement of radon in groundwater for earthquake prediction, *J. Geophys. Res.*, **83**, 1353–1357.

Ulomov, V. I. and B. Z. Mavashev, 1971: The Tashkent earthquake of 26 April, *Akad. Nauk Uzbek SSR FAN*, 188.

Usami, T., 1975: *Descriptive Table of Disastrous Earthquakes in Japan* (in Japanese), Univ. of Tokyo Press, 327 pp.

Vorhis, R.C., 1968: In the Great Alaska Earthquake of 1964, 140–236, National Academy of Sciences, Washington D. C.

Wakita, H., 1975: Water wells as possible indicators of tectonic strain, *Science*, **189**, 535–555.

———, 1978: Earthquake prediction and geochemical studies in China, *Chin. Geophys.* **1**, 443–457.

———, Y. Nakamura, K. Notsu, M. Noguchi and T. Asada, 1980: Radon anomaly—A possible precursor of the 1978 Izu-Oshima-Kinkai earthquake, *Science*, **207**, 882–883.

———, N. Fuji, S. Matsuo, K. Notsu, K. Nagao and N. Takaoka, 1978: Helium spots—Caused by a diapiric magma from the upper mantle, *Science*, **200**, 430–432.

Chapter 9 Earthquakes and Electromagnetic Phenomena

Hitoshi Mizutani

It has been known since the latter part of the 18th century that anomalous changes in the geomagnetic field and the earth's current appear before or during earthquakes. Records of earthquakes accompanied by earthquake luminescence and abnormal animal behavior go back to the middle of the 9th century in Japan. Scientific research on these phenomena began at about the same time as seismological research. Shida's paper [1886] is thought to be the world's first attempt to scientifically investigate the electromagnetic phenomena that accompany earthquakes. There is a long history of observations and research on electromagnetic phenomena, but progress in this area has been far from satisfactory. The research in this field has not contributed to studies of the earthquake source mechanism, nor have they played an important role in earthquake prediction, except for some cases in China. Recently, due to the availability of highly precise observation instruments, earthquake-related electromagnetic phenomena have become the center of much attention. In this chapter Sections 9.1 and 9.2 introduce examples of observations made in the U.S.A., USSR, China, and Japan. Section 9.3 summarizes the theory and experiments related to such phenomena, and Section 9.4 contemplates the future role and potential problems of research on electromagnetic phenomena.

There are already many good review papers and books available on this subject. Among them, Rikitake [1976] provides the most detailed explanation of this field. The reports by Sumitomo [1978] and Yukutake [1977] read in Tokyo at the Earthquake Prediction Symposium held in December 1976 also contain a great deal of information. Recently Yamazaki [1977] wrote a brief review paper on geoelectric phenomena. This chapter adds the latest data to these reports and attempts to incorporate these data with some of the author's views on the electromagnetic phenomena that accompany earthquakes. Another source is Noritomi's [1978] description of Chinese studies, which contains many interesting examples. A comparison of the contents of this chapter with these other reports would be most helpful.

9.1 Historical Observations

Many examples of earthquake-related anomalies in geomagnetism and the earth's current are listed in Shida [1886], Milne [1890], and Kato [1939]. For instance, Milne quotes this passage from *Ansei Kenbunroku* [Records of Personal Experiences during the Ansei Period (1854–60)]:

> An eyeglass shop called Osumi in Kaya-cho, Asakusa, had a magnet three feet long. On Nov. 11, 1855, about 8 p.m., all the old nails, locks, and other irons attached to the magnet fell off. The store owner was very disappointed and wanted to sell the magnet. Since it was unusually big, he had placed it in the window, hoping to attract the attention of a *daimvo*, or feudal lord; it had been the store's star attraction. But a magnet that no longer attracts iron is just a stone. The owner thought the magnet had lost its power because it was old, and regretted his loss. Then, at about 10 p.m., there was a great earthquake and the magnet again began to attract iron as it had in the past. The store owner's discovery that magnets lose their power to attract iron before great earthquakes was a topic of great interest to the people (quoted from K. Musya [1957]).

Judging from what we know now, this incident, in which the magnet lost its magnetism two hours before the Edo earthquake of 1855 (M 6.7), was probably not directly related to the earthquake. As Milne points out, the store owner was probably mistaken and the pieces of iron were in fact shaken off by the earthquake temblor. In any case, one has to be very wary of the possible inaccuracies in historical records and observations. In this regard, Rikitake's [1968] and Johnston et al.'s [1973] figures showing the change in the records of geomagnetic anomalies with time are quite interesting. As is shown in Fig. 9.1, there is a drastic decrease in the amplitude of earthquake-related geomagnetic anomalies reported after 1960. In comparison to the old data, which recorded anomalies as great as 100 nT ($T = $ tesla, $n = 10^{-9}$; $1\,nT = 1\gamma$), the new data, recorded since the invention of the proton magnetometer, which is stable and without drift, rarely show anomalies of more than 10 nT. This suggests that the measuring methods of the past were less than perfect. In order to distinguish the old from the new, observations made prior to 1960 will be called classical observations and observations made since 1960 will be called modern observations. One must be careful here, because not all modern observations are superior in precision to the classical observations, nor are all classical observations unreliable. Statistically speaking, however, modern observations are more accurate and dependable than classical ones.

In addition to changes in the intensity of geomagnetism, classical observations also included observations of variations in declination and

EARTHQUAKES AND ELECTROMAGNETIC PHENOMENA 219

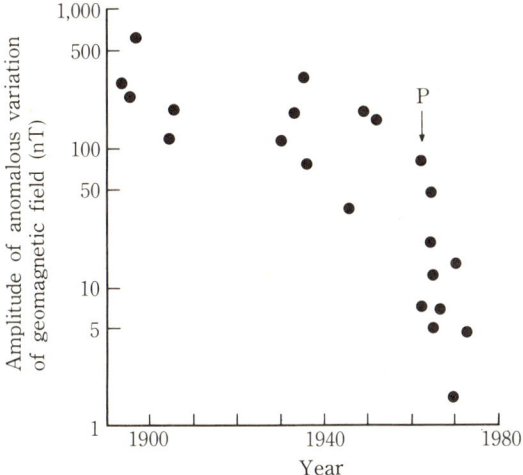

Fig. 9.1 Geomagnetic change reported associated with earthquakes since 1890 [Johnston et al., 1973].
P denotes the year the proton magnetometer was introduced.

inclination. According to Kato's summaries [1939, 1966] and Nagata's examples [1937], the change in declination and inclination accompanying earthquakes occurs within 10 minutes. The most celebrated example is the one that accompanied the Nankai earthquake of 1946, presented by Kato and Utashiro [1948], which is shown in Fig. 9.2: the difference between daily average of declination obtained at Katsuura (epicentral distance ≈ 60

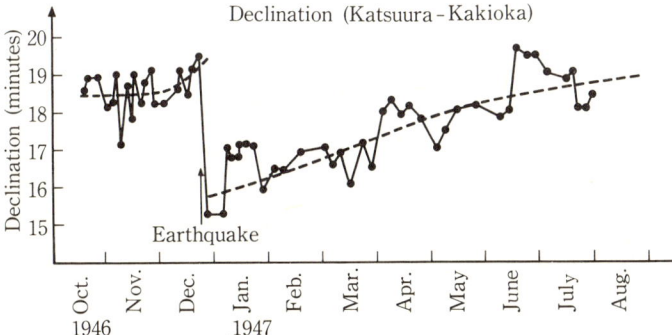

Fig. 9.2 Changes in the geomagnetic declination at the time of the Nankai earthquake (M 8.1) of 1946 [based on Kato and Utashiro, 1948].
The data represent the daily average difference in declination at Katsuura and Kakioka.

km) and that at Kakioka (epicentral distance ≈ 600 km) showed a slight increase prior to the earthquake, followed by a drastic coseismic decrease. Then it gradually returned to the original value during the next eight months or so. Judging from the degree of scatter of the data, it seems certain that the declination at Katsuura changed at the time of the earthquake. Nagata [1972] attempted to explain the observed coseismic change using the concept of piezo-magnetism. This simple concept, however, cannot easily explain the gradual change of declination that occurred after the earthquake. The gradual changes during the post-earthquake period can be explained as the result of post-earthquake crustal movement or, as Mizutani et al. [1976] suggest, as electrokinetic phenomena due to the diffusional process of groundwater.

In addition to geomagnetic changes, there have been many cases reported of changes in the earth's current and in its electric potential. The oldest of these are full of stories of anomalous electric currents running through communication lines and submarine cables either before or during earthquakes. It is difficult to determine the credibility of these reports, however. No substantial progress has been made recently in the methods used to observe the earth's current and its electric potential. Therefore, unlike the data on geomagnetism, older data do not necessarily indicate questionable dependability. The problem with the old data is not in the observation instruments, but in whether or not the effects on the earth's current caused by changes in the geomagnetic field, or by changes in other physical and chemical conditions, have been adequately taken into account. Milne [1890] and Yoshimatsu [1957] reported a great many classic examples of the observation of the earth's current and electric potential. One of these examples [Kato et al., 1950] is shown in Fig. 9.3. It shows the changes in the earth's current associated with an aftershock of the

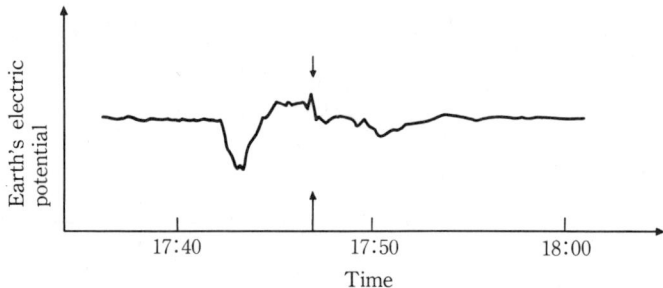

Fig. 9.3 Changes observed in the earth's electric potential during and after the July 19, 1948, Fukui earthquake [based on Kato et al., 1950].

The arrows indicate the aftershock of the earthquake. Absolute values of the changes in electric potential are not known.

Fukui earthquake (June 28, 1948). These observations were made in relation to an earthquake that occurred on July 19, 1949, at 17:48 hours. A few minutes before the earthquake, the earth's electric potential (between electrodes 100 m apart) began to show changes. It went back to normal just before the earthquake but continued to change for a few minutes afterwards. Although the absolute value of the change is not clear from the figure in the original paper, it is thought to be about 10mV/100m or more. Kato et al.'s paper includes some additional records of changes in the earth's electric potential that accompanied other aftershocks of the Fukui earthquake.

In addition to changes in geomagnetism and the earth's current, there is another electromagnetic phenomenon associated with earthquakes: earthquake lightning or luminescence. Whether luminescence is really earthquake-related or not has long been the subject of debate and discussion. Musya [1931, 1932] and Terada [1931] studied luminescence scientifically for the first time, and their results, combined with the photographs taken at the time of the Matsushiro earthquake, strongly suggest that luminescence is, indeed, an earthquake-related natural phenomenon. Like other electromagnetic phenomena, earthquake luminescence has been reported throughout Japanese history. The earliest record appears in *Sandai Jitsuroku* (True Records of Three Generations, 901) and concerns the Sanriku earthquake (M 8.6) of July 13, 869. According to this record:

> On that day, the land of Mutsu (presently Aomori and part of Iwate Prefecture, Northern Japan) was greatly shaken and light flashed like daylight and reflected . . .

This may be similar to the luminescence observed at the time of the Sanriku earthquake (M 8.3) of March 3, 1933 [Musya, 1934]. Such phenomena are not limited to Japan; Galli [1910], in a paper cited by Terada [1931], included examples of luminescence observed in Europe 800 years ago. According to Musya's [1932] list of earthquakes prior to the Meiji Era, 67 were accompanied by earthquake luminescence; between 1867 (the first year of Meiji) and 1932, there were 24 such earthquakes. Of the 60 "major earthquakes" since 1867 catalogued by Kaminuma et al. [1972], 15 were accompanied by luminescence. As these examples suggest, luminescence is observed in a comparatively large number of earthquakes.

Sometimes the luminescence takes place before the earthquake. For example, the record of the Kinai earthquake (M 6.4) of August 19, 1830, mentions:

> According to the people of Kyoto, the whole sky lit up the night before the earthquake and light also seemed to be coming from the ground. It was almost like daylight, and people were troubled and bewildered. Sure enough, a great earthquake struck the very next day [cited from Musya, 1957].

As in the case of other electromagnetic phenomena, understanding of the cause of earthquake luminescence is not sufficiently advanced to enable us to judge its exact scale or intensity. As our understanding of these phenomena deepens, however, earthquake-related electromagnetic phenomena will become a more useful tool for earthquake prediction.

9.2 Geomagnetic Changes Accompanying Earthquakes— Examples of Modern Observations

9.2.1 Cases in Japan

The observation of geomagnetism, as was noted, entered a new era with the invention of the proton magnetometer. Based on the Earthquake Prevention Plan, twelve geomagnetic observation sites equipped with proton magnetometers have been set up throughout Japan since 1968, and these are making continuous observations of the total geomagnetic force. The Geographic Institute of Japan and the Hydrographic Office of the Marine Safety Agency, at the same time, are carrying out magnetic surveys every one to five years at more than one hundred magnetic survey points throughout Japan. Although the intervals between surveys are long, the surveys are very effective at identifying extensive, long-range geomagnetic changes, thanks to the sheer number of observation sites. In addition to these government surveys, several university research centers, including the Earthquake Research Institute at the University of Tokyo, are undertaking temporary field observations.

Most geomagnetic changes thought to be related to earthquakes or volcanic activity were obtained from anomalies in the secular variation of geomagnetism as detected by magnetic surveys. The Matsushiro swarm earthquakes are only one example of an anomaly that was detected by continual geomagnetic observation.

An example of anomalous secular variation in geomagnetism is demonstrated by the observations made at Tanabe, Wakayama Prefecture, by Tazima [1968], which are shown in Fig. 9.4. From 1951 to 1960, the amount of secular variation in the horizontal component of the geomagnetic force at Tanabe was as great as 7 nT per year—much greater than that in other towns of Gojo (Nara Prefecture) and Taketoyo (Aichi Prefecture) farther from the epicenter. From 1966 on, however, the secular variation at these three observation points became equal. The reason for the change in the pattern of secular variation at Tanabe between 1960 and 1966 might be found in the two earthquakes that occurred in 1961 and 1962. Since there are no data available prior to 1951, however, it cannot be determined whether or not the large change in the secular variation between 1951 and 1960 was an earthquake precursor. Tazima *et al.* [1976] later extended their research to the data obtained at magnetic survey points throughout Japan, and pointed out that those regions with

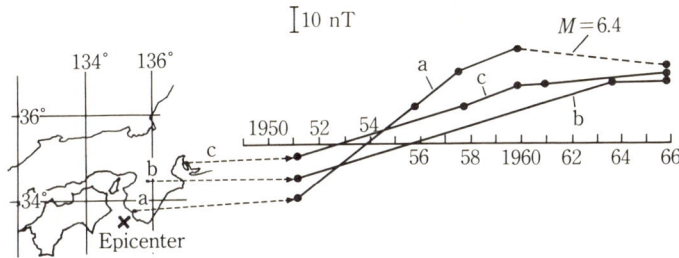

Fig. 9.4 Anomalies in the secular variation of geomagnetism at Tanabe, Wakayama Prefecture [based on Tazima, 1968]; a: Tanabe; b: Gojo; c: Taketoyo.
The anomalous variations at Tanabe, seen in conjunction with data from Gojo and Taketoyo, are thought to be related to the 1962 Kii Peninsula offshore earthquake ($M = 6.4$).

anomalous secular variation in geomagnetic force correspond to the regions where earthquake and volcanic activity are prevalent.

Figure 9.5 shows the regions where the secular variation in the total geomagnetic force is considered to be particularly large (data from 1970 ± 2.5 years). These regions include the sites of the Nemuro Peninsula offshore earthquake (M 7.4, 1973), the Eastern Akita earthquake (M 6.2, 1970), the Izu Peninsula offshore earthquake (M 6.9, 1974), and Mt. Iwate, an active volcano. Therefore, the researchers have concluded that the anomalies in Fig. 9.5 may reflect variations in crustal stress level in these regions. If this conclusion is correct, Fig. 9.5 will play an important role in earthquake prediction. Sumitomo [1978], on the other hand, made a detailed investigation of secular variation in the Chugoku and Kinki districts, and demonstrated that the region's pattern of secular variation was related to the region's crustal movement since the beginning of the Quaternary Period. These results do seem to indicate that geomagnetic anomalies in secular variation correspond to variations in crustal stress. Further evidence comes from the work of Fujita [1965], based on the results of magnetic surveys repeated during the one-year period after the Niigata earthquake (1964, M 7.5). Fujita found that the declination in the area adjacent to the epicenter had showed an anomalous variation of up to 2', compared to the surrounding areas, during the 10 years before the earthquake. After the earthquake this trend was reversed, and within a year conditions returned to normal. It has also been reported that the horizontal component of the geomagnetic force began to show an anomalous variation 10 years before the Niigata earthquake, and an anomaly of up to 20 nT was observed [Tazima, 1968].

Since there are intervals of up to five years in the data from magnetic

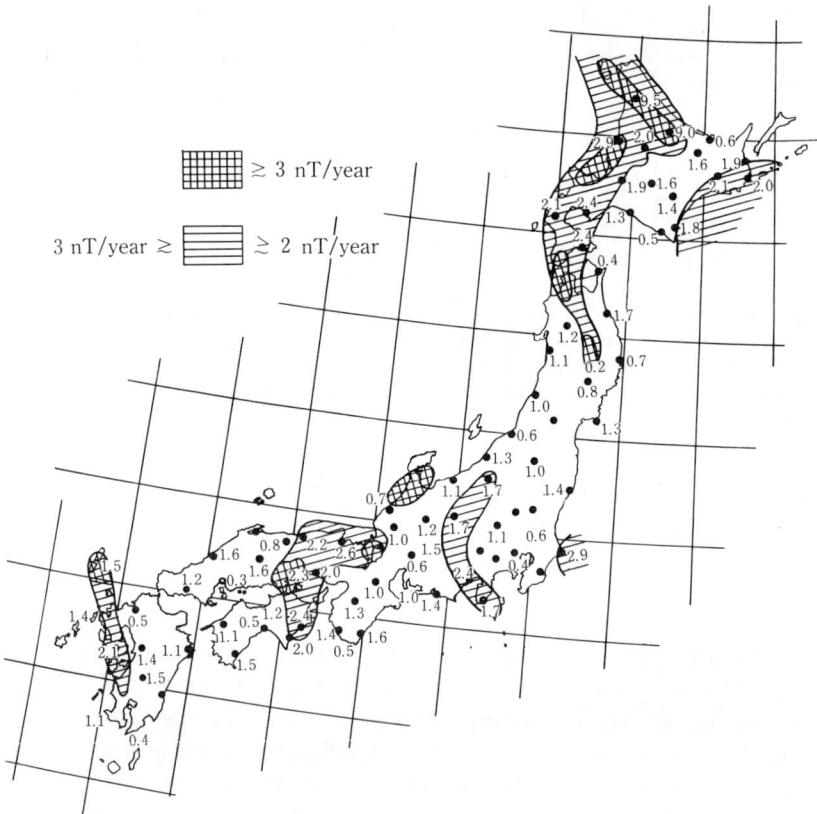

Fig. 9.5 Regions in Japan where the secular variation in total geomagnetic force is particularly great [based on Tazima et al., 1976].

surveys, more rapid changes cannot be detected. The most spectacular example of an earthquake-related anomaly discovered by continuous geomagnetic observation is the work of Rikitake et al. [1966] on the Matsushiro swarm earthquakes (1965 to 1967) (see also Yamazaki and Rikitake, [1970]). Figure 9.6 shows the total intensity values at Matsushiro (as compared to the values at Kanozan Observatory) as reported by Yamazaki and Rikitake [1970]. An anomaly of about 10 nT can be seen between August and October of 1966. The period of this anomaly, however, does not coincide with the period of high seismic activity in the area. Mizutani and Ishido [1976] point out that it should be attributed to electrokinetic phenomena—which will be discussed later in this chapter—since there is a good correlation between this anomaly and variations in the water flow that have been observed in the fault region of Matsushiro.

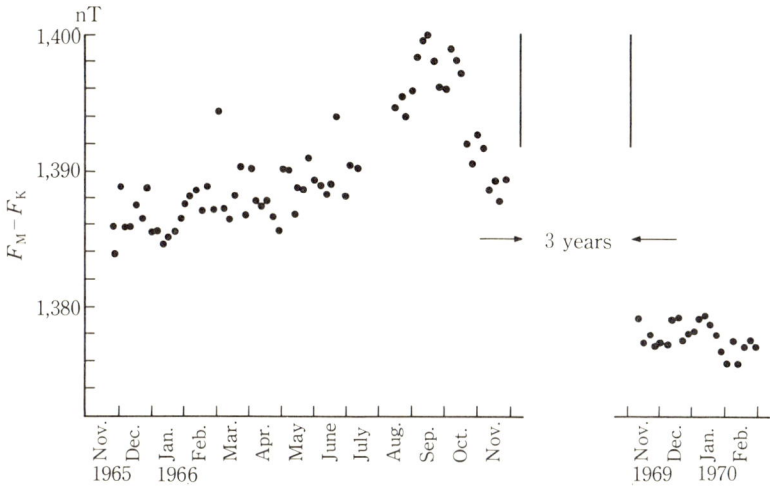

Fig. 9.6 Observed anomalies in the variations in total magnetic force accompanying the Matsushiro swarm earthquakes [based on Yamazaki and Rikitake, 1970]. The data show five-day average values of the difference between the observed values at Matsushiro and those at Kanozan. Observed values for the three-year period from Dec. 1966 to Oct. 1969 are not available.

9.2.2 Cases in the United States

Research on earthquake-related geomagnetic anomalies in the United States is being done by the Menlo Park Group of the U.S. Geological Survey. They have been engaged in the continuous observation of magnetic anomalies at seven fixed observation sites along the San Andreas Fault since 1973. They also perform periodic observations at more than 100 pairs of sites 8 to 12 km apart [Johnston et al., 1973]. Despite such an extensive observation network, discovering earthquake-related geomagnetic anomalies appears to be quite difficult: only a few such cases have been reported so far.

Data from the most remarkable case are shown in Fig. 9.7, which shows the precursory phenomena observed prior to the earthquake (M 5.2) that occurred on November 28, 1974, near Hollister, California. The epicenter and the positions of the observation sites are shown in Fig. 9.8. Figure 9.7 shows the five-day average of difference in the total magnetic force between observation site pairs 3 and 4 and between pairs 4 and 5. Data from site 2 were useless due to noise interference. The difference in the total magnetic force between sites 3 and 4 began to increase in October 1974 and reached 1.5 nT during the month of October. It stayed almost constant during the last two weeks in October, then dropped 0.3 nT on October 31 and by

226   SHORT-TERM PRECURSORY PHENOMENA

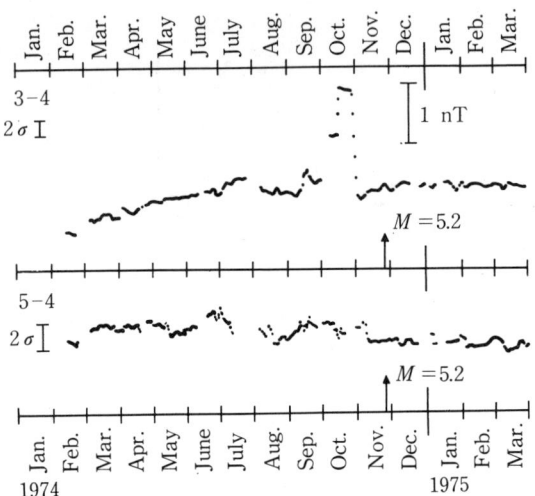

Fig. 9.7 Variations in total magnetic force observed at Observation Sites 3, 4, and 5 near Hollister, California, U.S.A. [Smith and Johnston, 1976].
The data show daily average differences in the values observed at sites 3 and 4 (top) at at sites 4 and 5 (bottom). The arrow denotes the time of the Nov. 28, 1974, earthquake (M 5.2). See Fig. 9.8 for the earthquake's epicenter and the positions of the observation sites.

another 1.3 nT on November 1. It had leveled off by November 28, when the earthquake finally occurred 1 km from site 3. Since it is difficult to imagine any other cause for the geomagnetic anomalies shown in Fig. 9.7, it seems likely that they are earthquake-related. As the change was quite rapid, it is thought to have been due to the piezo-magnetic effect on rocks (explained in Section 9.5.1 below) and not to electrokinetic phenomena.

In addition to the above example, Johnston *et al.* reported a case in which anomalies of about 1 nT were observed in relation to a M 4.2 earthquake. While investigating an earthquake of magnitude 7.1 which occurred in the sea near Sitka Island coast, Alaska, on July 20, 1972, Wyss [1975] discovered that the horizontal intensity of the geomagnetic field at Sitka had begun to decrease seven and a half years before the earthquake. The decrease totaled 20 nT shortly before the earthquake. This is the only example of an observed geomagnetic anomaly that exceeded 10 nT in recent years. Such an anomaly cannot be ruled out as impossible, however, since the magnitude of the earthquake itself was so great.

9.2.3 Cases in the USSR

The Soviet Union has made a concerted effort to use data on geomag-

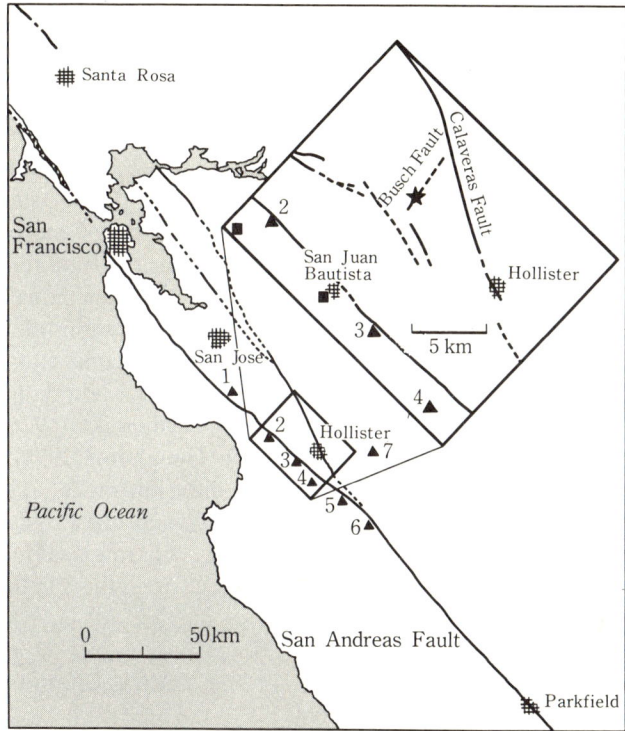

Fig. 9.8 Positions of geomagnetic observation sites and sites for observing the earth's electric potential and electric resistance near Hollister, California.
The star indicates the epicenter of the Nov. 28, 1974, earthquake ($M = 5.2$).

netism and the earth's current, collected by extensive observations, for purposes of earthquake prediction. So far, however, the results have been disappointing with regard to geomagnetic variation. In the district of Garm there are occasional geomagnetic anomalies of more than 10 nT, but it is very difficult to judge whether or not they are really earthquake-related [Abdullabekov et al., 1972]. The only case in which a geomagnetic anomaly seems to be certainly related to an earthquake is the one that accompanied an M 4.8 earthquake in the Tashkent Uzbek district. The observed anomaly was approximately 23 nT.

9.2.4 Cases in China

The observation of electromagnetic phenomena is used extensively in China for earthquake prediction. Noritomi [1978] reported in detail on these observations. The observation of declinations in geomagnetism, the earth's current, etc., are carried out at what are known as People's Observation Points. Looking at Noritomi's report [1978] and many others, one finds many geomagnetic anomalies being reported, just as in the rest of the world.

According to Zhu Fung-ming [1976], 15 months before the Haicheng earthquake (1975, M 7.3) the geomagnetism at Lüda showed an anomalous variation (compared with that at Beijing) of more than 20 nT. Oike [1977] and Noritomi [1978] reported that the total force began increasing at Beijing 3 months before the Tangshan earthquake (1976, M 7.8), reaching a maximum of 7 nT. Anomalous variations of 5' to 8' in the declination of geomagnetism were also reported in connection with the Yanyuan earthquake (1976, M 6.9), the Yanyuan-Ninglang Border earthquake (1976, M 6.8), and the Longling earthquake (1976, M 7.5) [Noritomi, 1978; Tang Chi-yang, 1978]. Qi Gui-zhong [1978] made a theoretical calculation based on a dilatancy diffusion model and electrokinetic effect similar to that in Mizutani et al. [1976] in order to explore the causes of earthquake-related anomalies. This calculation is said to explain the variation in the observed value of geomagnetism in the Yingtai district between 1968 and 1974 quite reasonably.

9.3 Earthquake-related Variation in the Earth's Current and the Earth's Electric Potential

9.3.1 Cases in Japan

Since 1960 there has been little observation of earthquake-related changes in the earth's current and its electric potential in Japan; thus good examples of anomalous variation are rare. Rikitake et al. [1966] did observe a variation in the earth's current that seemed to be related to one of the Matsushiro swarm earthquakes (Nov. 23, 1965; M 5.0), but it was a short-range variation that occurred during the earthquake, not a precursory variation like those that have been observed in the USSR, the U.S.A., and China. This may be due to the fact that artificial noise is so great during the daytime in Japan, and that observation of the earth's current and electric potential seemed almost impossible.

9.3.2 Cases in the United States

The San Andreas Fault is the site of observations of the earth's electric potential as well as of its geomagnetism. A group from the University of California, Berkeley, performed a 2 1/2-year-long continual observation and found two cases of anomalous variation that seemed to be earthquake-related [Corwin and Morrison, 1977]. One of these appeared to be related to the earthquake on Nov. 28, 1974 (M 5.2) that was cited earlier in the section on geomagnetism. It is shown in Fig. 9.9.

EARTHQUAKES AND ELECTROMAGNETIC PHENOMENA 229

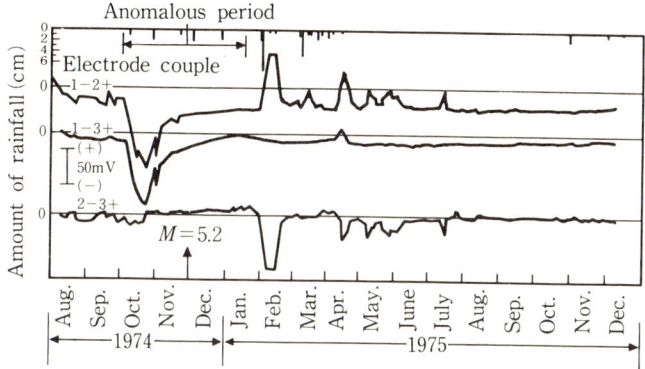

Fig. 9.9 Variations in the earth's electric potential observed near Hollister, California, in 1974 and 1975 [Corwin and Morrison, 1977].
The variations that appear in October and November 1974 at electrode couples $1-2^+$ and $1-3^+$ are thought to have been precursors of the Nov. 28 earthquake. The anomalous variations that appear at electrode couples $1-2^+$ and $2-3^+$ seem to have been a result of heavy rainfall in February 1975.

As can be seen in this figure, the voltage between the electrode couples 1–2 and 1–3 showed a decrease in electric potential 55 days before the earthquake. What is particularly remarkable about this variation is the way it decreases abruptly at the initial stage and recovers gradually after the decrease. The pattern of variation during February 6–13, 1975, on the other hand, is quite different and is attributable to rainfall. (Only electrode 2 must have been influenced by the rain.)

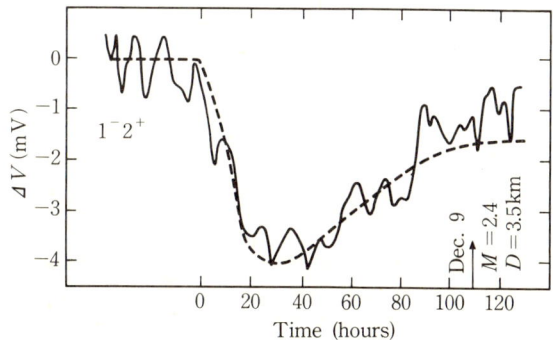

Fig. 9.10 Variations in the earth's electric potential observed at the same observation site as that used for data in Fig. 9.9 [Corwin and Morrison, 1977].
This variation pattern is thought to have been a precursor of the Dec. 9, 1975, earthquake in the area (M 5.4; epicentral distance 3.5 km), which is indicated by the vertical arrow. The dotted shows the calculated value based on the dilatancy dispersion model and on the electrokinetic effect.

The other example is seemingly related to the earthquake of December 9, 1975, and involves anomalous variation between a pair of electrodes approximately 2.5 km from the epicenter. This variation, shown in Fig. 9.10, has a pattern similar to those shown in Fig. 9.3 and Fig. 9.9. The variation began 110 hours before the earthquake. The dotted line in Fig. 9.10 is the calculated value of the variation in electric potential expected from the diffusion of groundwater and from electrokinetic effects—as advocated by Mizutani et al. [1976]. The agreement of the theoretical variation with the values actually observed seems convincing.

9.3.3 Cases in the USSR

In the USSR there have been numerous anomalous variations in the earth's electric potential observed in Kamchatka. Figure 9.11 demonstrates one of these, as reported by Myachkin et al. [1972]. It shows the time variation in the earth's electric potential after the short-range variation has been eliminated. In all these cases, the variation begins a few days before the earthquake and returns to normal just before the earthquake. The period during which the anomalous variation is sustained seems to be proportional to the magnitude of the earthquake, although this has not been determined for certain. At any rate, it is interesting to note that the cases in Japan (Fig. 9.3: the Fukui earthquake), the cases in the U.S.A. (Figs. 9.9 and 9.10), and the cases in the USSR (Fig. 9.11) all have similar variation patterns.

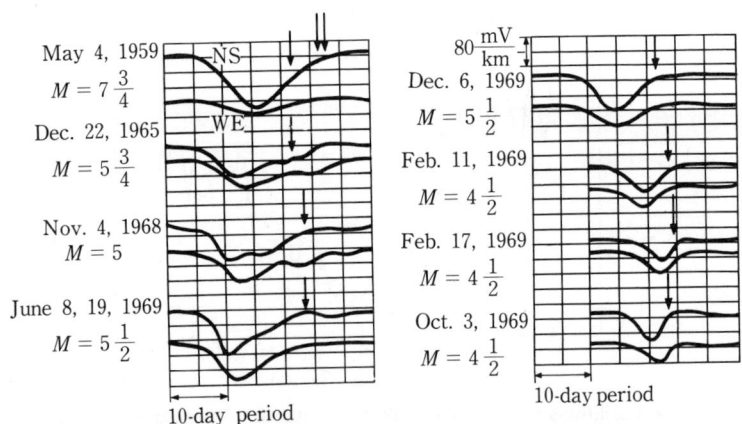

Fig. 9.11 Variations in the earth's electric potential observed prior to earthquakes in Kamchatka [Myachkin et al., 1972].
The arrows indicate the times of the earthquakes.

9.3.4 Cases in China

In China, as in the USSR, data on the earth's current and electric potential are used effectively to predict earthquakes. Figure 9.12 shows an anomalous variation in the earth's electric potential that is thought to have been a precursor of the Haicheng earthquake (Feb. 4, 1975; M 7.3). This variation was recorded on an instrument made by an amateur at Yingkou, approximately 45 km from the epicenter (distance between the electrodes: 60 m). In this case also the earth's electric potential began to change some time before the earthquake (in January 1975) and then returned to the original value 3 to 4 days prior to the earthquake. A similar variation was observed at many points at the time of the Songpan-Pingwu, Sichuan Province, earthquake (1976, M 7.2) [Noritomi, 1978].

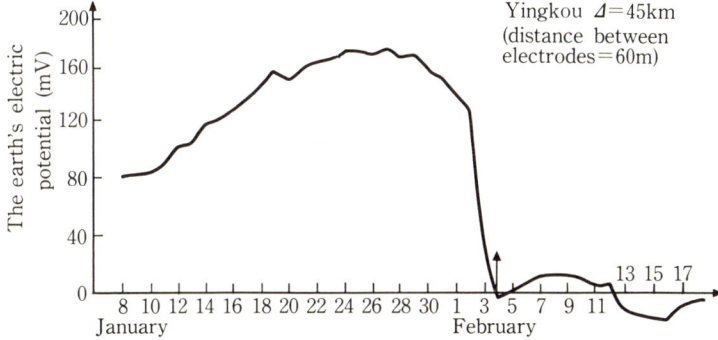

Fig. 9.12 Variations in the earth's electric potential observed at Yingkou at the time of the Haicheng earthquake (M 7.3), Feb. 4, 1975 [Zhu Fung-ming, 1976]. The arrow denotes the time of the earthquake.

From these examples it can be seen that anomalous variations of 10 to 100mV between two electrodes 100 m apart were observed as earthquake precursors.

9.4 Variations in Electric Resistance Accompanying Earthquakes

9.4.1 Cases in Japan

Yamazaki et al. of the Earthquake Research Institute, University of Tokyo, discovered, as a result of their lengthy experience at Aburatsubo, Kanagawa Prefecture, that the electric resistance in this area (which is largely Lapilli-tuff) can be used as a very sensitive strain gauge. As is demonstrated in Fig. 9.13, electric resistance shows a step-wise variation at the time of an earthquake [Rikitake and Yamazaki, 1970]. The variation

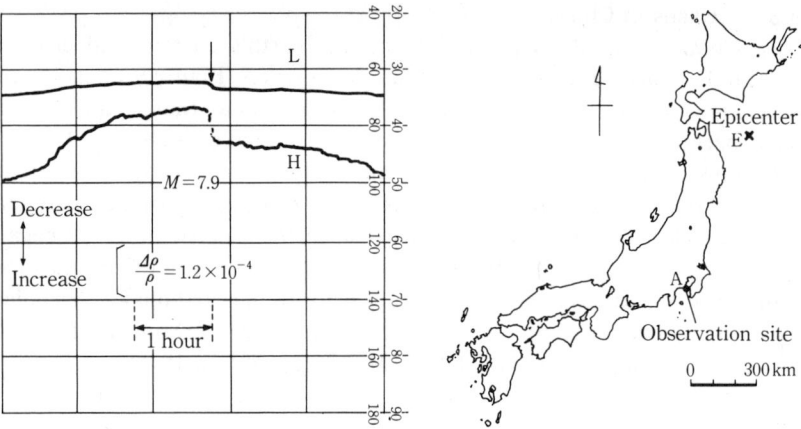

Fig. 9.13 Variations in specific reactivity observed at Aburatsubo at the time of the Tokachi offshore earthquake (M 7.9), May 16, 1968 [based on Rikitake and Yamazaki, 1970].

H denotes a high sensitivity record; L denotes a low sensitivity record. The arrow indicates the exact time of occurrence of the earthquake.

does not occur instantly, however, but seems to take several minutes to complete. This may mean that the progress of the strain in the direct current components caused by earthquakes is relatively slow. In addition to the variation that occurs at the same time as the earthquake, there is also anomalous variation in electric resistance that seems to begin several hours before the earthquake, as shown in Fig. 9.13. The period of precursory variation does not seem to be related to the magnitude of the earthquake. Rikitake and Yamazaki [1977] suggest that this variation may reflect the progress of non-elastic creep strain near the epicenter prior to the earthquake. Although the cause of this variation has not yet been determined, it may prove to be an effective element in earthquake prediction, provided that such variations occur frequently before earthquakes. According to Yamazaki [1975], during 1968–1974 there were 20 cases in which such precursory variations were observed.

There have been many attempts in Japan to use the time variation of the $\Delta Z/\Delta H$ ratio for geomagnetic bays. These attempts are based on the fact that this ratio of the vertical component ΔZ and the horizontal component ΔH for geomagnetic bays may change with the underground electric resistance. Such a ratio is one of the means by which the variation in underground electric resistance associated with earthquakes can be examined.

9.4.2 Cases in the United States

The observation of earth resistivity is also being made to monitor seismic activity. In the observation made by a team at the University of

California, Berkeley, an electric current of square waves with an amplitude of 200 A and a period of 10 sec was transmitted through the ground using a gigantic 85 kW generator, and the ratio of the voltage to the current between two electrodes 1.5 km apart was measured [Mazzella and Morrison, 1974]. The results of this measurement for the period January 1973 through February 1974 are shown in Fig. 9.14. The variation that began about 60 days prior to an earthquake of magnitude 3.9 that occurred about 80 km from the electrodes on June 22, 1973, can be interpreted as a precursor of that particular earthquake. In this case, as in many others, there was a decrease in resistivity prior to the earthquake. The electric resistance went down to 10–15% of the normal value. It was also pointed out that the variation at observation point 2 from Sept. 15 to Oct. 15, and the variation at observation point 3 during mid-December, may have been associated with earthquakes in these areas, but the variations here are too slight to be significant.

9.4.3 Cases in the USSR

Extensive investigations are under way in the district of Garm and other areas in the USSR to monitor underground electric resistance using a method similar to that used by the U.S. team. The results of these investigations are presented in Fig. 9.15. Almost without exception there is a 10 to 15% decrease in electric resistance prior to earthquakes of more than a certain magnitude. Although these results are similar to those obtained by Yamazaki [1975], the decreases begin several months before the earthquake. Thus the precursory periods here are more than 100 times longer than the several hours observed by Yamazaki. The difference may be due to the distance between the epicenter and the observation point, or to the geological characteristics of the observation point. This is a point that warrants further investigation.

9.4.4 Cases in China

Many precursory variations in electric resistance have also been observed in China. Noritomi [1978] reports that the Chinese regard the measurement of electric resistance as much more dependable than the measurement of geomagnetism or of the earth's current and electrical potential. The specific resistivity of the ground associated with the Songpan-Pingwu earthquake (1976, M 7.2) is shown in Fig. 9.16. Among the data in Fig. 9.16, the data at Songpan are noteworthy because the resistivity there increases before the earthquake rather than decreases, unlike the cases previously discussed. It is also interesting to note that a 17% decrease in electric resistance was recorded in Tongwei, 300 km away from the Songpan earthquake's epicenter.

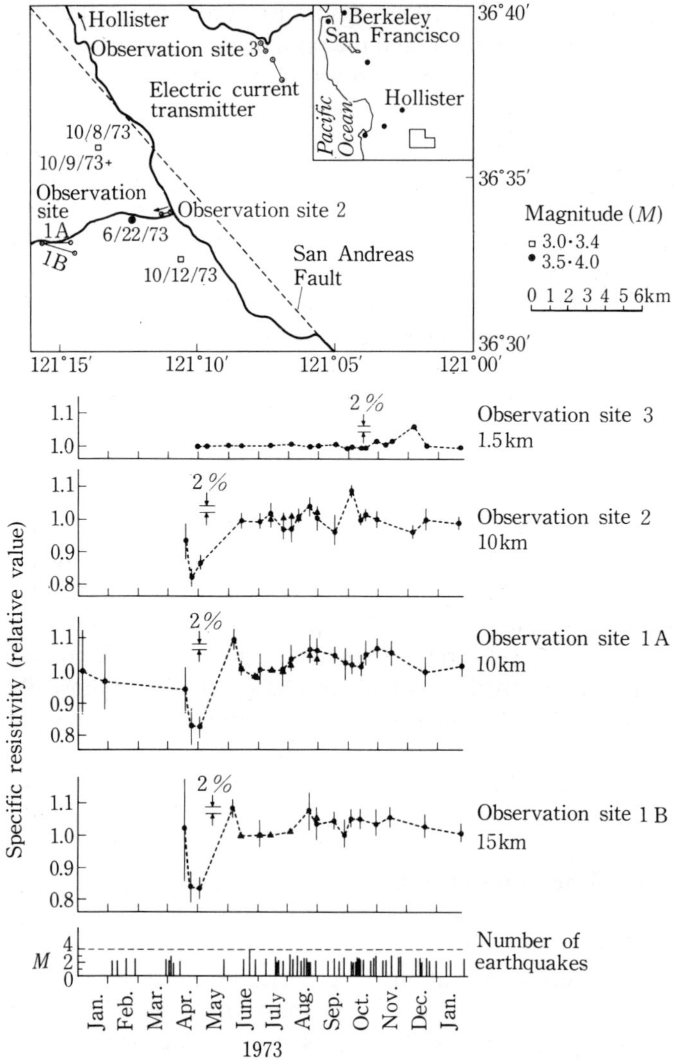

234 SHORT-TERM PRECURSORY PHENOMENA

Fig. 9.14 Variations in the ground's specific resistivity observed in the vicinity of Hollister, California, in 1973 [Mazzella and Morrison, 1974].

The map at the top shows the position of the transmitter and the positions of four specific resistivity measurement dipoles (1A, 1B, 2, and 3).

EARTHQUAKES AND ELECTROMAGNETIC PHENOMENA   235

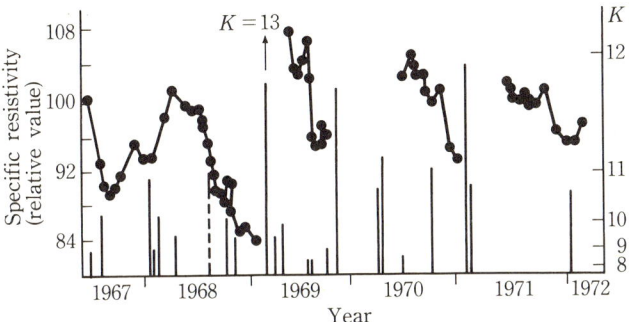

Fig. 9.15 Variations in specific reactivity observed in the Garm district of the USSR between 1967 and 1972 [Barsukov and Sorokin, 1973].
The vertical lines indicate the scales of the earthquakes; K is the exponent of 10 when earthquake energy is measured in ergs.

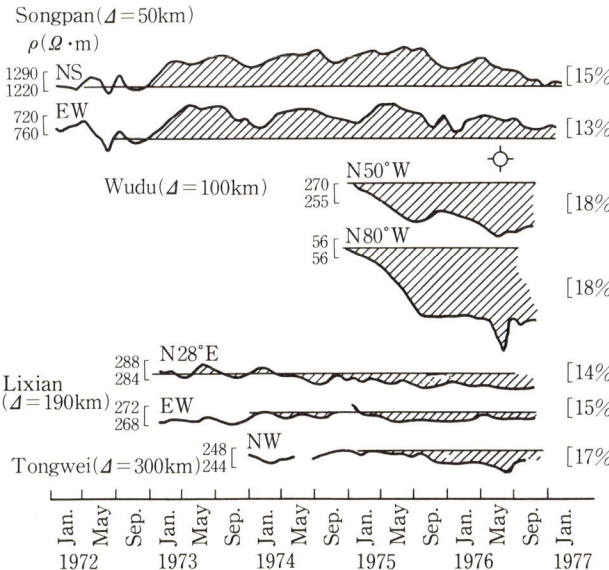

Fig. 9.16 Variations in the specific resistivity of various areas in relation to the Songpan-Pingwu earthquake of Aug. 16, 1976 (M 7.2).
On the left side are listed the four observation points and their distances from the epicenter. The percentages on the right side are the relative variation ratios [Noritomi, 1978]. The time of the earthquake is indicated by the circular symbol in the upper right quarter of the figure.

9.5 The Etiology of the Electromagnetic Phenomena Associated with Earthquakes

9.5.1 Piezo-Magnetic Effect

Rocks contain magnetic minerals such as titano-magnetite and are themselves magnetic as a whole, although their magnetism is quite weak. This rock magnetism affects the distribution of the magnetic field on the earth's surface. Consequently, any variation in the magnetism of rocks preceding an earthquake will cause a variation in the observed value of the earth's magnetic field on the surface.

It is both theoretically and experimentally proven that the magnetism of rocks changes when the rock is stressed. This variation due to stress is called the piezo-magnetic effect. The term "tectonomagnetism" coined by Nagata [1969] refers to the study of the variations in geomagnetism caused by the piezo-effect that accompanies large-scale geological activities such as earthquakes and volcanoes.

There are two kinds of piezo-magnetic effect. In volcanic rocks, such as basalt and andesite, rock magnetism is governed by the remanent magnetism of the minerals contained in the rocks; the effect of stress on the remanent magnetism is important for volcanic rocks. In sedimentary rocks, however, remanent magnetism is weak, so that the rock magnetism is governed by the magnetism induced by the earth's magnetic field. Therefore the effect of stress on the magnetic susceptibility is important for sedimentary rocks. A great many studies have been done on these two types of stress effects, including experiments by Kapista [1955], Ohnaka and Kinoshita [1968], Nagata [1966], and Martin and Wyss [1975], and theories by Kern [1961], Stacey [1962], and Nagata [1970] among others.

The effect of stress (uniaxial compression: confining pressure $= 0$) on remanent magnetism and on magnetic susceptibility can be summarized in the following equations, based on Nagata [1970]:

$$\chi = \chi_0 \{1 - 1/4(1 + 3\cos2\theta) \beta \cdot \sigma\} \quad (9.1)$$

$$J_R = J_{R0} \{1 - 1/4(1 + 3\cos2\theta) \gamma \cdot \sigma\} \quad (9.2)$$

where χ and J_R represent magnetic susceptibility and the strength of the natural remanent magnetism, respectively. The suffix 0 represents the value at stress $= 0$. β and γ are constants to be determined by experiment. σ is the stress applied to rocks. θ is the angle between the directions of magnetization and stress. When σ is the compressive stress it is assumed to be positive. Experiments show that the value of the constants β and γ for most rocks are as follows:

$$\beta = (0.5 \sim 5.0) \, 10^{-4} \, \text{cm}^2/\text{kg} \quad (9.3)$$

$$\gamma = (0.3 \sim 1.2)\ 10^{-4}\ cm^2/kg. \qquad (9.4)$$

Thus, both χ and J_R change about 10%, corresponding to a stress change of about $\sigma = 1\text{kbar} = 10^3 \text{kg/cm}^2$. These experiments and equations only apply to the rocks which are stressed for the first time. If stress is applied repeatedly, β and γ become much smaller. An example of the effect of stress cycles on remanent magnetism is shown in Fig. 9.17 [Martin and Wyss, 1975]. Future research will determine whether rocks *in situ* actually respond in the manner suggested by these equations or by Fig. 9.17. For the moment, however, the values in equations (9.3) and (9.4) will necessarily be used.

Variations in crustal stress cause variation in the magnetism of underground rocks, thus causing variation in the intensity of the magnetic field and in the direction of magnetism on the earth's surface. Now that variations in the magnetic susceptibility or remanent magnetism of rocks and

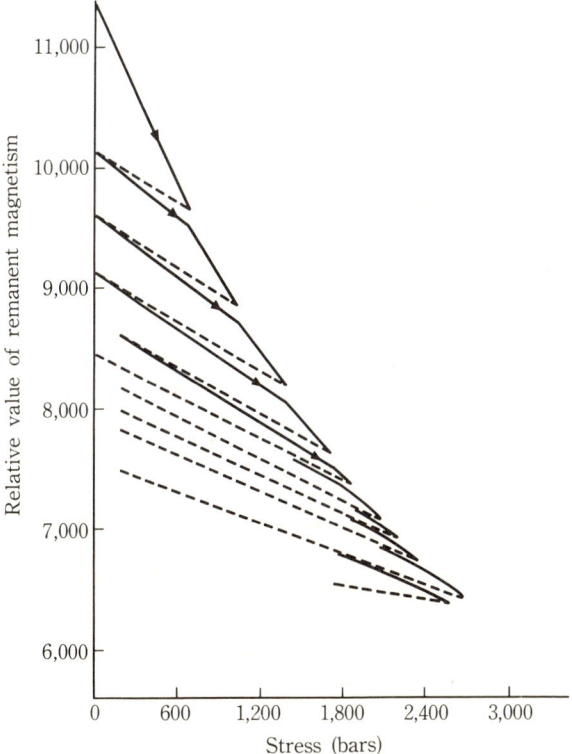

Fig. 9.17 Variations in the remanent magnetism of rocks (gabbro) caused by the repeated application of stress [Martin and Wyss, 1975].

variations in crustal stress have been combined in such equations as (9.1) and (9.2), variations in the magnetic field on the earth's surface can be calculated quite easily. These calculations were made by Stacey [1964] and many others for various seismic fault models. Figure 9.18 is typical of such calculations [Shamsi and Stacey, 1969].

This calculation assumes that the San Francisco earthquake (M 8.3) of 1906 occurred on a horizontal displacement fault 5 km deep and of infinite length, and that the stress drop at the fault plane was 100 bar. As is apparent in the figure, the variation in total geomagnetic force, even on the fault line, is only 2 nT or so. At 25 km away from the fault the horizontal and vertical components also drop below 2 nT. If the fault's length is regarded as limited, then stress will be concentrated at both ends. Magnetic field variations would then be greater by several nT in these areas. The variations would decrease, of course, as the distance from the fault increased. Geomagnetic variation observed prior to the earthquake should be the same as Fig. 9.18, but with the positive and negative signs reversed. Thus precursory geomagnetic variations may also be only several nT or less. Detection of such geomagnetic variations is possible but it is not simple, given the noise caused by factors,

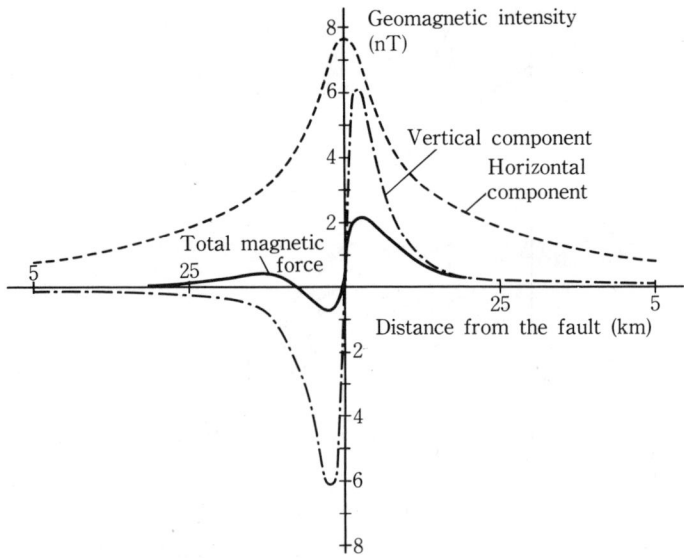

Fig. 9.18 Geomagnetic variation on the earth's surface based on the horizontal displacement fault model for the San Francisco earthquake of 1906 [Shamsi and Stacey, 1969].
The magnetic susceptibility of rocks is assumed to be 10^{-3} emu.

such as magnetic storms and secular variation, that are unrelated to seismic activity. There have been cases in which geomagnetism varied by 2 to 8 nT due to the increased stress created by water in a dam [Davis and Stacey, 1972]; underground nuclear explosions and other explosions have been known to cause geomagnetic variations of 10 nT [Hasbrouck and Allen, 1972; Abdullabekov *et al.*, 1972]. Thus earthquake-related geomagnetic variation may not be as small as the present simple piezo-magnetism theory indicates.

9.5.2 Electrokinetic Effect

According to the dilatancy-diffusion model, dilatancy takes place in rocks prior to an earthquake. This causes a decrease in the water pressure (pore pressure) in the rocks near the epicenter, thus triggering the flow of groundwater into the epicentral zone. Mizutani *et al.* [1976] pointed out that observable electromagnetic phenomena should occur due to the "electrokinetic effect" of the rock and water system if such groundwater movement takes place before an earthquake.

The electrokinetic phenomenon is caused by the uneven distribution of ions in flowing liquids. The uneven distribution is due to the double electric layer that forms at the solid–liquid interface. The schematic diagram of ion distribution near the interface between rock and water is shown in Fig. 9.19. If some force moves the water in the capillary, an electric current will be transported by the water. This current will correspond to the amount of unevenness in the ion distribution (represented in the figure as zeta potential). In other words, the water that flows within rocks is accompanied by an electric current, establishing a correlation between them. The

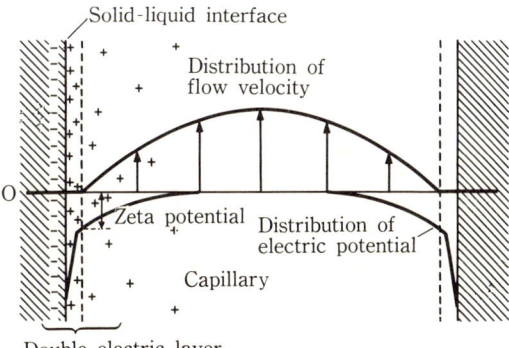

Fig. 9.19 Schematic diagram of the distribution of electric potential and the distribution of flow velocity of fluid in a capillary, and of the double electric layer [Mizutani *et al.*, 1976].

electrokinetic phenomenon can be represented simply in the following equations:

$$I = -\phi\sigma \operatorname{grad} E + \frac{\phi\varepsilon\zeta}{\eta} \operatorname{grad} P \qquad (9.5)$$

$$J = \frac{\phi\varepsilon\zeta}{\eta} \operatorname{grad} E - \frac{k}{\eta} \operatorname{grad} P \qquad (9.6)$$

where I represents the earth's current; J, the flow rate of groundwater; E, the earth's electric potential; P, the pressure (pore pressure) of groundwater; σ, the electric conductivity; k, the permeability; η, the coefficient of viscosity; ζ, the zeta potential; ε, the dielectric constant of water; and ϕ, the porosity. The first term on the right side of equation (9.5) represents Ohm's law. The second term on the right side of equation (9.6) represents Darcy's law. The second term on the right side of equation (9.5) and the first term on the right side of equation (9.6) represent the electrokinetic effect.

As is apparent in equation (9.5), electric current is generated even if there is no electric potential gradient when the gradient of the pore pressure, grad P, exists. This current generates a magnetic field, thus causing variations in the magnetic field on the surface. Mizutani and Ishido [1976] indicated that the geomagnetic anomalies observed during the Matsushiro swarm earthquakes could be explained by this effect. If there is not a sudden temporal change in grad P (the gradient of pore pressure) or if there is no heterogeneity or anisotropy, however, electric potential E is generated so as to suppress the current I. Thus the total current I becomes zero ($I = 0$) under stationary conditions, and the gradient of the earth's electric potential, grad E, as represented in the following equation, results:

$$\operatorname{grad} E = \frac{\varepsilon\zeta}{\eta\sigma} \cdot \operatorname{grad} P. \qquad (9.7)$$

By inserting the appropriate numbers in the above, the following formula is obtained:

$$\operatorname{grad} E = (10^2 \sim 10^3)\,(V/\text{kbar}) \cdot \operatorname{grad} P. \qquad (9.8)$$

As the variation of grad P before earthquakes can be estimated at grad $P \approx 10^{-3}\,10^{-1}$ kbar/km, the following variation of grad E can be expected:

$$\operatorname{grad} E = (10^{-1} \sim 10^2)\,\text{V/km}. \qquad (9.9)$$

Consequently, the following variation of the gradient should be expected near the epicenter between two electrodes 100m apart:

$$\Delta E \approx (10 \sim 10^4)\,\text{mV}. \qquad (9.10)$$

EARTHQUAKES AND ELECTROMAGNETIC PHENOMENA 241

This is in good agreement with the observed values mentioned in the preceding section. As the water flow diffuses over time, the pressure gradient of groundwater becomes smaller. Since the temporal change in water pressure, P, is represented in the diffusion equation, the earth's electric potential will also show a diffusion-type temporal change in accordance with equation (9.7). An example of such a calculation is shown in Fig. 9.10. Temporal change in the earth's electric potential can thus be successfully explained by the electrokinetic effect.

Although the electrokinetic effect was first suggested as a purely theoretical possibility, recent rock mechanics experiments and field experiments at geothermal zones and hot springs seem to provide evidence that this electrokinetic effect may really appear as natural phenomena [Ishido, 1977]. Fitterman [1979a, b] has recently made a theoretical calculation of the geomagnetic anomalies and the variations in the earth's electric potential caused by the electrokinetic effect that demonstrates the significance of these anomalous variations for the purposes of earthquake prediction.

9.5.3 Dilatancy and Electric Resistance

When stress is more than 60% of the failure strength, a phenomenon called dilatancy takes place in a rock mechanics experiment. The dilatancy is the volume increase of the rock due to crack extension or crack opening. The formation of cracks and the increase in volume greatly affect the elastic wave velocity and other physical properties of rocks. The rock's electric resistance is one such property. Relations between axial stress and electric resistance of rocks are demonstrated in Fig. 9.20. As is shown in

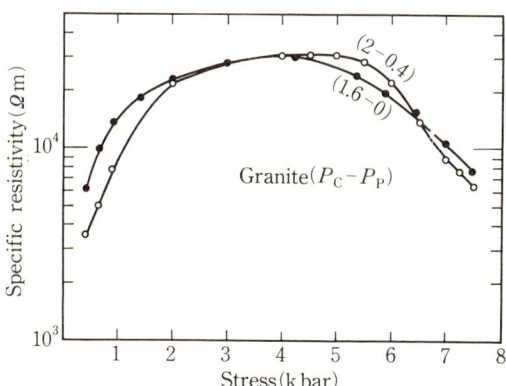

Fig. 9.20 The relationship between electric resistance and stress in rocks [Brace and Orange, 1968].

the figure, whether the rock is completely saturated with water (open circles) or not (solid circles), the relationship between electric resistance and axial stress remains similar as long as the effective pressure ($P_C - P_P$: P_C = confining pressure; P_P = pore pressure) is constant. Most notable in this figure is the decrease in resistance at the point where axial stress exceeds 5 to 6 kbar. Immediately before the rupture, the resistance drops to a figure one order smaller than the original one. Fujii and Hamano [1976] investigated the effect of stress cycling on rocks' electric resistance and its anisotropy. They showed that, as in the case of remanent magnetism, the electric resistance of rocks also shows a large hysteresis when subjected to stress. This is because the water that moistened the crack surface remains even when the stress has been released. The hysteresis, however, has not been observationally identified as the stress cycle associated with actual earthquakes. The explanation for the discrepancy between laboratory results and field data must await the results of future research.

9.6 Problems in Earthquake Prediction Research Based on Electromagnetic Phenomena

Many cases have already been discussed in which observations of geomagnetism and of the earth's electric potential and electric resistance have led to the detection of precursory phenomena. On the other hand, there are also many cases where no significant electromagnetic phenomena have been observed before and/or during earthquakes. This does not necessarily mean that such phenomena did not occur: the signals of the electromagnetic phenomena might have been just outside the detection capability.

The approximate amplitude of the signals of electromagnetic phenomena associated with earthquakes is estimated on the basis of the theoretical and experimental research discussed in Section 9.5. A summary follows. If, for example, the observation point was 50 km from the epicenter of an M 7 earthquake:

(1) Variations of about 1 to 2 nT in the total magnetic force of geomagnetism, and 1 to 10 nT in the horizontal and vertical components, will result.

(2) If there is a movement of groundwater that can be estimated by the dilatancy diffusion model, variation of an electric potential gradient of $10^{-1} \sim 10^2$ mV/m will be observed.

(3) If dilatancy such as that observed in laboratory experiments occurs, a $10 \sim 20\%$ decrease would be seen in electric resistance.

These variations in geomagnetism, in the earth's electric potential, and in specific resistivity can be detected by present-day methods of measurement. The problem in industrialized countries like Japan is the presence of artificial noise that hinders the measurement of these phenomena. The difficulty might be partly avoided by using a suitable filter, because the spectrum of the artificial noise is of much shorter period than that of signals associated with earthquakes.

The above estimate of the electromagnetic signals associated with earthquakes, however, is very much dependent on the physical processes that occur in the source area of earthquakes. Therefore, it is not even certain whether or not such phenomena accompany all earthquakes. Finding out which earthquakes are accompanied by such phenomena and which earthquakes are not may be an effective means of finding out about the processes themselves. The most important thing is to eliminate as much noise as possible and to systematically and comprehensively observe electromagnetic phenomena. Unfortunately, Japanese research in the field has been rather sporadic. In order to demonstrate the usefulness of electromagnetic observation for earthquake prediction, a much more extensive and thorough observation system should be devised. There is much to be learned along these lines from the U.S. Geological Survey's magnetic survey of the San Andreas Fault and the University of California team's continuous observation of the earth's electric potential and resistance. Observations in China and the USSR are also much more extensive and have more continuity than those in Japan.

In addition, various phenomena that accompany earthquakes such as earthquake luminescence, abnormal behavior of animals, etc., have not been investigated scientifically, and they may represent another facet of electromagnetic phenomena. Objective observational data are limited here, and there have been no hypotheses that successfully explain them. In order to explore their usefulness for earthquake prediction, basic research on electromagnetic phenomena should be strongly promoted. Not only will such research enrich the content of earthquake-related electromagnetic research, but it will also provide an effective way of better understanding the earthquake phenomenon itself.

References

Abdullabekov, K. N., L.S. Bezuglaya, V.P. Golovkov and Y.P. Skovorodkin, 1972: On the possibility of using magnetic method, *Tectonophys.*, **14**, 257–262.

Barsukov, O.M. and O.N. Sorokin, 1973: Variation in apparent resistivity of rocks in the seismically active Garm region (English edition), *Izv. Acad. Sci. USSR* (Phys. Solid Earth), **8**, 685–687.

Brace, W.F. and A.S. Orange, 1968: Electrical resistivity changes in saturated rocks during fracture and frictional sliding, *J. Geophys. Res.*, **73**, 1433–1445.

Corwin, R.F. and H.F. Morrison, 1977: Self-potential variations preceding earthquakes in central California, *Geophys. Res. Lett.*, **4**, 171–174.

Davis, P.M. and F.D. Stacey, 1972: Geomagnetic anomalies caused by a man-made lake, *Nature*, **240**, 348.
Fitterman, D.V., 1979a: Theory of electrokinetic-magnetic anomalies in a faulted half-space, *J. Geophys. Res.*, **84**, 6031-6040.
——, 1979b: Calculations of self-potential anomalies near vertical contacts, *Geophys.*, **44**, 195-205.
Fujii, N. and Y. Hamano, 1976: Anisotropic changes in resistivity and velocity during rock deformation, *High Pressure Research*, edited by M.H. Manghnani and S. Akimoto, 53-64.
Fujita, N., 1965: The magnetic disturbances accompanying the Niigata earthquakes, *J. Geod. Soc. Japan*, **11**, 8-25.
Galli, I., 1910: Raccolta e classificazione di fenomeni luminosi osservati nei terremoti, *Bolletino della Società Italiana*, **14**, 221.
Hasbrouck, W.P. and J.H. Allen, 1972: Quasistatic magnetic field changes associated with the CANNIKIN nuclear explosion, *Bull. Seism. Soc. Amer.*, **62**, 1479-1487.
Ishido, T., 1977: A study of electrokinetic phenomena in rock-water systems and its applications to geophysics, Ph. D. thesis, Univ. of Tokyo, Tokyo.
Johnston, M.J.S., B.E. Smith and R. Mueller, 1976: Tectonomagnetic experiments and observations in Western U.S.A., *J. Geomag. Geoelectr.*, **28**, 85-97.
——, B.E. Smith, J.R. Johnston and F.J. Williams, 1973: A search for tectonomagnetic effects in California and Western Nevada, *Proc. Conf. Tectonic Problems of the San Andreas Fault System*, edited by R.L. Kovach and A. Nur, Geological Sciences, **XIII**, 225-238.
Kaminuma, K., T. Iwata, I. Kayano and M. Ohtake, 1972: *Illustrated Japanese Earthquakes* (in Japanese), Special Report of Earthquake Research Inst., Univ. of Tokyo.
Kapista, S.P., 1955: Magnetic properties of eruptive rocks under mechanical stresses, *Izvestia Akad. Nauk. USSR*, Ser. Geophys., **6**, 489-504.
Kato, Y., 1939: Investigation of the changes in the earth's magnetic field accompanying earthquakes or volcanic eruption, *Sci. Rep. Tohoku Univ.*, Ser. 1, **27**, 1-100.
——, 1966: Recent studies on geomagnetic changes accompanied by earthquakes, *Proc. Symp. Geomagnetic Changes Associated with Earthquakes and Volcanic Activities*, edited by T. Nagata, Geophys. Inst., Univ. of Tokyo, Tokyo, 1-20 (in Japanese).
—— and S. Utashiro, 1948: On the changes of the terrestrial magnetic field accompanying the Great Nankaido earthquake of 1946, *Sci. Rep. Tohoku Univ.*, Ser. 5, **1**, 40-41.
——, S. Utashiro, R. Shoji, J. Ossaka, M. Hayashi and F. Inaba, 1950: On the changes of the earth-current and the earth's magnetic field accompanying the Fukui earthquake, *Sci. Rep. Tohoku. Univ.*, Ser, 5, **2**, 53-57
Kern, J.W., 1961: The effect of stress on the susceptibility and magnetization of a partially magnetized multidomain system, *J. Geophys. Res.*, **66**, 3807-3816.
Martin, R. J. and M. Wyss, 1975: Magnetism of rocks and volumetric strain in uniaxial failure tests, *Pure Appl. Geophys.*, **113**, 51-61.
Mazzella, A. and H. F. Morrison, 1974: Electrical resistivity variations associated with earthquakes on the San Andreas Fault, *Science*, **185**, 855-857.
Milne, J., 1890: Seismic, magnetic, and electric phenomena, Trans. Seis. Soc. Japan, **XIX**, 23-33.
Mizutani, H. and T. Ishido, 1976: A new interpretation of magnetic field variation associated with the Matsushiro earthquakes, *J. Geomag. Geoelctr.*, **28**, 179-188.
——, T. Ishido, T. Yokokura and S. Ohnishi, 1976: Electrokinetic phenomena associated with earthquakes, *Geophys. Res. Lett.*, **3**, 365-368.
Musya, K., 1931: On the luminous phenomenon that attended the Idu earthquakes, *Bull.*

Earthq. Res. Inst., **9**, 214–215.

———, 1932: Investigations into the luminous phenomena accompanying earthquakes, *Bull. Earthq. Res. Inst.*, **10**, 649–673.

———, 1934: On the luminous phenomena that accompanied the great Sanriku tunami in 1988 (Part I), *Bull. Earthq. Res. Inst.*, Suppl., **1**., 87–111.

———, 1957: *Nihon Jishin Shiryo* (Japanese historical records relevant to earthquakes), Mainichi Press, Tokyo, 1019 (in Japanese).

Myachkin, V.I., G.A. Sobolev, N.A. Dolbikina, V.N. Morozow, and V.B. Preobrazensky, 1972: The study of variations in geophysical fields near focal zones of Kamchatka, *Tectonophys.*, **14**, 287–293.

Nagata, T., 1937: A comparison of the results of magnetic surveys before and after the earthquake in Niisima, December 27, 1936. *Bull. Earthq. Res. Inst.*, **15**, 497–505.

———, 1966: Main characteristics of piezo-magnetism and their qualitative interpretation, *J. Geomag. Geoelctr.*, **18**, 81.

———, 1966: Magnetic susceptibility of compressed rocks, *J. Geomag. Geoelct.*, **18**, 73.

———, 1969: Tectonomagnetism, *International Association of Geomagnetism and Aeronomy Bulletin*, **27**, 12.

———, 1970: Basic magnetic properties of rocks under the effect of mechanical stresses, *Tectonophys.*, **9**, 167–195.

———, 1972: Application of tectonomagnetism to earthquake phenomena, *Tectonophys.*, **14**, 263–271.

———, 1976: Tectonomagnetism in relation to seismic activities of the earth's crust: Seismo-magnetic effect in a possible association with the Niigata earthquake in 1964, *J. Geomag. Geoelectr.*, **28**, 99–111.

Noritomi, K., 1978: Application of precursory geoelectric and geomagnetic phenomena to earthquake prediction in China, *Report by Japanese Seismological Society Delegation to the People's Republic of China*, Seism. Soc. Japan., 57–87 (in Japanese; for English translation, see *Chinese Geophys.*, **1**, No. 2, 377–391, Amer. Geophys. Union).

Ohnaka, M. and H. Kinoshita, 1968. Effects of uniaxial compression on remanent magnetization, *J. Geomag. Geoelectr.*, **20**, 93–99,

Oike, K., R. Shizi and T. Asada, 1975 : Earthquake prediction in China, *Zisin* (J. Seismol. Soc. Japan.), Ser. 2, **28**, 75–94.

Oike, K., 1977: Lessons from Tangshan earthquake and short-term forcast, *Shizen*, **381**, 37–47 (in Japanese).

Qi, Gui-zhong, 1978: On the dilatancy-magnetic effect, *Acta Geophysica Sinica*, **21**, 18–33.

Rikitake, T., 1968: Geomagnetism and earthquake prediction, *Tectonophys.*, **6**, 59–68.

———, 1976: *Earthquake Prediction*, Elsevier, Amsterdam, 357 pp.

———, T. Yukutake, Y. Yamazaki, M. Sawada, Y. Sasai, Y. Hagiwara, K. Kawada, T. Yoshino and T. Shimomura, 1966: Geomagnetic and geoelectric studies of the Matsushiro earthquake swarm, 3. *Bull. Earthq. Res. Inst.*, **44**, 1335–1370.

———, T. Yamazaki, M. Sawada, Y. Sasai, T. Yoshino, S. Uzawa and T. Shimomura, 1967: Geomagnetic and geoelectric studies of the Matsushiro earthquake swarm, 4. *Bull. Earthq. Res. Inst.*, **45**, 89–107.

Rikitake, T. and Y. Yamazaki, 1970: Strain steps as observed by a resistivity variometer, *Tectonophys.*, **9**, 197–203.

———, 1977: Precursory and coseismic changes in ground resistivity, *J. Phys. Earth.*, **25**, Suppl., S161–S173.

Shamsi, S. and F. D. Stacey, 1969: Dislocation models and seismomagnetic calculations for California 1906 and Alaska 1964 earthquakes, *Bull. Seis. Soc. Amer.*, **59**, 1435–1448.

Shida, J., 1886: On earth current, *Trans. Seis. Soc. Japan*, **IX**, 32–50.

Smith, B.E. and M.J.S. Johnston, 1976: A tectonomagnetic effect observed before a magnitude 5.2 earthquake near Hollister, California, *J. Geophys. Res.*, **81**, 3556–3560.

Stacey, F.D., 1962: Theory of the magnetic susceptibility of stressed rocks, *Phil. Mag.*, **7**, 551.

——, 1964: The seismomagnetic effect, *Pure Appl. Geophys.*, **58**, 5–22.

Sumitomo, N., 1978: Geomagnetism in relation to tectonic activities of the earth's crust in Japan, *Earthquake Precursors, Advance in Earth and Planetary Sciences,* 2 , edited by C. Kisslinger and Z. Suzuki, Center for Academic Publications Japan, 147–160.

Tang, Chi-yang, 1978: Basis for prediction of Lungliao Earthquake and its space-time characteristics of precursor phenomena, *Report by Japanese Seismological Society Delegation to the People's Republic of China,* Seism. Soc. Japan, 13–32 (in Japanese).

Tazima, M., 1968: Accuracy of recent magnetic survey and a locally anomalous behaviour of the geomagnetic secular variation in Japan, *Bull. Geograph. Sur. Inst.*, **13**, 1–78.

——, H. Mizuno and M. Tanaka, 1976: Geomagnetic secular change anomaly in Japan, *J. Geomag. Geoelectr.*, **28**, 69–84.

Terada, T., 1931: On luminous phenomena accompanying earthquakes, *Bull. Earthq. Res. Inst.*, **9**, 225–255.

Wyss, M., 1975: A search for precursors to the Sitka, 1972, earthquake:Sea level, magnetic field, and P-sesiduals, *Pure Appl. Geophys.*, **113**, 297–309.

Yamazaki, Y., 1975: Precursory and coseismic resistivity changes, *Pure Appl. Geophys.*, **113**, 219–227.

——, 1977: Tectonoelectricity, *Geophys. Surveys,* **3**, 123–142.

—— and T. Rikitake, 1970: Local anomalous changes in the geomagnetic field at Matsushiro, *Bull. Earthq. Res. Inst.*, **48**, 637–643.

Yoshimatsu, T., 1957: Universal earth-currents and their local characteristics, *Mem. Kakioka Magn. Obs.,* Suppl. **1**, 1–76.

Yukutake, T., 1977: The earth resistivity studies for the earthquake prediction, *Proc. Symp. Earthq. Prediction Studies,* 1976 (in Japanese).

Zhu, F. M., 1976: Prediction, warning, and disaster prevention related to the Haicheng earthquake of magnitude 7.3, *Proc. Lectures by the Seismological Delegation of the People's Republic of China,* Seism. Soc. Japan Special Publication, 15–26 (in Japanese).

PART IV   THE ROAD TO ACTUAL EARTHQUAKE PREDICTION

Earthquake prediction research was launched as a special project more than a decade ago; as a result, great progress has been made in basic seismology and much more is known about precursory phenomena. And yet only a few minor earthquakes have been successfully predicted in Japan. In fact, earthquake prediction has been successful only in China so far. Precursors are often identified in post-earthquake observations—"post facto" predictions, as it were—when what is truly needed, of course, is advance prediction. The data at our disposal for judging precursory phenomena in advance, however, are minimal. Most of the phenomena that are definitely identified as precursory after an earthquake do not look so definite before the earthquake. Based on the available data, it is beginning to be possible to study phenomena and determine whether or not they are precursory. In a comprehensive sense, however, there is still much to learn about earthquake prediction.

Making do with the available information may be inherent to the nature of earthquake prediction, for truly accurate predictions still seem a long way off. In Japan, scientists are focusing on detecting precursors to a Tokai earthquake which seems imminent, temporarily setting aside prediction attempts for the rest of the country. How will the public react if precursors are successfully detected and publicly announced? It is important to remember that earthquake prediction not only has a profound effect on the public but also is closely related to a country's politics, administration, and economics.

PART IV: THE ROAD TO ACTUAL EARTHQUAKE PREDICTION

Chapter 10   A Practical Strategy for Earthquake Prediction

Katsuhiko Ishibashi

Following the establishment of the Earthquake Assessment Committee for the Tokai Area in April 1977, the Large-Scale Earthquake Countermeasures Act was passed in June 1978. This act is based on the premise that earthquakes can and will be predicted successfully.

During the past several years considerably optimistic views on earthquake prediction have been circulated; the public seems to believe, as a result, that the scientific strategy for successful prediction is already an accomplished fact, and that routine earthquake prediction will be simple once the necessary organizations and observation networks have been consolidated. Not a few scientists who have taken the lead in grappling with the actual earthquake phenomena, however, feel that there is a certain danger in the discrepancy between the public's optimism and their knowledge of the prediction techniques that are in fact available. The recent surprise attacks of the damaging Izu Oshima offshore earthquake (Jan. 14, 1978; M 7.0; 25 dead) and the Miyagi offshore earthquake (June 12, 1978; M 7.4; 27 dead) are further reminders of the difficulties that lie ahead before successful earthquake prediction can be realized.

In this chapter I will point out that Japan's present earthquake prediction strategy is not necessarily the best, and propose an alternative which is based on the true nature of earthquake phenomena. I will also point out the extreme difficulties that still beset earthquake prediction generally. As an example of the proposed strategy I will examine the Tokai earthquake prediction that has attracted so much public notice in Japan.

10.1   Present Earthquake Prediction Strategy

The guiding principle of earthquake prediction in Japan has been as follows: as an earthquake is a rupture that takes place in the rocks within the earth's surface layer, by detecting precursory phenomena preceding the main rupture, it is possible to predict the probable location, size, and time of the earthquake. Precursory phenomena, especially in the case of large earthquakes, have been categorized as long-term (with a lead time of years) and short-term (with a lead time of days or hours). This concept was formalized in a proposal made by the Geodetic Council, Ministry of Education, in July 1968 as "nationwide basic observations, specific observations→intensified observations→concentrated observations (→realization of prediction)." This formula was adopted by the Coordinating Committee for Earthquake Prediction (CCEP) that was established in April 1969, and has been in effect ever since.

The content of this formula was rather abstract and phenomenological at first. However, since the dilatancy-diffusion model ("DD model" for short) concerning the rupture process in the earthquake source region was introduced in the U.S.A. [e.g., Nur, 1972; Scholz et al., 1973], maintaining that premonitory effects occur at a characteristic time before earthquakes which increases with the earthquake's magnitude (M) (such phenomena will be called "M-dependent precursors" in this chapter), a rather definite idea on prediction procedure has become predominant: that it consists of long-term prediction based on M-dependent precursors and short-term prediction based on those premonitory phenomena which appear shortly before an earthquake. Recently there have been considerable doubts about the applicability of the DD model to actual earthquakes. The main consideration remains unchanged, however, in that the detection of precursors is still considered to be supreme not only in short-term but also in long-term prediction. This is also the basic principle of the proposal for the Fourth Five-year Earthquake Prediction Project (for 1979–1983) made by the Geodetic Council, Ministry of Education, in July 1978.

M-dependent precursors, however, are not very definite; basing long-term prediction on such phenomena, therefore, may not be particularly practical.

10.1.1 The Validity of M-Dependent Precursors

For M-dependent precursors, the following expressions of the relationship between precursor time (T) (in days) and earthquake magnitude (M) have been proposed or referred to thus far:

$\log_{10} T = 0.79M - 1.88$ [Tsubokawa, 1969]   (1)

$\log_{10} T = 0.80M - 1.92$ [Whitcomb et al., 1973]   (2)

$\log_{10} T = 0.65M - 1.2$ [Scholz et al., 1973]*   (3)

---

* Interpreted from a figure as there is no equation in the original paper. According to Rikitake [1975], $\log_{10} T = 0.685M - 1.57$.

$\log_{10} T = 0.52M - 0.24$ [Fujii, 1974]   (4)

$\log_{10} T = 0.76M - 1.83$ [Rikitake, 1975]   (5)

$\log_{10} T = 0.60M - 1.01$ [Rikitake, 1978]   (5')

$\log_{10} T = 0.77M - 1.65$ [Sekiya, 1976, 1977]   (6)

Equation (1) was proposed as an expression of the relationship between duration of crustal movement that accompanies strain accumulation (*i.e.*, earthquake "preparation time") and M, in the original paper published before the DD model emerged. Thus it is a grave error to refer to it (as has frequently been done) as the equation for M-dependent precursors. The bases of this equation are only three pieces of data, among which is the 12-year preparation time for an intraplate earthquake of magnitude 7.0 (the 1961 Northern Mino earthquake), whose recurrence time is ordinarily considered several hundred or several thousand years. This would seem to cast doubt on the logic of the original paper itself. Furthermore, the paper maintains that the durations of preseismic crustal deformations of six other earthquakes apply to equation (1); but all data including the first three are quite uncertain. The contents of this paper were introduced overseas inaccurately [Rikitake, 1969] and exerted considerable influence on equations (2) and (3).

The equation that has no logical problem in the relationship between M and duration of anomalous crustal movement as an earthquake precursor is (4). The eight pieces of data that were used to derive this equation, however, are subject to controversy. Table 10.1 lists several Japanese earthquakes of magnitude 7 or greater whose long-term precursory crustal movements are frequently debated. From this table, the uncertainty of most of the data can be comprehended. The only reasonable long-term precursory crustal movement thus far may be the crustal uplift that began about ten years before the 1964 Niigata earthquake [Dambara, 1973]. (This, too, does not seem to be a simple phenomenon.)

One of the bases of the DD model was the result of observations of anomalous changes in seismic wave velocity that preceded many earthquakes (M from 1 to 6.4) in the USSR and the U.S.A. Equations (2) and (3) are mostly based on these data. Since then many results of analyses of seismic wave velocity changes that harmonize with equation (3), etc., have been reported in Japan and the United States. Equations (5) and (5') use a great deal of these reported data. As has already been discussed in Chapter 5, however, the detection of seismic wave velocity change presents a great many technical problems. It has even been suggested that the greater the analytical precision, the harder it is to detect changes. Consequently, at present many seismologists doubt most research results, including the

Table 10.1 Duration Time of Long-term Precursory Crustal Movement of Several Japanese Earthquakes of Magnitude 7 or More

Earthquakes	Tsubokawa* [1969]	Fujii [1974]	Rikitake [1975]	Remarks
Nankai (1946, M 8.1)	92 years	13 years	Not accepted	Time since the 1854 Ansei earthquake. Tidal range between Hosojima and Tosashimizu; 4 to 5 years according to Geographical Survey Institute [1971] (S/N ratio poor).
Tonankai (1944, M 8.0)	Not accepted	Not accepted	10 years	Vertical movement of Kakegawa relative to Numazu; existing measurement data very few, subjective interpretation.
Kanto (1923, M 7.9)	Not accepted	13 years		Tide at Aburatsubo; approx. 10 years according to Geographical Survey Institute [1971].
			4 years	Dilatation of Mitaka rhombus base-line net; there may have been a survey error, according to H. Sato [personal communication]. Imamura reported anomalous tidal level at Aburatsubo 4 years prior to the earthquake, but Tsumura [1970] contended that the influence of the cold water mass is most probable.
Niigata (1964, M 7.5)	40–50 years	12 years	10 years	Leveling.
			1 year	Tidal range between Kashiwazaki and Nezugaseki.
Northern Mino (1961, M 7.0)	12 years	Detection is questionable	12 years	Leveling.

* Tsubokawa [1969] gave these numbers in relation to crustal movement accompanying strain accumulation. Included here as reference.

early discoveries in the USSR and the U.S.A. Equation (2) was obtained from these seismic wave velocity change data first recorded in the U.S.A. and the USSR, and from Tsubokawa's crustal movement equation (1). Equation (3) was obtained by adding data on electrical resistivity, radon emission, and b value. Judging from the above-mentioned conditions, it is very doubtful that these equations objectively reflect natural phenomena.

Equations (5) and (5') are the result of classifying and consolidating data on precursory phenomena that have been reported by researchers and research institutes throughout the world. (5) is based on approximately 90 pieces of data, and (5') on nearly 200 pieces of data (70 of these are data on premonitory changes in seismic wave velocity). Rikitake [1975] calls the precursory phenomena that fit into these equations "the A_2-type precursors." As has been suggested thus far, however, it is doubtful that all these reports contain true information on natural phenomena. At present, therefore, these two equations are questionable as representations of natural laws.

Equation (6) is an expression on the anomalous seismic activity that preceded ten Japanese earthquakes (M from 4.1 to 7.9). At present, the reliability and true meaning of this equation is not clear. The estimate of T as 82 years for the 1923 Kanto earthquake (M 7.9) seems groundless and open to question.

The truth is that there are very few available examples of reliable long-term precursory phenomena (which can last more than several years in the case of large earthquakes). And whether or not such long-term precursors can be applied to equations such as (2) through (6) is much less clear. Thus it is dangerous to use them as the basis for earthquake prediction at this time.

One of the reasons that the idea of M-dependent precursors was so popular was that they could be physically explained by the DD model. There are many scientists who question the applicability of the DD model to interplate earthquakes with magnitudes around 8—even though the model may be quite appropriate. Those concerned with earthquake prediction in Japan, however, still say sometimes that "since the precursor time interval of M-dependent precursors is 10 years (for example) for magnitude 7, that for magnitude 8 should be 20 to 30 years (for example)." Such mechanical logic, by ignoring the tectonic meaning of earthquake occurrence, results in contradictions such as this: The recurrence time of large earthquakes in Japan is estimated roughly at 100 to 200 years for M 8-class interplate earthquakes on the Pacific coast, and at several hundred to several thousand years for M 7-class intraplate earthquakes in inland Japan. If M-dependent precursor time intervals due to dilatancy are 10 years (for example) for magnitude 7 and 30 years (for example) for magnitude 8, then—assuming (A) that the strain accumulation velocity remains constant during interseismic periods—dilatancy occurs at 70 to 85% stress level of the breaking strength in the case of M 8 interplate earthquakes and more than 98% in the case of M 7 intraplate earthquakes. Aside from the problem inherent in assumption (A), this example demonstrates the limitations of the DD model that ignores the strain velocity which is actually an important factor.

10.1.2 Rikitake's Earthquake Prediction Strategy Based on M-Dependent Precursors

Rikitake [1975; 1978] has proposed an earthquake prediction strategy that emphasizes M-dependent precursors or A_2-type precursors. He maintains that it is possible to make step-by-step earthquake predictions that proceed from long-term to short-term (and immediate) on the basis of observations of precursory phenomena alone, aside from very long-term prediction (which is achieved by crustal strain monitoring and statistics on recurrence times of large earthquakes). Especially during the stage of long-term prediction, according to him, the size and location of a coming earthquake can be forecast by detecting A_2-type precursors, and the time of its occurrence can be predicted by using equations such as (1) through (6) (ideally including a guaranteed period during which the earthquake will not take place).

It is clear from the preceding section, however, that there are still many questions about the A_2-type precursors themselves. Hence Rikitake's earthquake prediction strategy hardly seems practical. Even if A_2-type precursors do exist, actual long-term earthquake prediction still has the following difficulties.

First of all, there seems to be only a slim chance that various kinds of A_2-type precursors will be observed simultaneously and clearly over an extensive area which will be the epicentral region of a coming large earthquake. In reality it is quite possible to overlook precursors, or to make errors concerning their spatial extent or the times of their appearance. This is easily suggested by the experts' disagreement on long-term precursors to past large earthquakes as demonstrated in Table 10.1, and by the many failures to detect A_2-type precursors despite the geodetic surveys and tide-gauge observations that were being performed.

Secondly, even when some A_2-type precursors are successfully detected, it is very difficult to pre-estimate the magnitude of the coming earthquake based on them. If this cannot be done with certainty, then equations (1) through (6) are useless. Rikitake [1978] maintains that Dambara's [1966] equation, which obtains M from the mean radius of the anomalous crustal movement area, is effective for this purpose. However, this equation was originally an expression of the relationship between earthquake magnitude and the extent of vertical crustal movement "caused by" that earthquake. Therefore M is not attainable properly unless anomalous vertical crustal movement appears throughout the future epicentral area.

Such a case, however, would be a rarity in itself, and even if it did occur, successful detection of the extent of anomalous crustal movement might not be so easy.

Thirdly, the estimated value of precursor time interval T, and hence earthquake occurrence time, can vary so broadly, due to scatters in phenomena and in estimation of M, that it can become practically meaningless. For instance, if the pre-estimated value of M is 7.0 to 8.0, T is 8.5 to 48.7 years according to (5), and 4.2 to 16.9 years according to (5'). And this is not all. A range of error must be added to these values. The result is said to be expressed by probability, but its calculation inevitably depends on a formal assumption. At present such assumptions are far from the true nature of a coming earthquake.

In the first place, precursory phenomena do not appear with a label proclaiming them to be "signs of a forthcoming earthquake." Long-term precursors, in particular, are often only recognized as such after the earthquake has occurred, using that earthquake as the basis for analysis. It is obvious, then, that an adequate forecast of the location and size of a coming earthquake and definite recognition of the possibility of its occurrence at the stage of long-term prediction, both of which are indispensable for accurate short-term prediction, are almost impossible on the basis of precursory phenomena alone.

After all, Rikitake's earthquake prediction strategy is, at present, an idealized one which is far from being realized. Our inadequate research on earthquake phenomena is not necessarily to blame. Rather, in this strategy earthquake phenomena have been considered so abstractly that some of their most crucial aspects may have been overlooked. It may be necessary to return to the essence of the "live" earthquake and begin afresh.

10.2 Earthquake Prediction in China

It is now well known that the Chinese have put earthquake prediction to practical use and have had success in disaster prevention. Prediction in China is reported to be done in stages, based on various observations, *i.e.*, long-term, intermediate-term, short-term, and just before the earthquake; a step-by-step earthquake prediction strategy based on observations of precursory phenomena seems to be enormously successful. However, in the case of the Haicheng earthquake (Feb. 4, 1975; M 7.3) in Liaoning Province, which is considered the best example of successful earthquake prediction, a closer examination of Zhu Fung-ming's [1976] detailed report suggests that the classification of its predictions into long-term, intermediate-term, short-term, and immediate ones were made after the earthquake. Probably at the time of the issue of each prediction seismologists were not clearly conscious of the class of the prediction. Consequently, disaster prevention measures based on earthquake prediction were not necessarily carried out in an orderly sequence. This is not surprising, but too often reports have given the impression that the prediction was made in an organized, step-by-step fashion by seeing through the course of the earthquake occurrence in advance. If this impression were to remain with the public, it could be damaging to the cooperative effort necessary for the successful realization of earthquake prediction.

What the Chinese call a long-term prediction of the Haicheng earthquake was issued in mid-1973. But it was not a prediction of the location, size, or time of the earthquake based on something like A_2-type precursors. Rather, it simply pointed out that Liaoning Province was earthquake-prone, based on extensive and detailed research on tectonic movement and the characteristics of seismic activity in and around this region. The area cited as earthquake-prone was the size of Southern California, and, as Zhu Fung-ming indicates, it was uncertain in which part of this vast area the earthquake was going to occur or whether it would strike within several years or not—even though it occurred only a year and a half later.

From late 1973 on, observation efforts were increased throughout this extensive area in order to detect various precursory phenomena. By June 1974, when a conference was called by the State Seismological Bureau, various anomalous phenomena had been discovered, including a rapid increase in tilting at the crustal movement observatory in Jinxian (approximately 200 km south by southwest of Haicheng), a large increase in the intensity of the geomagnetic field in Lüda (approximately 250 km south by southwest of Haicheng), a considerable rise in sea level in Liaodong Bay, and a remarkable increase in microseismic activity in Liaoning Province. The conference examined these data carefully and arrived at the conclusion that an earthquake of magnitude 5 to 6 might occur in the next year or two on the north of Bohai (which covers approximately the area of the Liaonan and Liaodong Peninsula districts). This can be called a quasi-long-term prediction. Though it eliminated the vagueness of the first long-term prediction except for the location, the bases for the estimates of magnitude 5 to 6 and a 1- to 2-year time interval are not clear. Possibly the Chinese, rather than basing their estimate on quantitative laws, made an overall judgment of the data based on their determination to attain success by all means in earthquake prediction and disaster prevention. If it had been Japan, such phenomena as mentioned above would have been regarded as insufficient to be linked to earthquake occurrence, and a prediction would never been made.

Another aspect of this prediction that is worthy of our attention is that it must have played substantially the same role as a short-term prediction

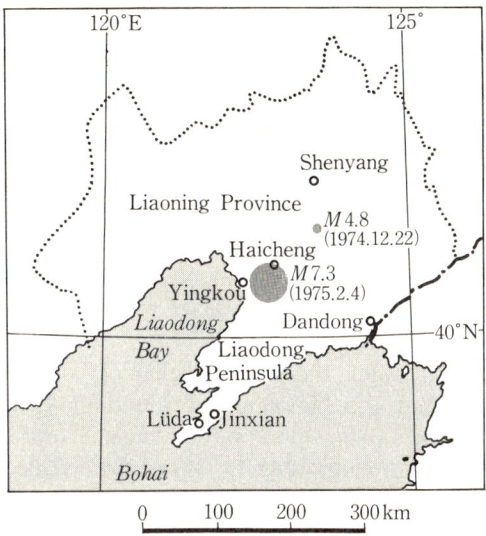

Fig. 10.1 Map of the area around Haicheng, with the epicenters of an earthquake on Dec. 22, 1974 (M 4.8) and a second, major one (M 7.3) on Feb. 4, 1975.

at the time it was issued. In fact the conference requested that prediction, warning, and disaster prevention efforts should be reinforced north of Bohai. And in response to this request the Revolutionary Committee of Liaoning Province acted quickly and forcefully as described at length by Zhu Fung-ming in his report. At that time, they must have thought that a large crowd of rats might run out and well water might bubble over even the following day. Thus, for the next seven months, until the immediate warning was issued, an area as large as Southern California was probably placed in a tense atmosphere, and strong efforts to detect immediate precursors were made on a continuing basis.

Another conference was called in November 1974. By that time, the tilting of the whole Liaodong Peninsula to the northwest had been discovered. The conference announced that there was a comparatively strong possibility that a destructive earthquake would occur in the near future in Yingkou (approximately 50 km west by southwest of Haicheng), Lüda, etc. This was the so-called intermediate-term prediction. This prediction, however, rather than narrowing down the location, size, and time of the earthquake quantitatively based on distinct precursors, just increased the conviction that an earthquake was imminent under tense circumstances.

In mid-December of 1974 there appeared numerous anomalies in well water and in animal behavior in the Dandong district (approximately 150 km southeast of Haicheng). Also noted were drastic changes in ground tilt and anomalous changes in radon emission in several areas in the Liaonan district. An emergency conference was called, and it concluded that there was a possibility that an earthquake of magnitude 4 to 5 would occur in the Liaonan district in the near future. On Dec. 22 an earthquake of magnitude 4.8 took place 70 km northeast of Haicheng. Then, according to Zhu Fung-ming, the Revolutionary Committee of Liaoning Province contacted various areas by telephone, urging them to take emergency disaster prevention measures, and called an emergency earthquake disaster prevention conference. This was a very significant step, akin to the issuance of an immediate warning, and it took place prior to what is now categorized as the "short-term" prediction.

Anomalies in well water and animal behavior continued and spread extensively throughout the Liaonan and Liaodong Peninsula districts. Crustal movement at Jinxian Observatory became unstable. The distribution of earthquakes of magnitude 3 to 4 in 1974 revealed an anomalous seismicity gap which covered the entire area from Jinxian to Yingkou. On the basis of these anomalous conditions, the conference called by the State Seismological Bureau in mid-January 1975 pointed out definitely that the area from Yingkou to the Jinxian and Dandong districts was prone to an earthquake of magnitude 5.5 to 6 during the first half of 1975. This is what is called the short-term prediction.

Anomalies in well water and animal behavior became increasingly pronounced in the beginning of February. Radical changes appeared in the earth's current in Yingkou and in the ground tilt at Shenyang (approximately 130 km north by northeast of Haicheng). What ultimately strongly suggested the imminence of an earthquake was the microearthquake activity that began on Feb. 1 in the Haicheng area, where seismicity had been very low. The activity increased rapidly in a limited area and was remarkably characteristic of foreshock activity. On Feb. 3, felt earthquakes began to occur. This was the decisive factor in convincing the Revolutionary Committee of Liaoning Province to make emergency phone calls throughout the province at 10 a.m. on Feb. 4 to warn of the possibility of a comparatively large earthquake in Haicheng, Yingkou district. This is the immediate earthquake prediction and warning. Nine and one-half hours later, at 19:36 (Beijing time), the main shock hit.

Although the main shock occurred by chance half a month after the issuance of the short-term prediction, the prediction assigned a too large area of an earthquake-prone zone and a too long time span, for it to have been meant as a short-term prediction. Therefore, the prediction seems to have been a kind of reconfirmation of the possibility of a future earthquake under circumstances where a large earthquake was very late in occurring in spite of continuing anomalies in well water and animal behavior, rather than the prediction of a specific earthquake that would occur at a specific time and place. "The entire area from Yingkou to Jinxian and the Dandong district" was the smallest area that the prediction could define, until foreshocks began to occur four days before the main shock. The predicted magnitude was no more than 6 as compared to the actual magnitude of 7.3.

Although the Haicheng earthquake prediction was certainly a great success, as analyzed so far, it can hardly be termed a totally successful step-by-step prediction based on the observation of precursory phenomena. The only predictions that were substantially significant as triggers of concrete actions were the quasi-long-term prediction which aroused a tense vigilance program, and the immediate prediction, which urged the evacuation of the public. This analysis is not intended to minimize the achievement of the Chinese, but rather to reveal the realities of earthquake prediction itself.

10.3 The Intrinsic Nature of Earthquake Phenomena

Let us return to the essence of "live" earthquakes.

According to the evidence accumulated by modern seismology, the intrinsic nature of shallow large earthquakes—the subject of earthquake prediction efforts—is a grand-scale rupture of underground rocks in the form of "seismic faulting" to release strain that has accumulated in the lithosphere of the earth's surface layer. It is the sudden occurrence and spreading of "dislocation" within the rocks which have been critically strained, forming a "dislocation surface (fault plane)." The rocks on the two sides of the fault plane move in opposite directions along it during a very short time. As a result, seismic waves are generated, which cause ground vibration, and surface deformations are brought about. Today it is possible to know what kind of faulting took place under the ground when a large earthquake occurs: the location, dimension, and configuration of the fault plane, and the direction, amount, and time function of the relative displacement between the rocks on both sides of the fault plane can be determined objectively from various observational data. The results of this determination can be demonstrated in a "fault model" of the earthquake.

The size of an earthquake actually has to do with the scale of its seismic faulting (in a magnitude 8 earthquake around Japan, the length of the fault plane is usually about 100 km, the width, about 50 km, and the amount of dislocation, several meters). The location of an earthquake is the extent of the underground fault plane (source region). Consequently, in the last analysis, earthquake prediction is estimating the most probable fault model to occur and predicting the time at which the faulting will take place.

The faulting of large earthquakes is controlled by two essential factors. One is tectonic and individual; the other, physical and universal. The first is based on the fact that the faulting is a radical manifestation of tectonic movement in a certain area; it is mainly concerned with the "spatial" aspect of faulting—*i.e.*, the tangible image of the earthquake apart from the time of occurrence. The second is the rupture generating process within the source region; it is concerned with the "temporal" aspect of faulting—*i.e.*, when the rupture actually takes place.

The occurrence of faultings of large earthquakes is never arbitrary. As one link in a chain of specific tectonic movement that has been taking place in each individual area over several hundred thousand years, a large-scale seismic faulting possesses a certain inevitability. Since the places that can be deformed easily by tectonic forces and the pattern and velocity of strain in any particular place are generally predetermined, and since the ultimate strain, or breaking strength, is thought to be almost constant, the seismic faulting that is to release the accumulated strain recurs in the same place with some regularity and in the same pattern, to the first order of approximation. This is a definite fact manifested by wide-ranging research of earth sciences. The basis for this phenomenon is most clearly explained by plate tectonics theory. Within this framework, the correlation that has often been recognized between seismic activities over a rather vast area can be understood as the inevitable result of the dynamic causal relationship of the whole.

The type of seismic faulting that is likely to occur in a specific area and its average recurrence interval can now be estimated with considerable accuracy based on the careful and comprehensive analysis of the available data in the framework of the theory of seismic faulting and plate tectonics, if the area is in a convenient location, such as the converging plate boundary at the continental margin. The data come from earthquake observations, geodetic surveys, and the results of research on historical earthquakes, topography, geology, active faults, crustal structure, etc. This kind of study, which argues earthquake generation in a specific region concretely on the basis of regional neotectonism, is known as "seismotectonics."

As for the actual timing of seismic faulting, some earth scientists claim that it cannot be definitely predicted since rupture is in general a probabilistic phenomenon. For this reason they judge earthquake prediction to be an essentially impossible task [e.g., Takeuchi, 1973]. Still, there are many instances in which remarkable precursory phenomena such as foreshocks and acute crustal movements have been recognized several hours or several days prior to the main shocks, even with the less perfect observation techniques of the past. Many seismologists today consider that within the heterogeneous rocks of the earth's surface layer an irreversible process probably goes on in the source region shortly before the main rupture. This process would determine the occurrence of seismic faulting and be responsible for short-term and immediate precursors. It is possible that these rupture generating processes are universal and governed to some extent by some decisive law.

If this is the case, by elucidating the true nature of the rupture-generating process, an accurate judgment can be made when anomalous phenomena are detected, and the timing of the faulting can be predicted concretely and with high precision. The DD model—although it concerns itself with a much longer period of time than that which is presently under discussion—was in any case one hypothesis for such a process. In addition, there is a postulation that the main rupture is preceded by an extremely slow preliminary faulting (pre-slip) that is too slow to generate seismic waves.

At present, however, no definite facts are known about the rupture-generating process, and there is, therefore, no guarantee whatsoever that decisive short-term and immediate earthquake prediction is possible in principle. Research in this field leaves a great deal to be desired in Japan. Whether or not highly reliable earthquake prediction can be made on a scientific basis hinges on theoretical and experimental basic research on physics in earthquake source regions as well as on the steady accumulation of observational data.

In short, a large-scale earthquake that is to be the object of prediction is an event that is both unique and individual, in that it is a part of the tectonic movement of a particular area, and general and universal, in that it is a rock-breaking phenomenon. Furthermore, it has the general inevitability of being a part of tectonic movement and the contingency of being a rupture phenomenon. The former almost determines the pattern, size, and rough timing of the faulting in long-range terms, while the latter affects largely the actual timing of its occurrence. When an earthquake is imminent, however, there is a possibility that the actual timing of the occurrence is determined by universal decisive laws that apply to the pre-failure process.

10.4 The Practical Procedure of Earthquake Prediction

10.4.1 A Two-Stage Earthquake Prediction Strategy

If earthquake prediction is possible in principle, it would naturally occur in two stages, given the intrinsic nature of the earthquake phenomenon (Fig. 10.2). The first stage is to provide a realistic picture of a coming seismic faulting and to predict the degree of its likelihood of occurrence, in a broad sense, based on its general inevitability. The second stage is a period

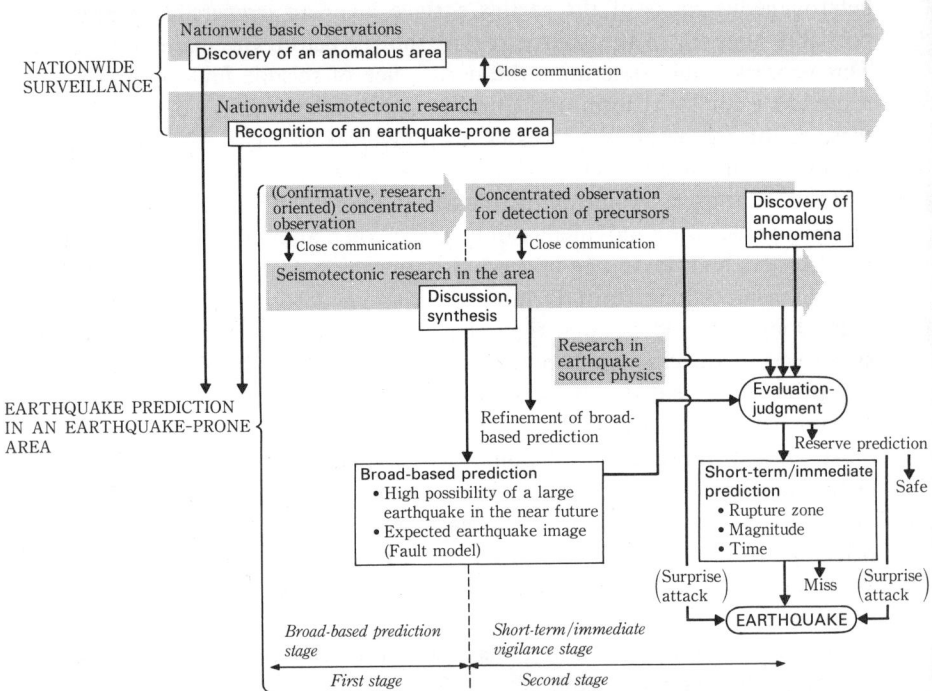

Fig. 10.2 Practical flow chart of earthquake prediction.

of short-term vigilance which ultimately leads to a prediction of the time of the earthquake based on short-term decisive laws that apply to the pre-failure process. The first stage would be based mainly on seismotectonics supported by various geophysical observations and geoscientific surveys. The second stage would be based mainly on the detection of short-term precursors and their seismophysical interpretation. Neither of these stages is phenomenological or statistical, but rather substantial and realistic.

The kernel of earthquake prediction is in the second stage. However, it is indispensable to recognize clearly the near-future possibility of a large earthquake and to draw a picture of its expected faulting as clearly as possible during the first-stage effort, since it is not known for certain whether or not the pre-failure process provides any clues to the scale and pattern of faulting.* It will then be also possible to carry out the effective observation

---

* The impossibility of basing earthquake prediction on the detection of short-term precursory phenomena alone was clearly analyzed and demonstrated by Tsumura [1973] using foreshocks as an example.

to detect short-term precursors. The first stage effort would include such means so far recognized for long-term prediction as the investigation of crustal strain, recurrence intervals of large earthquakes, seismic gaps, migration of earthquakes, and seismic activity in neighboring areas. The important point is, however, incorporating these investigations into a seismotectonic analysis, rather than analyzing them separately from a phenomenological point of view. Statistical prediction is included, of course, in the first-stage analysis, but it should never be left at the elementary level of phenomenology.

In essence, the most crucial aspect of the first stage is to try to draw up a comprehensive and realistic sketch map of the anticipated earthquake and the characteristics of the surrounding crustal activity, paying special attention to the individual phenomena in that particular area as well as following the general laws of seismology and physical geology. In this sense, the work in the first stage resembles more than anything a criminal investigation. It is, above all, grounded in the principles of basic research. The fact that the first-stage effort as above forms one of the bases of earthquake prediction is clearly demonstrated by the two successful (or quasi-successful) predictions of non-damaging small earthquakes in Japan in 1977. One of these earthquakes took place in the vicinity of Wakayama City on Aug. 7 (M 4.7) [Mizoue et al., 1977]; the other occurred on the Yamasaki fault in Hyogo Prefecture on Sept. 30 (M 4) [The Research Group for the Yamasaki Fault, 1978]. These are areas in which close-knit microearthquake observation networks were first established in Japan and these elaborate networks have been studying the characteristics of seismic activity in each area in great detail for more than 10 years. The area around the Yamasaki fault in particular has been subjected to various kinds of comprehensive observation and investigation as an "earthquake prediction test field."*

Under present conditions, rough prediction of intraplate earthquakes—those that occur in inland Japan where tectonic movement is slow and recurrence intervals long—based on seismotectonics is extremely difficult compared to the prediction of interplate earthquakes on the Pacific coast. This does not necessarily discount seismotectonic considerations in inland earthquake prediction, however. As one of the problems in predicting inland earthquakes is the difficulty of pinpointing a prediction target, it is

---

* So far, a "test field" has been regarded as a sort of open laboratory where basic research on the method of earthquake prediction is carried out by means of comprehensive observation and study employing every conceivable technique. Since both the actual characteristics of seismic activity and the means to survey them effectively are unique to each region, however, all earthquake-prone areas must become "test fields" in order to watch for actual seismic activity during the first-stage prediction procedure.

very important to develop a seismotectonic research method that will be effective in studying intraplate activities so as to make drawing a clear picture of an anticipated earthquake possible.

The factor of time, in the strict sense of the word, is not included in the first stage of earthquake prediction. And there are no objective guidelines or definite standards for when the second-stage prediction work, or the short-term surveillance, should begin. Although M-dependent precursors were expected to quantitatively fill the gap between the two stages, they do not seem to be useful at present as was discussed earlier. Probability of earthquake occurrence as mentioned by Rikitake [1978] is mostly a reflection of a perfunctory assumption for calculation, does not necessarily represent actual future earthquakes, and thus does not serve as a substantial standard. Ideally of course there should be a dependable standard for when to switch from broad prediction of the first stage to short-term surveillance of the second stage. In developing an actual prediction strategy for the present, however, it is vital to pay attention to the gap that exists between the two stages.

Practically, it seems better for us to return to the viewpoint that the primary purpose of earthquake prediction is to minimize earthquake disasters, and to base our work on the idea that our recognition of earthquake occurrence is always in danger of being a step behind what is actually happening. Namely, when the most probable fault model with few contradictions has been obtained for the expected earthquake, and many researchers have agreed on the possibility of its near future occurrence, we should consider that the irreversible pre-failure process may start "even tomorrow," and jump immediately into the second stage. That this is not an exaggeration was proven by the occurrence of the 1978 Miyagi offshore earthquake (M 7.4).

For some time before the earthquake, a few researchers had pointed out the existence of a seismic gap off Miyagi Prefecture. In particular, Seno [1979] published a broad prediction, based on comprehensive seismotectonic considerations, of an imminent interplate earthquake of M 7.7 or thereabouts, at the 1977 Fall Meeting of the Seismological Society of Japan. No one paid much attention at the time, and Seno himself had a later date in mind. The Sendai area was therefore completely taken by surprise when the earthquake hit in June 1978. This earthquake apparently was not the one that was predicted, but seems to have a close connection with it. Given the example of 1897, it is feared that the great earthquake predicted by Seno may follow.

The second-stage prediction—establishing short-term surveillance—centers around observations to detect short-term precursors. In the future, when prediction techniques are more advanced, it will be possible to concentrate on those observations that have proved to be most effective. Without knowing which these are, however, it is necessary at present to deploy all conceivable kinds of instruments as densely as possible, concentrate all records on one place, and continue surveillance around the clock.

When anomalous phenomena appear, it must be determined whether or not they are earthquake precursors, what kind of earthquake they will lead to, and how soon the earthquake will strike. Such a judgment must be based on a thorough understanding of the pre-failure process. Since this understanding is still far from thorough, however, scientists must depend on a combination of experience and intuition, aided by the predicted fault model and the limited knowledge available on source physics. Consequently the potential for error, especially in the prediction of when the earthquake will occur, is quite high. To put it another way, this process of judgment is, at this time, a research activity that is highly original and creative every time.

In addition, there is no guarantee that a short-term surveillance setup will, indeed, be short-range. This raises the question of how the public will be affected. What is important at this time, however, is to realize that this is the nature of earthquake prediction and to devise some ways to minimize social inconvenience. Broad prediction activities based on seismotectonic research will continue to be carried out in the second stage, ever correcting and improving their precision, until an earthquake occurs.

The foundation for nationwide surveillance in Japan is of course the basic observations for detecting anomalous phenomena throughout the country, which have already been in effect: repeated geodetic surveys, tide observation, gravity surveys, geomagnetic surveys, and observations of large, intermediate, and small earthquakes. In addition, however, there is one more cornerstone that should be formally incorporated into the foundation, and that is basic seismotectonic research, *i.e.*, various investigations of earth sciences directly relating to earthquake generation in and around Japan. Since many researchers are already pursuing such investigations on their own initiative—often outside the framework of the earthquake prediction program—it may not be necessary to set up a new organization for this purpose. And yet it is necessary to establish a method to consolidate these research results, as was seen earlier in the case of the prediction of the Miyagi offshore earthquake, from the viewpoint of picking up earthquake-prone areas. Nationwide basic observations and seismotectonic research are like the two wheels of a vehicle. Even if the results of the former do not reveal any anomalous phenomena, they should be continuously analyzed and interpreted from the viewpoint of seismotectonics. It is also vital that an opportunity be provided for thorough academic discussion and broad participation when earthquake-prone areas are being determined and the first-stage prediction of a specific earthquake examined.

10.4.2 Realizing Earthquake Prediction

Based on the above strategy, what are the chances of realization of earthquake prediction now? Where the first-stage prediction is concerned, there is a sharp division between those areas where predictions in a broad sense can actually be made and those for which prediction is extremely difficult. When future earthquakes are being considered, more areas are in the latter category than in the former. Especially in the case of inland earthquakes with long recurrence intervals, even if an image of the anticipated earthquake could be conjectured, determining the possible imminence of an earthquake that would compel the adoption of the second-stage prediction procedures is a task of utmost difficulty. As for the second stage, there seem to be some cases in which key short-term precursors can be anticipated to a considerable degree and some in which they cannot, depending on the region and the type of faulting. I have already mentioned that present techniques for judging urgency leave a lot to be desired. Under these circumstances, therefore, it must be admitted that a technical system of earthquake prediction is far from complete. The realization of earthquake prediction, in the sense that the rate of successful predictions throughout the country would exceed a certain level, cannot be achieved for the time being.

On the other hand, however, the chances are good of achieving successful prediction by detecting clear precursors if a short-term surveillance system has been set up in the earthquake-prone area that has been designated as such by a rather accurate first-stage prediction procedure. Considering the great loss of life and damage to property that occurs when an unpredicted large earthquake strikes in a highly civilized and densely populated society, it is obvious that every possible lead to successful prediction should be pursued. Then, the best practical plan now, as already proposed by Tsumura [1973], seems to be to divide the country's areas into four categories—those in which large earthquakes are highly possible and successful prediction likely, those in which the threat of large earthquakes is real but prediction is difficult, those in which the threat of large earthquakes is minimal, and those for which no definition is possible—and to concentrate our efforts in the first area for purposes of practical short-term prediction. Undoubtedly, the region covering Sagami Bay, the Izu Peninsula, Suruga Bay, and the Enshu Sea would be the first candidate at present.

If it is to be of service to the public, earthquake prediction must be an established, day-to-day undertaking. The major work in attaining successful prediction, however, is original research works, as is clear from the discussion so far. Then, routinizing earthquake prediction in the following sense is impossible in itself; predictions can be made mechanically based on some sort of programmed handbook only if an observation network is set up for constant surveillance.

These realities of earthquake prediction must be recognized as steps are considered which will prepare society for disaster prevention.

10.5 Prediction of a Tokai Earthquake

The Tokai earthquake expected in the Tokai district, on the Philippine Sea coast of central Japan, is the most important target in the current Japanese earthquake prediction program as well as a big social problem because of the scale of the pre-estimated catastrophic disaster [Ishibashi, 1981]. Its longterm and short-term prediction efforts are among the most advanced in the world. The prediction work on this earthquake is a good example of the two-stage earthquake prediction strategy discussed above.

The Tokai district facing the Enshu Sea and Suruga Bay was already mentioned as an example of the "area of special observation" for M 8 class earthquakes in "A Long-Term Yearly Plan of Earthquake Prediction Research (for 1970 Fiscal Year)" published in June 1969 for the reason that, despite a history of large earthquakes, there has been no event there in recent years. Mogi [1970] reiterated definitely the possibility of a large earthquake in the district based on an interpretation of geodetic data by means of his model for the occurrence of offshore great earthquakes. In November 1969, the CCEP officially designated this district as an area of special observation, taking into account the remarkable secular subsidence of the west coast of Suruga Bay and the history of great earthquakes in the region. In May 1973, with rapidly-developed plate tectonics around this region as background, Fujii [1973] discussed the nature of the expected Tokai earthquake. In November of the same year Ando [1975] proposed a fault model for the Tokai earthquake, which occupied almost the entire Enshu Sea, based on full-scale seismotectonic discussions. He stated that the possibility of its occurrence was very high. In February 1974, the CCEP upgraded the area to the area of intensified observation, based on a seismicity gap in the Enshu Sea and crustal movements along the coast.

Thus the idea of the "Enshu Sea earthquake" was established. Here, we should pay attention to the strong influence of a seismotectonic argument that a great earthquake will not occur in Suruga Bay where plates are colliding [Ando, 1975]. Although Rikitake [1974] had calculated the cumulative probability of the occurrence of a Tokai earthquake as more than 90% based on a simple assumption, many researchers who had been making precise studies on the area were plagued by inconsistencies in the image of the expected earthquake [e.g., Ando and Fukao, 1975]. The major problems were the strain release that appeared to have taken place in most of the Enshu Sea at the time of the Tonankai earthquake of 1944, and disagreement between the crustal movement calculated from Ando's fault model and the actual observation data. As a result, some seismologists even began to think that a Tokai earthquake would not take place for 100 years.

In essence, up to 1976, the major part of the Tokai earthquake prediction was a broad prediction based on seismotectonic discussions. The significant point here is that the possibility of the imminent earthquake itself was neither certain nor clear because a convincing earthquake image could not be produced.

The "Suruga Bay earthquake hypothesis" which I proposed in 1976 [Ishibashi, 1976; 1977; 1981] presented a fault model of the Tokai earthquake that spans the eastern half of the Enshu Sea and the innermost part of Suruga Bay, based on reexaminations of the fault models of the 1944 Tonankai earthquake and of the 1854 Ansei Tokai earthquake, and on reevaluations of geodetic and topographic data. It maintained that the time of recurrence might be near due to the amount of time that has elapsed since 1854 and to the estimated amount of accumulated strain. Although this hypothesis was not unique academically, it did delineate certain circumstances in the past which had made it difficult to determine the possibility of an imminent earthquake due to the lack of a convincing earthquake image. The author insisted that "it would be no surprise if precursory phenomena were to begin today. Since there is no guarantee that such phenomena would continue for a long time, a concentrated observation effort must begin immediately" [Ishibashi, 1977]. By this the author meant nothing more than to suggest, based on the two-stage earthquake prediction strategy explained above, that short-term surveillance of the second stage should now be set up, given his broad-based prediction.

In November 1976, the CCEP announced its official view that the most likely rupture zone of the Tokai earthquake was the area from south off Omaezaki to the interior of Suruga Bay, but that no precursor indicating its occurrence time had been detected yet.* On the other hand, however, the reinforcement of observation, and the concentration of records and their continuous watching at Japan Meteorological Agency (JMA) have been under way, providing against an emergence of short-term precursors. In April 1977, the Earthquake Assessment Committee for the Tokai Area

---

*It is a fact, however, that anomalous crustal movement has been observed in an extensive area of the Chubu district since the end of 1973, and that the crustal subsidence at Omaezaki was accelerated at about the same time.

was established, although there are still many unresolved problems. Viewing the progress made thus far, the Tokai earthquake prediction operation is basically proceeding along the lines represented in Fig. 10.2. It is important to recognize that the operation has now entered the short-term surveillance stage.

In addition, I would like to point out that serious attention must be paid from now on to the coseismic and postseismic observation of the Tokai earthquake itself, although this is unrelated directly to its prediction. An M 8 class interplate earthquake occurring right in the middle of a highly industrialized country is a very rare event. We in Japan have a duty to the world to carry out complete observations of such an earthquake for the purposes of basic geophysical research as well as earthquake prediction research. It is very important to prepare, for example, a special near-field seismological observation network with broad-band and large dynamic range and a concrete program for quick geodetic re-survey so as not to miss details of postseismic crustal movements.

References

Ando, M., 1975: Possibility of a major earthquake in the Tokai district, Japan, and its pre-estimated seismotectonic effects, *Tectonophysics*, 25, 69–85.
——— and Y. Fukao (eds.), 1975: *Off-Tokai Earthquake* (in Japanese), 61pp.
Dambara, T., 1966: Vertical movements of the earth's crust in relation to the Matsushiro earthquake (in Japanese), *J. Geod. Soc. Japan*, 12, 18–45.
———, 1973: Crustal movements before, at, and after the Niigata earthquake (in Japanese), *Rep. Coord. Comm. Earthq. Predict.*, 9, 93–96.
Fujii, Y., 1973: Seismic crustal movements in the Tokai district (in Japanese), *Abstracts, Geod. Soc. Japan*, No. 39, 72.
———, 1974: Relation between duration period of the precursory crustal movement and magnitude of the earthquake (in Japanese), *Zisin (J. Seismol. Soc. Japan)*, Ser. 2, 27, 197–214.
Geographical Survey Institute, 1971: On tidal records before and after earthquakes (in Japanese), *Rep. Coord. Comm. Earthq. Predict.*, 5, 67–71.
Ishibashi, K., 1976: Re-examination of a great earthquake expected in the Tokai district—Possibility of the 'Suruga Bay earthquake' (in Japanese), *Abstracts, Seismol. Soc. Japan*, 1976 No. 2, 30–34.
———, 1977: Re-examination of a great earthquake expected in the Tokai district, central Japan—Possibility of the 'Suruga Bay earthquake' (in Japanese), *Rep. Coord. Comm. Earthq. Predict.*, 17, 126–132.
———, 1981: Specification of a soon-to-occur seismic faulting in the Tokai district, central Japan, based upon seismotectonics, *Earthquake Prediction: An International Review*, D. W. Simpson and P. G. Richards (eds.), Maurice Ewing Ser., IV, AGU, Washington, D.C., 297–332.
Mizoue, M., M. Nakamura, Y. Ishigeta and N. Seto, 1977: Change of the seismicity pattern around Wakayama City and its regularities (in Japanese), *Abstracts, Seismol. Soc. Japan*, 1977 No. 2, 90–91.

Mogi, K., 1970: Recent horizontal deformation of the earth's crust and tectonic activity in Japan (1), *Bull. Earthq. Res. Inst., Univ. Tokyo*, **48**, 413–430.

Nur, A., 1972: Dilatancy, pore fluids, and premonitory variations of t_S/t_P travel times, *Bull. Seismol. Soc. Amer.*, **62**, 1217–1222.

Rikitake, T., 1969: An approach to prediction of magnitude and occurrence time of earthquakes, *Tectonophysics*, **8**, 81–95.

――――, 1974: Probability of earthquake occurrence as estimated from crustal strain, *Tectonophysics*, **23**, 299–312.

――――, 1975: Earthquake precursors, *Bull. Seismol. Soc. Amer.*, **65**, 1133–1162.

――――, 1978: Earthquake precursors and earthquake prediction (in Japanese), *Kagaku*, Iwanami Shoten, Tokyo, **48**, 27–32.

Scholz, C. H., L. R. Sykes and Y. P. Aggarwal, 1973: Earthquake prediction: A physical basis, *Science*, **181**, 803–810.

Sekiya, H., 1976: The seismicity preceding earthquakes and its significance to earthquake prediction (in Japanese), *Zisin (J. Seismol. Soc. Japan)*, Ser. 2, **29**, 299–311.

――――, 1977: Anomalous seismic activity and earthquake prediction, *J. Phys. Earth.*, **25**, Suppl., S85–S93.

Seno, T., 1979: Intraplate seismicity in Tohoku and Hokkaido and large interplate earthquakes: A possibility of a large interplate earthquake off the southern Sanriku coast, northern Japan, *J. Phys. Earth*, **27**, 21–51.

Takeuchi, H., 1973: *Science of Earthquake* (in Japanese), Nippon Hoso Shuppan Kyokai, Tokyo, 225pp.

The Research Group for the Yamasaki Fault, 1978: On some abnormal observations accompanied by an earthquake ($M = 4$) at the Yamasaki fault, September 30, 1977 (in Japanese), *Rep. Coord. Comm. Earthq. Predict.*, **19**, 122–128.

Tsubokawa, I., 1969: On relation between duration of crustal movement and magnitude of earthquake expected (in Japanese), *J. Geod. Soc. Japan*, **15**, 75–88.

Tsumura, K., 1970: Investigation of mean sea level and its variation along the coast of Japan (Part 2), *J. Geod. Soc. Japan*, **16**, 239–275.

――――, 1973: Microearthquake observation and problems of earthquake prediction (in Japanese), *Proc. Symp. on Earthq. Predict.* (1972), 81–89.

Whitcomb, J. H., J. D. Garmany and D. L. Anderson, 1973: Earthquake prediction: Variation of seismic velocities before the San Fernando earthquake, *Science*, **180**, 632–635.

Zhu Fung-ming, 1976: An outline of prediction and disaster prevention of the Haicheng earthquake of magnitude 7.3 (in Japanese), *Proceedings of the Lectures by the Seismological Delegation of the People's Republic of China*, Seismol. Soc. Japan, 15–26 (Available in English in *Proceedings of the Lectures by the People's Republic of China*, J.P. Muller (ed.), Spec. Publ. 43–32, Jet Propulsion Laboratory, California Inst. of Tech., Pasadena, 1976, 11–19).

Chapter 11 The Evaluation of Short-Term Precursory Earthquake Phenomena

Akio Takagi

The generative process and recurrence cycles of great earthquakes are varied and complex. Some earthquakes recur after several decades, while others do not recur for more than a thousand years. It would be impossible to try to observe every single earthquake, analyze its precursors, and devise an objective, quantitative prediction method. Although no two earthquakes have identical rupture sequences, some similarities in immediate earthquake precursors have been found that can be explained by basic research.

These similarities were the basis for beginning earthquake prediction research. The strategic objectives of this research are the establishment of a method of making long-range predictions of the position and size of future earthquakes based on anomalous crustal movement and seismic activity and a method of making short-range predictions of when earthquakes will occur based on the analysis of immediate precursory phenomena. The road to earthquake prediction does not already exist—it is being created by those who travel it.

Short-range prediction is based on assessing radical changes in the continuous observation data and judging the imminence of an earthquake on that basis. This assessment is the first step toward the realization of earthquake prediction. Since earthquake phenomena are extremely varied and complex, however, there are many basic problems that must be resolved before earthquake prediction can be realized. The evaluation of precursors must be based on experience. The knowledge that results from that experience is, literally, the substance of earthquake prediction research.

Figure 11.1 is a flow chart that shows how research begins with long-range prediction and ends with a short-range prediction. Much of the information gained from basic research must be applied throughout this process, as it would be dangerous to make a decision that relied solely on models or on the interpretation of anomalous phenomena.

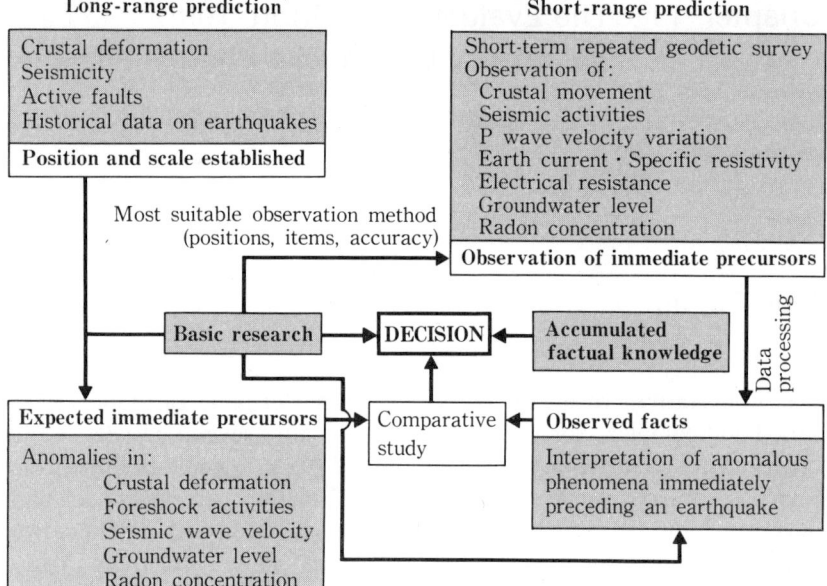

Fig. 11.1 An earthquake determination flow chart, from long-range to short-range prediction.

This process is now being applied in the Tokai district of Eastern Japan. In 1969 survey results indicated a pronounced accumulation of strain in a northwesterly, inland direction in the coastal region of Shizuoka Prefecture. At the same time, it was reported that seismic activity off the coast of Tokai was remarkably low. These results prompted scientists to believe that the Tokai district was in the process of accumulating seismic energy. The imminence of another Tokai earthquake was confirmed when, in 1976, Ishibashi pointed out that, based on historical and other research data, the focal zone of the Ansei Tokai earthquake of 1854 extended well into Suruga Bay. There are various points of view on the focal zone and timing of a future Tokai earthquake, but a majority of researchers agree on its probability. The area has been designated a reinforced observation area and an intensive observation network established. Figure 11.2 shows the positions of the observation sites. An Earthquake Assessment Committee for the Tokai Area has been established within the Coordinating Committee for Earthquake Prediction. Its function is to determine quickly whether any sudden anomalous variations in the continuous observation data portend an imminent earthquake. The decision will be made in accordance with the short-term prediction method illustrated in Fig. 11.1. The following factors will be central to this decision.

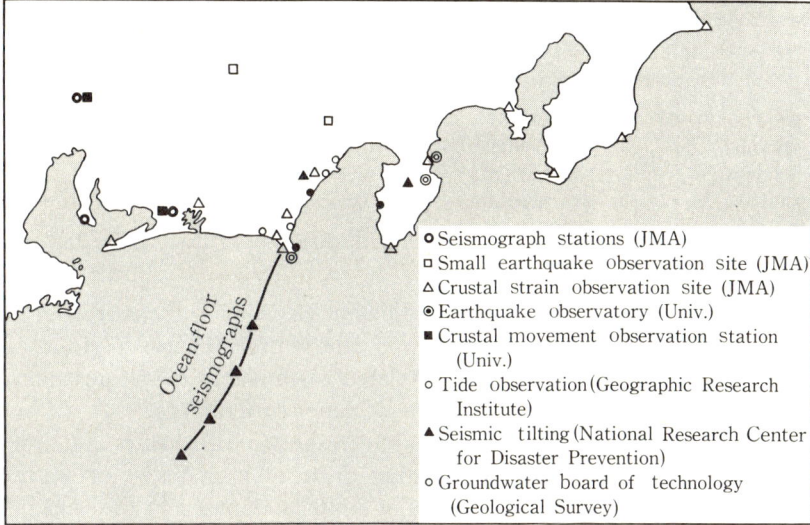

Fig. 11.2 Observation stations in the Tokai district for the observation of earthquakes, crustal movement, strain in rocks, tide level, tilting, and groundwater.

11.1 Foreshock Activity

There is a seismic gap off the coast of Tokai, but does this mean that foreshock activity can be expected? Principal earthquakes are sometimes preceded by small earthquakes called foreshocks that occur anywhere from several days to 30 to 40 days before the major earthquake. Earthquakes in Japan since 1860 of M 6.5 or more that have been preceded by foreshocks are listed in Table 11.1.

Rikitake [1979] selected 73 earthquakes with reported foreshocks and investigated the relationship between the magnitude (M) and the time (T)— the time between the onset of foreshocks and the occurrence of the principal earthquake. He concluded that there was no regularity in the relationship. The examples in Table 11.1 confirm this conclusion. Mogi, who studied the spatial distribution of earthquakes accompanied by foreshocks, pointed out that they occur in a specific area, as is shown in Fig. 4.15 of Chapter 4.

Examples of earthquakes preceded by foreshocks are rare, but if detailed research based on highly accurate observation is continued, some regional regularities, or some regularity between M and T in each region, may be discovered. If even a handful of foreshocks could be observed, they could

Table 11.1 $M \geq 6.5$ Earthquakes in Japan That Have Been Preceded by Foreshocks

Date of principal earthquake	Name of earthquake	M	Time elapsed between the first foreshock and the principal earthquake
1872 Mar. 14	Hamada	7.1	4 to 5 days
1896 Aug. 31	Rikuu	7.5	8 days
1905 Jun. 7	Izu Oshima	6.5	10 days
1906 Apr. 21	Central Gifu	6.6	7 hours
1930 Nov. 26	Northern Izu	7.0	19 days
1945 Jan. 13	Mikawa	7.1	2 days

contribute to effective earthquake prediction, as will be discussed later in this chapter. If a method can be found that will distinguish a group of swarm earthquakes or a group of unrelated earthquakes from foreshocks, it will be very helpful in the prediction of major earthquakes.

There have been numerous reports of foreshocks with values of b or m (see Chapter 4) that are smaller than those of aftershocks or swarm earthquakes. There are also examples of a radical decrease in the value of b immediately before the principal earthquake occurs. Mogi carried out a series of experiments on rock fracture that indicated that foreshocks do not occur easily in an extremely homogeneous crust. Rather, this sort of crust is prone to principal earthquake→aftershock type earthquakes. The slightly heterogeneous crust is prone to a series of foreshock→ principal earthquake→aftershock type earthquakes. In a totally heterogeneous crust, swarm earthquakes tend to occur without a principal earthquake. The values of b or m, however, are not necessarily smaller than those of regular earthquakes. Nor can it be said that a decrease in b always brings about a great earthquake. It is difficult, therefore, to decide whether or not a tremor is a foreshock based solely on the value of b. Furthermore, to obtain b and m, data processing is necessary. Data on foreshocks are scarce, however, making it difficult to arrive at meaningful values or else leading to erroneous values.

Hirasawa and others [1980] in an experiments on rock fracture, detected a minute acoustic emission that accompanies minute rupture and obtained the relationship between the decrease in m that accompanies an increase in axial stress and the amount of stress drop due to acoustic emission. Figure 11.3 demonstrates the results of this experiment. The x-axis in the figure represents axial stress as standardized by the fracture stress of a granitic sample. The open circles denote the value of m (the y-axis on the right side), which is obtained from the frequency distribution of the acoustic emission classified by amplitude. The solid circles represent the amount of local stress drop (y-axis on the left side; scale can be arbitrary) that accompanies micro cracks. As is obvious from the figure, the values of m and the amount of local stress drop change as the axial stress increases: m in particular decreases drastically, and the amount of local stress drop shows a radical increase as the axial stress of the sample goes beyond 85% of its fracture stress. This experiment also confirmed that the high-frequency component of the wave pattern of acoustic emission becomes predominant as axial stress increases. These results indicate that not only are variations in m an immediate precursor of rupture, but that an increase in the amount of local stress drop and variations in the seismic wave spectrum can also function as precursory phenomena.

At present the Tokai district is thought to be a seismic gap zone. In the event that a great earthquake does occur, anomalous seismic activity may be noted in adjacent or distant areas shortly before the earthquake, and foreshock activity, in the narrow sense, may be minimal. From the minimal foreshocks that do occur, it may be necessary to find out whether or not the stress drop is greater than that of ordinary earthquakes, or to examine the seismic wave spectrum, since it will be impossible to obtain the value of b or m from minimal foreshocks. In order to do this, it will be necessary to improve the observation system for smaller earthquakes so that the amount of stress drop and any changes in the seismic wave spectrum can be detected immediately under a continuous surveillance program. An observation system must also be devised that can obtain this information for microearthquakes as easily as the degree of seismic activity is now obtained.

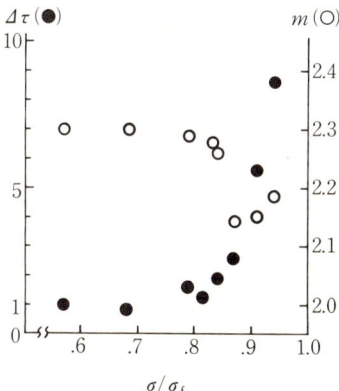

Fig. 11.3 The m value in Ishimoto–Iida's equation and the amount of stress drop by acoustic emission versus the applied stress normalized by fracture stress in a granitic sample.

11.2 Continuous Observation of Crustal Movement

At the time of the 1943 Tottori earthquake, a horizontal pendulum tiltmeter belonging to the Faculty of Science of Kyoto University was positioned in a vault in Ikuno Mine, which is about 60 km from the epicenter. Six hours prior to the earthquake an anomalous tilting (0.25 ″) was recorded in which the epicentral region uplifted. More recently, a water tube tiltmeter positioned in an observation vault at Matsushiro Observatory recorded frequent anomalous tiltings which began several hours before the M 5 Matsushiro earthquake (1965–).

There have been cases in which anomalous tilting has been detected by instruments other than tiltmeters. When the Tonankai earthquake took place in 1944, leveling was underway near Kakegawa. Recent investigations have revealed that an anomalous tilting of 1.3″ upward toward the south began on December 6, one day prior to the earthquake, and continued until just before the earthquake began on December 7. Due to these reports, anomalous tilting is anticipated as an earthquake precursor, thus making the observation of crustal movement vital to short-term as well as long-term earthquake prediction.

In fact, a compressive strain was successfully detected by a volume strainmeter belonging to the JMA, one month before the Izu Oshima offshore earthquake of 1978. Figure 11.4 (a) (b) represents the information registered on the volume strainmeters in the Tokai district, as compiled by the JMA. (a) shows the record for a two-year period beginning in April 1976, and (b) shows that from November 1977 to January 1978. Anomalous variations were clearly noted in the records from Irozaki and Ajiro—the areas closest to the epicenter—40 and 25 days, respectively, prior to the principal earthquake.

There are two important lessons in these facts which must be learned. The first is that the time span of precursory phenomena is longer than expected. Why does the crust begin to change as much as one month before the earthquake? The results of rock mechanics experiments can be applied here as a possible interpretation. Within a heterogeneous medium, local stress concentration occurs as axial stress increases, and when minor rupture takes place, the concentrated stress is redistributed. As the stress increases again, local stress concentration recurs until the principal rupture finally takes place. The duration of precursory phenomena may reflect the duration of this repeated process. At any rate, as far as the duration of anomalous precursory phenomena is concerned, the heterogeneity of the medium must be taken into account.

The second lesson is that it is important to look at the records simultaneously in order to get an overview of the progress of crustal stress.

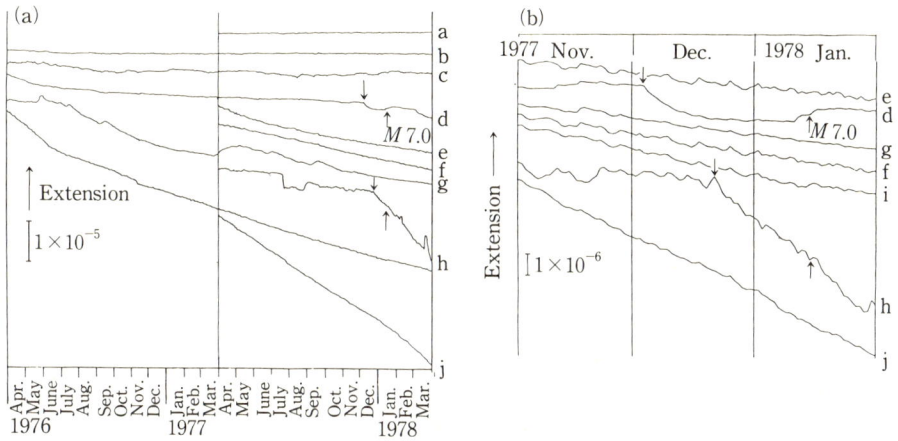

Fig. 11.4 Secular variation in strain measured by embedded strain gauges.
a: Choshi, b: Irago, c: Shizuosa, d: Irozaki, e: Yokosuka, f: Tateyama, g: Mikkabi, h: Ajiro, i: Omaezaki, j: Katsuura.

This is made possible in the Tokai district by the fact that records are sent from Choshi, Irago, Shizuoka, Yokosuka, Tateyama, Mikkabi, Omaezaki, and Katsuura, as well as from the places mentioned earlier, by telemeter, thus making it easy to catch anomalous phenomena in their early stages. Crustal movement should be grasped as not only a recorded phenomenon at one point, but as part of a pattern as well. Looking at an array of network observations, in other words, is helpful in discounting the influence of sources of disturbance.

As far as the Tokai district is concerned, volume strainmeter observations, complemented by a network providing an array of tilting observations, are desirable. As the epicenter is likely to parallel the Nankai Trough, dual observation by volume strainmeters and tiltmeters along the coast is an absolute must if the earthquake process is to be monitored more accurately.

The function of constant surveillance is to detect anomalous change in the continuous records of crustal movement. In terms of short-term earthquake prediction, the automation of this function is desirable. Ishii [1976] has already attempted to detect anomalies in the past records of crustal movement that accompanied earthquakes, using Chebychev's approximate function. Figure 11.5 shows an example of this method. The solid line in (a) represents the observed value of two components of the extensometer observations made at Erimo Observatory. The dotted line represents the predicted value of these components 30 days later, as obtained and plotted from data gathered in the past year. This predicted value assumes that there have been no anomalous changes. The two dotted lines at the bottom of the figure were arrived at by subtracting the observed values from the predicted values of the two components. This is called the prediction error.

Approximately three months before the Nemuro Peninsula offshore earthquake (June 17, 1973; M 7.4, $h = 40$ km), anomalous phenomena were noted which did not fall within the normal value on the appropriate scale. (b) represents the results of an analysis of observations made at Mase of the north-south component of the water tube tiltmeter both before and after the Niigata earthquake (June 16, 1964; M 7.5, $h = 40$ km). The solid line below represents the observed value, while the broken line represents the predicted value 2.5 months later, calculated and plotted

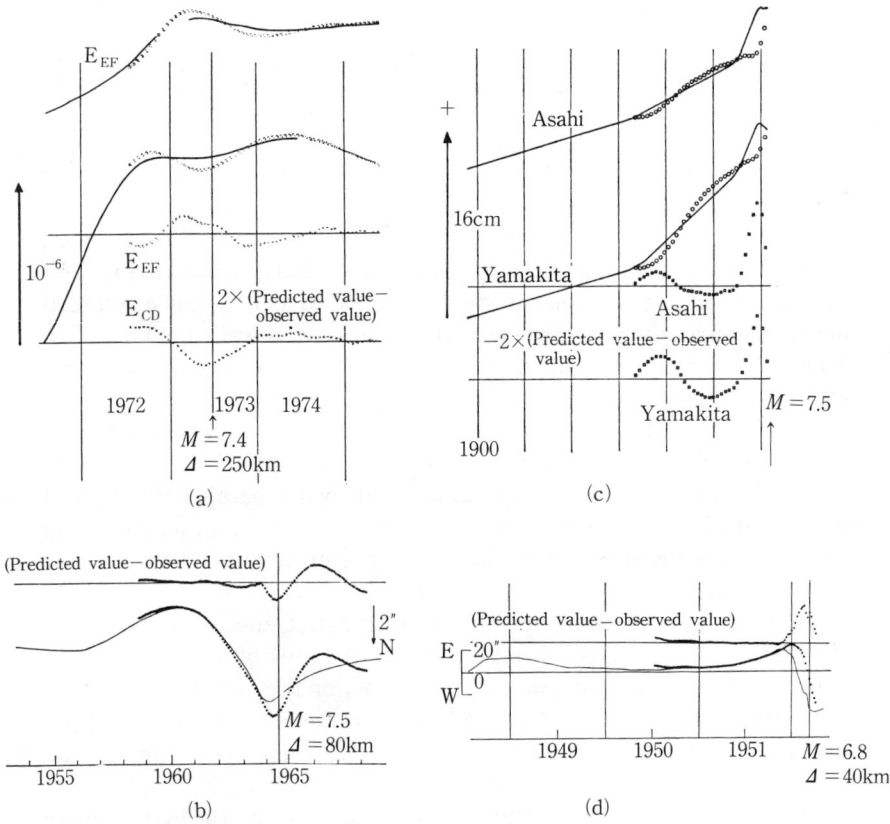

Fig. 11.5 Prediction errors recorded at Erimo Crustal Movement Observatory (Southern Hokkaido).
The arrows indicate the occurrences of earthquakes.

from data gathered over the past five years. When the difference between the two values was obtained, anomalous changes were noted approximately three months before the earthquake that were different from normal variations in value. (c) represents the results of an identical calculation using changes in levels at Asahi and Yamakita, with Kashiwazaki as the standard, at the time of the Niigata earthquake. The calculation is identical except that the predicted value is for two years later and the calculation is based on data collected over the past thirty years. The anomalous change in this case was noted approximately three years prior to the earthquake. (d) shows the results obtained by using the same method to analyze the EW component of the horizontal tiltmeter at the time of the Daishoji offshore earthquake (March 7, 1952; M 6.8, $h = 20$ km), as observed at the Ogoya Observatory of Kyoto University. The predicted value for one month later was based on data from the last two years. The anomalous phenomena in this case appeared three months before the earthquake.

Thus the predicted value can be obtained for a designated time in the future by calculating Chebychev's approximate function from data already accumulated. An abnormal change, if it appears, will be objectively detected. The function is the best approximation of the data for the entire period and is considered to be quite accurate. It will be useful for detecting anomalous changes for the purposes of short-range prediction.

11.3 Groundwater Level and Radon Concentration

It is important that observed changes in groundwater level directly reflect changes in the earth's crustal stress. (See Chapter 8 for an extensive discussion of groundwater level as an earthquake precursor.) The volume strainmeters of the JMA continue to make observations from twelve locations. They are embedded at depths ranging from 50 to 250 m. It is interesting to note that the deepest of these is still affected by atmospheric pressure, but those at a depth of 200 m or more cannot be influenced by rain, regardless of the surrounding rock type. This seems to be an indication of how great the rain's influence is on the upper crust of the earth. Consequently it is necessary to take outside disturbances, such as rain, into consideration when measuring the groundwater level, since these disturbances can cause far greater changes than do precursory anomalies. Given these circumstances, groundwater level measurements should also be performed at an array of different observation stations. If a determination is going to be made based on the observation data from one position, one needs to know the influence of outside disturbances in advance.

During the Izu Oshima offshore earthquake (1978), precursory phenomena were detected by the continuous radon observation program of the University of Tokyo's Faculty of Science at nearly the same time that anomalous changes were registered on the JMA's volume strainmeters. This seems to indicate that there is a correspondence between changes in radon and changes in crustal stress—a very intriguing finding indeed.

11.4 Changes in the Specific Resistivity of Rocks and in Crustal Electric Resistance

When porous rocks that contain a certain amount of water are compressed, their electric resistance changes. This phenomenon is thought to be due to compression which collapses the pores, thus consolidating the water within the pores and decreasing the electric resistance of the rocks. Continuous observation of precursory changes in the electric resistance of volcanic tuff—a type of rock that possesses the above prerequisites—is being carried out at Aburatsubo in Kanagawa Prefecture. Since precursory changes in specific resistivity are often observed in connection with great earthquakes, it is considered to be one of the most effective short-range prediction aids. Specific resistivity acts as a highly sensitive strain gauge and is useful to research on the earthquake process as it relates to the period just before the earthquake strikes. An instrument is needed that will make multiple observations of specific resistivity easier.

Changes in the earth's exterior magnetism induce changes in the electromagnetic field according to the scale and distribution of the specific resistivity of the earth's interior. The longer the period of the exterior magnetic field, the more information these changes will provide on the scale and distribution of electric resistance in the depths of the earth. According to research being done by the Chinese, there is a layer of abnormally low (several Ω-m) electric resistance, several kilometers thick, at a depth of about 20 km. Research is being done to examine the relationship between the changes in specific resistivity in that layer and crustal stress. At the same time, the structure of specific resistivity in each region is being examined in detail. Research on changes in the specific resistivity of the earth's crust will become more and more necessary to short-range earthquake prediction in the future.

11.5 Identification of Anomalous Changes

If anomalies are discovered in the observation records forwarded by the extensive observation network set up for short-range prediction purposes, it must be determined whether or not these anomalies are pre-earthquake changes related to a great earthquake. As illustrated in Fig. 11.1, that determination will be based in part on whether the scale of the recorded anomalous change is in quantitative agreement with the expected precursors of the anticipated earthquake. The more items there are for comparison, the better the chances of making an accurate determination. When various anomalous phenomena begin to concentrate more and more in time and place, then highly accurate determination will become possible. At present, however, this is an impossible task. Realistically speaking, determinations will often have to be made on the basis of a limited number of anomalous phenomena, or from records with a low S/N ratio. Under these circumstances, the most important task becomes the interpretation rather than the determination of anomalous precursory phenomena. In order to interpret accurately, it is important to know in advance the particular characteristics of the observation site at which the anomalies were detected. In other words, do anomalous changes appear easily as a result of other external forces?

For example, when an earthquake is imminent, local stress is bound to concentrate in the area where the earthquake will eventually occur. If the crust is homogeneous, the anomalous changes will not reflect local differences. If the crust is heterogeneous and the stress field is uniform, as elucidated by the GDP traverse survey mentioned in Chapter 4, the amount of strain change will not be spatially uniform. The various precursory phenomena accompanying the strain change become spatially quite complex as a result. This is an example of a local characteristic. It is important to know about such characteristics in advance—an area's response to rainfall, for instance, and the amount of stress drop at the time of past earthquakes. The accumulation of such knowledge will be of great help in the future when determinations regarding anomalous changes must be made.

References

Ishii, H., 1976: Application of prediction method for analysis of crustal movement, *J. Geod. Soc. Japan*, **22**, 299–301.

⸻, T. Sato and K. Tachibana, 1978: Observation of crustal movements at the Akita Geophysical Observatory (3)—Application of Chebychev approximation function for data observed by extensometers and tiltmeters (in Japanese), *J. Geod. Soc. Japan*, **24**, 122–131.

Kusunose, K., K. Yamamoto and T. Hirasawa, 1980: Source process of microfracture in granite with reference to earthquake prediction, *Sci. Rep. Tohoku Univ.*, Ser. 5, Geophysics, **26**, 111–121.

Rikitake, T., 1979: Classification of earthquake precursors, *Tectonophysics*, **54**, 293–309.

Chapter 12 The Development of the Earthquake Prediction Project and Its Problems

Tatsuo Usami

12.1 The Blueprint

Full-scale earthquake prediction research in Japan began with the publication of "The Prediction of Earthquakes: Progress to Date and Plans for Future Development"—the so-called "Blueprint"—in January 1962. This 32-page pamphlet published by the Earthquake Prediction Project Research Group (C. Tsuboi, K. Wadati, T. Hagiwara, representatives) attempted to clarify the concept of earthquake prediction at that time. The development of earthquake prediction in Japan since then has been based solely on this Blueprint. In October 1963, the Japan Science Council, following the Blueprint, recommended that the Japanese Government "establish the necessary research facilities, and reinforce and enlarge the geophysical observation work of the various organizations that supply the basic data." In the meantime, in June of the same year, the Earthquake Prediction Standing Committee was established within the Japanese Geodetic Council, the Ministry of Education. The purpose of the Committee was to determine how to expedite the recommendation. Then, in 1965, the first phase of the Five-year Project was budgeted and inaugurated.

Stimulated by the Japanese Blueprint, American scientists submitted a report to the U.S. Government, entitled "Earthquake Prediction: A Proposal for a Ten-year Research Program," in September 1965. Although the proposal was never implemented, it did spur on rapid progress in earthquake prediction research in the United States.

In China, a destructive earthquake of M 6.8 occurred in the Yingtai District of Hebei Province (approximately 300 km southwest of Beijing) on March 8, 1966, causing considerable damage. This was followed, on March 22, by another earthquake of M 7.2. Prime Minister Zhou En-lai immediately visited the area to comfort the victims and inspect the damage. The national earthquake prediction project in China is said to have begun immediately after Zhou's visit.

Even now, on re-reading, the Blueprint seems simple and very much to the point. Although the timing of some of its proposals, and some of its contents, may have to be revised due to recent research results, the Blueprint's fundamental concept does not need any alteration. Rather, it was necessary to return to the original concept of the Blueprint before the Fourth Project, which commenced in 1979, could proceed.

One characteristic of the Blueprint is that it was based on research. There were some experts who were somewhat contemptuous of earthquake prediction; yet the spirit and the objectives of the Blueprint succeeded in satisfying even the skeptics. The tone of the Blueprint can be seen clearly in the following statements, taken from the foreword and conclusion:

> These points were agreed on by all:
> "What kinds of measurements should be made, using which methods, if measurements are to be made for the purpose of earthquake prediction?"
> "Measurements must be made on a grand scale and cannot be adequately handled by a single organization. An effective national project to undertake these measurements will require a radical re-examination of the present research set-up, as it is unsatisfactory."
> This is a kind of blueprint, which briefly summarizes the items that are indispensable to the efficient and conscientious achievement of the original goal.
> It is highly probable that significant correlations will be found between earthquake events and observed phenomena, based on the data accumulated during the several years of this Project. At this time it is impossible to say when earthquake prediction will become a reality or, in other words, when earthquake warning will be issued as part of a routine operation. But if all phase of this Project can be started within the next ten years, these questions can be answered with confidence.

Although the Blueprint was obviously research-oriented, the authors did not forget to add the following:

> The importance of establishing a large, new organization, such as a Ministry of Land Management, to coordinate disaster measures has been recognized already. If such a national organization were to be established, most of the work proposed here could be handled most effectively by it.

Reflecting the Blueprint's concept and the conditions of the time, the First Five-year Project was called the Earthquake Prediction Research Project. From the Second Project on, however, the word "research" has been deleted.

Another characteristic of the Blueprint is its scientific nature. It does not include engineering and sociology, which are deeply related to disaster relief. This was an oversight that became painfully obvious at the time of the Kawasaki earthquake of 1974. It became clear at that time that earthquake prediction would only be supported by the Japanese public if disaster prevention technology and the social consequences of forecast announcements were taken into consideration. Spurred on by the Tokai earthquake scare of 1976, the Earthquake Assessment Committee for the Tokai Area was established in April 1977, to determine the possibility of an earthquake. From the point of view of the predictive sciences, this was not a very clear-cut beginning. It is important to reflect on the fact that the original plan was unable to predict growing social demands for disaster prevention and forecasts and did not include socially oriented research.

The third characteristic of the Blueprint concerns the distribution of responsibilities, the organization of communication among prediction-related organizations, and the introduction of routine prediction work to educational institutions such as universities. Since great deal of research is still needed before the most effective earthquake prediction method is found, these tasks could not be assigned entirely to government bureaus as routine operations. In recognition of these circumstances, the plan recommended several proposals for the actual implementation of the project. Some of these proposals are still worthy of examination today. And yet they have never been implemented—probably due to a certain inflexibility on the part of Japan's administration, particularly where organizational reforms are concerned.

The research observation topics recommended in the Blueprint were:
- Examination of crustal movement by geodetic survey;
- Establishment of tide gauge stations to detect crustal movement;
- Continuous observation of crustal movement;
- Examination of seismic activities (great, intermediate, minor, micro- and ultra-microearthquakes);
- Observation of seismic wave velocity by way of explosion seismology;
- Examination of active faults;
- Examination of geomagnetism and the earth's current.

The above items were divided into numerous subtopics.

Ten years have passed since the Earthquake Prediction Project went into effect. In addition to the items included in the Blueprint as needing continuing research, some new topics have entered the picture. The following fields of research are currently recognized as relevant to earthquake prediction and are therefore part of the Project—whether or not they were formally included in its original mandate:
- Research and the collection of historical material on ancient earthquakes;
- Examination of crustal movement by strain measurement using laterallation surveys;
- Observation by strain gauge;
- Examination of crustal movement by gravity changes;
- Development of ocean floor seismographs and observation by instruments;

Gathering information by telemeters and data processing;
Observation of changes in electric resistance;
Rock fracture experiments;
Changes in the crust's micro structure using explosion seismology;
Observation of Tokyo district (earthquake observation using deep wells);
Changes in radon content, etc.;
Observation of groundwater, etc.;
Anomalies in animals and plants;
Test fields;
Theory of earthquake prediction;
Influence of prediction techniques on society.

Especially noteworthy is the development of hypotheses regarding plate tectonics, dilatancy, and earthquake faults. None of these were included in the Blueprint. And yet they are the three basic fields of research that provide the theoretical grounds for today's earthquake prediction. Without research in these areas, the results of past earthquake prediction research would exist in a vacuum. These hypotheses were developed outside the financial framework of the Earthquake Prediction Project. The National Research Center for Disaster Prevention, which is presently performing seismic tilt observations using deep wells, did not even exist when the Blueprint was published.

The list of topics alone is evidence of the spectacular results achieved by the Earthquake Prediction Project in the last ten years. Thanks to what seemed a rather unglamorous research effort, a part of the Project can now become a routine, technical observation operation. This progress was made despite the fact that the Project was severely criticized at the beginning for ignoring many important problems.

12.2 Outline of Earthquake Prediction Progress

Two years after the Blueprint was published, the Niigata earthquake occurred (June 16, 1964). It caused heavy damage, including oil tank fires and collapsed reinforced concrete apartment buildings, the most serious of which were triggered by liquefaction. With the government's financial support, the Earthquake Prediction Project was put into effect the very next fiscal year (1965). Immediately afterwards the Matsushiro swarm earthquakes began. As a result, the Geodetic Council recommended a partial revision of the Project in order to "further advance earthquake prediction research" in July 1966. Then, two years later, on May 16, 1968, the Tokachi offshore earthquake occurred, causing damage that extended from Aomori Prefecture to Hokkaido. The most serious damage was that to reinforced concrete buildings, especially school structures. This prompted a proposal for a Second Earthquake Prediction Project starting in 1969. The First Earthquake Prediction Project concluded after four years. The Second Project was in effect for five years, from 1969 to 1973, followed by the Third Earthquake Prediction Project from 1974 to 1978. The Fourth Earthquake Prediction Project began in 1979.

The Second Earthquake Prediction Project emphasized a new need "to establish a comprehensive system to advance the Project." As a result, the Coordinating Committee for Earthquake Prediction was formed in April 1969 within the Geographic Research Institute of the Ministry of Construction, as the headquarters for the comprehensive development of earthquake prediction. The Committee's purpose was to accelerate the work of the Earthquake Prediction Project and to realize, eventually, an actual earthquake prediction operation. Three detection and observation Centers were established for this purpose, namely, the Crustal Activity Detection Center of the Geodetic Survey Institute, the Seismic Activity Measurement Center of the Japan Meteorological Agency, and the Earthquake Prediction Observation Center of the Earthquake Research Institute of the University of Tokyo, each of which sends its relevant data to the Coordinating Committee for Earthquake Prediction. In addition, nine locations were designated as Special Observation Districts, or districts where earthquake prediction efforts would be concentrated. One of these, the Southern Kanto district, was further designated as a Reinforced Observation District in 1970, following the discovery of an anomalous uplift. Tokai district was designated in 1974 by the speculation that strain was accumulating off the coast of the Tokai district.

During the Second Project, the National Research Center for Disaster Prevention began to observe tilting in relationship to earthquakes using wells over 3,000 m deep. This is known as the "Tokyo observation." In addition, the Japan Meteorological Agency and the various universities involved began the observation of ocean floor earthquakes, mobile observations, and rock fracture experiments.

During the Third Project, data transmission and processing using telemeters and observation by strainmeters was initiated. During this period the "Kawasaki earthquake scare" took place, and the difficult task of formulating actual earthquake prediction practices fell to the Coordinating Committee for Earthquake Prediction. In addition, the heated debate over the possibility of a Tokai earthquake, which began in 1976, involved the entire nation and prompted the formation of the Earthquake Prediction Development Headquarters in October 1976. This organization monitors communication between government agencies and ministries. On April 1, 1977, a six-member Earthquake Assessment Committee for the Tokai Area

Table 12.1 The Budget of the Earthquake Prediction Project [figures through 1976 from T. Rikitake, *Introduction to the Theory of Earthquake Prediction*, Kyoritsu Zensho, 1976, p. 18].

Fiscal year	Budget (yen)
1965	212,539,000
1966	290,035,000
1967	334,362,000
1968	328,559,000
1969	496,116,000
1970	596,120,000
1971	805,720,000
1972	898,900,000
1973	761,164,000
1974	1,552,674,000
1975	2,007,638,000
1976	2,313,676,000
Total	10,596,503,000

Table 12.2 Changes in Facilities, Special Instruments, and Mobile Groups

	Prior to Blueprint (1962)	Newly established for Earthquake Prediction Project			
		1st	2nd	3rd	Total
Tide gauge stations	66	14			90
Crustal movement observatories	3	8	4	1	
Microearthquake observatories	3	9	6		
Ultra-microearthquake observation mobile groups	0	5	2		7
Geodetic survey mobile groups	0	1		1	2
Geomagnetism observatories	0		1		1
Local earthquake prediction centers (Telemeter)	0		2	3	5
Proton magnetometers					11
Deep-wells	0		1	1	2
Ocean-floor seismographs	0			1	1
Rock crushing facilities	0				6
Strainmeters	0			12	12

was established to determine the need for forecasts. The accumulation and centralization of data for the purpose of timely evaluation was achieved at the end of fiscal year 1978. From an academic point of view, there is a problem in the very formation of the Earthquake Assessment Committee itself, since Japan has little experience in how to determine potential earthquake danger from what data.

The Earthquake Prediction Project's budget is shown in Table 12.1.

This budget does not include construction or personnel expenses. Although the original budget proposed in the Blueprint was revised and cut, the budget's total of 10 billion yen is still rather impressive. From 1965 to 1976 some 80 persons were added to the Project, thanks to the efforts of those already involved.

Table 12.2 shows changes in the approximate number of facilities, special instruments, groups, etc. from the beginning of the earthquake prediction effort to the end of the Third Earthquake Prediction Project.

12.3 Earthquake Prediction Knowledge at the Time of the Blueprint and at Present

When the Blueprint was published, the examples of various phenomena that could be tied into scientific prediction were scarce. The accumulated knowledge regarding earthquake prediction was painfully limited. The special issue *Seismology in Japan* published by the Seismological Society of Japan two years after the Earthquake Prediction Project began describes the progress of earthquake prediction efforts up to 1966. Taking this as representative of the condition of seismological study at that time, it is possible to make a comparison of the past with the present.

Although the theory of plate tectonics does not predict the timing of earthquakes, it is fundamental to a consideration of Japan's great Pacific coast earthquakes. There are some arguments as to exactly when this theory was conceived, but it was sometime in the early 1960s—just about the same time as the Blueprint. The special issue *Seismology in Japan*, however, did not even mention it—probably because no one had any idea it would develop so rapidly. The theory of plate tectonics has provided a rough idea of why there are repeated great earthquakes on the Pacific coast. It also explains why reverse fault-type earthquakes occur on the inner side of oceanic trenches, while normal fault-type earthquakes occur on the outside. It also gave a clue to the link between great earthquakes off the coast of Kanto and minor seismic activities in the vicinity of Wakayama City. The tectonic analysis of the area around the northern Izu Peninsula where the Pacific Plate, the Philippine Sea Plate, and the Continental Plate meet in a most complex way profoundly affects the prediction of Tokai coastal earthquakes. Plate tectonics makes it easy for any layman to visualize the earth's movements and understand the repetition of great earthquakes. Nevertheless, plate tectonics has yet to explain certain local variations, such as why the area off the coast of Ibaragi Prefecture has fewer great earthquakes than the areas to its north and south, why the intervals between the great earthquakes on the Tokai or Nankai coasts fluctuate by as much as 100 to 200 years, or why earthquake recurrence cycles get longer and longer as they move south along the Japan Trench from the Hokkaido coast to the Sanriku coast.

In comparison, the dilatancy diffusion theory that is thought to apply to shallow inland earthquakes shows that there are typical variation patterns in such precursory phenomena as seismic activity, crustal movement, radon content, the crust's specific resistivity, the V_P/V_S ratio, and so forth. These patterns indicate the existence of a relationship between the duration of these phenomena and the size of the main earthquake. This seemed to be a new tool for earthquake prediction. A dry model of this theory was added later. In Japan, however, no one could find convenient examples to which the theory applied. Thus the dilatancy diffusion theory has entered a period of reevaluation. When Scholz, one of the originators of the theory, visited the Earthquake Research Institute in 1976, he seemed to be confident that he could explain Japanese earthquakes successfully. When he left Japan early the following year, however, it seemed to be with the realization that Japanese earthquakes were too complex to be explained by a single theory.

Tsubokawa [1973], Scholz *et al.* [1973], Rikitake [1975], and others have all been examining the various relationships between the duration of precursory phenomena and the magnitude of main earthquake. Their work has pushed earthquake prediction knowledge to the point where the drawing in Fig. 12.1 is now possible.

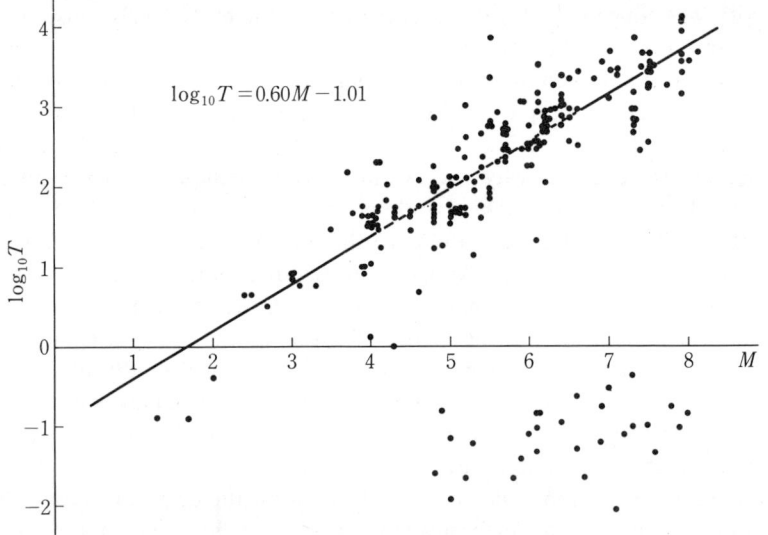

Fig. 12.1 Relationship between the common logarithm of the duration of precursors T (unit: days) and earthquake magnitude M [from T. Rikitake in *Kagaku* 43–1, p. 38].

One of the indispensable tools in current earthquake prediction research involves a fault model theory. The crustal movement that would accompany a model fault, and the pattern of elastic waves that movement would generate, are then calculated. This is another method that was not mentioned in *Seismology in Japan*. This method, like plate tectonics theory, does not predict the timing of earthquakes; rather it is one of the theoretical underpinnings of earthquake prediction research. It is particularly helpful in postulating a fault model with which the results of observation can be compared, and in estimating the damage that would be caused by imminent earthquakes.

Some of the research mentioned above can be undertaken without a special budget while some can be financed by money not allocated specifically for earthquake prediction. There must be a change in the attitude that earthquake prediction is a budget-buster, or that money should only be spent on projects that relate directly to earthquake prediction. Prediction science is meaningful only when it stands on a foundation of balanced development in various other related fields.

The Blueprint pointed out: "A number of scientists maintain that some crustal deformation may be detectable before a large earthquake. This supposition is supported by only a few examples, however." The examples given were the Ajigasawa earthquake of 1793 and the Hamada earthquake of 1872. The Sekihara earthquake of 1927, the Futatsui earthquake of 1955, and the Nagaoka earthquake of 1961 were also listed as examples of pronounced pre-earthquake crustal movement detected by observation and leveling. In addition, the changes discovered in the Mitaka rhomboid (base line) prior to the Kanto earthquake of 1923 were mentioned. Since the beginning of the Earthquake Prediction Project, however, the number of examples of precursory phenomena have increased rapidly, thanks to the accumulation of new data and the re-examination of old data.

An outline of the developments based on this new information follows. The re-examination of the results of leveling near Kakegawa in Shizuoka Prefecture led to the discovery of precursory crustal movement prior to the Tonankai earthquake of 1944 (detailed in Chapter 11).

Slow changes in the leveling data which had been collected since 1900 in the coastal area that formed the epicenter of the Niigata earthquake were found to agree with the precursory change process as explained by the dilatancy diffusion model.

Active attempts are being made to improve the instruments used for the continuous observation of crustal movement, and to introduce and explore the possibilities of new instruments. A geodimeter has made it possible to measure distances of up to 40 to 50 km with an accuracy of 10^{-6} with ease. This prompted a change in emphasis from triangulation to laterallation surveys.

Test observations are being performed using bore-hole type tiltmeters instead of water tube tiltmeters which require a larger site for observation activities. The Sachs-Evertson volume strainmeters stationed along the coast of the Tokai and Kanto districts have been making great contributions in this area.

Examples of precursory phenomena continue to accumulate. Tilting was observed shortly before the Matsushiro earthquake. About a year before the Central Gifu earthquake of 1969 and the Atsumi Peninsula earthquake of 1971, tilting began in Inuyama and Kamitakara. These are but a few of the increasing number of examples of intermediate, short, and extra-short-term precursory phenomena. There is cause for optimism in the fact that some kind of precursory crustal strain can always be observed prior to great earthquakes.

The Blueprint maintains that "the main objective is to obtain information regarding the earthquake process by clarifying the relationship between strain accumulation in the crust and earthquake occurrences. This can be accomplished by monitoring crustal deformation from time to time. Thus our efforts need not be confined to the detection of direct earthquake precursors." There have already been some achievements made along these lines—studies in the visco-elastic migration phenomenon of crustal movement, for instance, and the elucidation of the distribution of strain step that accompanies earthquakes, as well as the observation of the anomalous movements that occurred extensively in various areas of Japan between 1969 and 1973.

True achievement in observation over the past decade would mean that, in addition to the accumulation of data on precursory phenomena, a firm basis had been established from which to clarify and logically pursue the actual process of crustal movement that accompanies earthquakes.

The key to earthquake prediction is generally thought to lie in the observation of earthquakes. But it is difficult to look for precursory phenomena in earthquake observation. When the Blueprint was written, microearthquake observation was carried out only intermittently. Now, however, microearthquakes are observed on a daily basis. This has made it possible to construct a rough picture of microearthquake activity throughout Japan. As a result, many basic facts have come to light: earthquakes in the Kanto and Tohoku districts take place in two planes along the Wadati zone; microearthquakes in inland Japan occur only in the layer with a P wave velocity of 6 km/sec; and earthquakes are concentrated near faults.

Since microearthquake observation began, seven swarm microearthquakes have occurred in the Kanto district. Only one of these developed into an intermediate-size main earthquake. At the time of the Central Gifu earthquake only one foreshock (M 0.5) was observed. There was also only one foreshock (M 1.0) registered prior to the earthquake near Hiroo on January 21, 1970. Suyehiro discovered that the b value of foreshocks is smaller than that of aftershocks [Suyehiro, 1966]. Since that time, however, some foreshocks have been observed in which the b value is not significantly smaller. It is obvious from this example that even if we try prediction by foreshocks, there is so much variation that the difficulty of accurately predicting main earthquakes is very great.

Since 1970, seismic gaps have been more widely studied abroad than in Japan. In 1972 Utsu pointed out the existence of a seismic gap off the coast of the Nemuro Peninsula. The area was struck by an M 7.4 earthquake on June 17, 1973. Since then the importance of examining seismic gaps has been increasingly recognized. Basic research in this area is still in its infancy, however. When a seismic gap is discovered, it is necessary to determine whether it is a region where aftershocks have ceased and an earthquake is now imminent, whether it is an innately aseismic area, or whether it is an area in which strain has been released by aseismic crustal movement. In some cases this determination can be made on the basis of the area's geological properties or on its record of ancient earthquakes. In other cases such a determination is extremely difficult.

Many cases in which the V_P/V_S ratio decreased before an earthquake were reported in the USSR in 1962. In the 1970s this topic has received a great deal of attention. Ever since a drop in the V_P/V_S ratio was discovered at the time of the Nagano earthquake of 1968, many cases of V_P/V_S decrease have also been reported in Japan (see Chapter 5). Variations in the V_P/V_S ratio are thought to be due primarily to changes in the V_P. Since 1968, annual explosion seismology experiments have been undertaken at Izu Oshima to study velocity changes in P waves as they are transmitted from the Izu area to the Kanto district. Nothing noteworthy has been found so far, however.

Efforts are being made to detect the precursory activities that precede great, intermediate, and minor earthquakes, and further efforts are being proposed. Such proposals should be examined carefully. There has been a remarkable increase in observational data and knowledge. The time has come to utilize these data and eventually arrive at an effective earthquake prediction method. Since there are innumerable microearthquakes, microearthquakes are being examined in the hope that data can be gathered quickly that will elucidate the principles of seismic activity. There is a contradiction here, however, since a great earthquake must be experienced before the relationship between microearthquakes and great earthquakes can be clarified.

Rapid progress has been made in the investigation of active faults. The distribution of active faults throughout Japan, with the exception of Hokkaido, has been outlined, and they have been ranked in three categories, A, B, and C, depending on their average displacement velocity. Where there are many active faults, it becomes important to be able to determine which one is going to cause an earthquake next. Otherwise there is not much practical meaning in the study of fault distribution. Thus the direction of today's research is to clarify the degrees of danger presented by these faults.

Although geomagnetism and the earth's current have long been under observation, their relationship to earthquakes has remained rather vague. On the other hand, variations in the earth's specific resistivity seem quite promising where earthquake prediction is concerned. Observation in this area began at the same time as the Earthquake Prediction Project, but it did not occur within the framework of the Project. Now the number of cases in which precursory variation in the earth's specific resistivity has been observed before an earthquake has climbed to 20 out of 29 cases. At this point the record should be examined in detail to find out whether or not this particular phenomenon can be utilized in earthquake prediction.

Research on the physical properties of rocks has also progressed. Basic facts have been discovered concerning the nature of foreshocks, variations in the value of b both before and after a main earthquake, the temporal and spatial vicissitudes of foreshock activity, variations in electric resistance before rupture, the influence of water on rupture, and so forth.

Compared to the time the Blueprint was issued, the examples of earthquake precursors have multiplied tenfold, and a remarkable amount of knowledge on the fundamental nature of earthquakes has been accumulated. Based on this accumulation of data and knowledge, it seems that the point has been reached where the search for the substance of earthquake phenomena can be undertaken so as to find effective means of prediction, at the same time that the relationship between precursors and earthquakes can be pursued from a phenomenological point of view. The accumulated data and knowledge must be sorted out and organized as soon as possible so that the Earthquake Assessment Committee for the Tokai Area can function effectively.

12.4 The Rise of Amateurism

Amateurs in the field of seismology have focused on earthquake prediction from the very beginning. The oldest example in Japan is probably that of the Mukuhira Rainbow, which has been reported to appear before earthquakes since the Taisho Era. Another example is the use of the astatic magnetometer for observations. Shirashoji has been observing geomagnetism in the Kansai district for a long time, while in Kanto Ozawa

THE DEVELOPMENT OF THE EARTHQUAKE PREDICTION PROJECT   295

has been observing the groundwater level for 25 years. Stimulated by the success of earthquake prediction in China and by the participation of the Chinese public in prediction activities, several amateur groups have recently formed—the Namazu (Mudfish) Group, the Earthquake Prediction Research Group, the Japan Earthquake Prediction Club, etc. These groups exchange information and observation data and make earthquake prediction attempts. Some of these groups call me whenever there are precursors; others send me their newsletters. The city of Tokyo has also begun to observe and investigate abnormal animal behavior at zoos and to study mudfish* in earnest.

Amateurs are interested in geomagnetism, the earth's current, rainbows, clouds, groundwater levels, abnormal behavior of animals and movements of celestial bodies, statistics, and many other subjects. Often the amateur seems to specialize in one subject.

The characteristics of amateur observations can be summarized as follows:

(1) The accuracy of the observations leaves something to be desired. In some cases the subject has only been observed by one particular person. In other words, the observations often lack universal applicability.
(2) The statistical processing of observation results is often inadequate.
(3) Rivalry over who can make the first prediction may get in the way of cooperative efforts.

Since the accuracy of observation can be partially improved by increasing the number of observation sites, (2) would seem to present the more significant problem. This problem will be examined in terms of the data collected by Ozawa, who makes predictions based on his long-range observations.

Ozawa measures the groundwater level, the atmospheric pressure, and the underground temperature. With some assistants who have recently joined him, he studies anomalies in the water level variations at 16 sites as shown in Fig.12.2. The groundwater level changes according to the amount of rainfall, the atmospheric pressure, the amount of water drawn to the surface, etc. Ozawa has determined a coefficient for each site, and he multiplies the amount of rainfall by the coefficient. If the groundwater level is higher than the figure obtained by this calculation, after eliminating obvious changes caused by atmospheric pressure, Ozawa labels it anomalous. He claims that 23 ± 3 days after such an anomaly appears, an earthquake takes place. Recently he also discovered an anomalous drop in the ground-

---

* *Translator's note:* In Japanese folk tradition, mudfish are said to cause earthquakes by moving their whiskers.

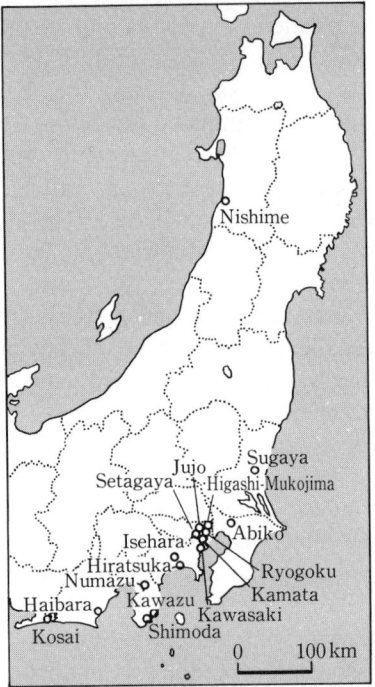

Fig. 12.2 Groundwater observation sites in Northern Japan.

water level 1 to 2 days prior to an earthquake. He maintains that the combination of these two factors makes for effective earthquake prediction.

Let us focus our attention on Ozawa's data from August 1976 to March 1977. Since there are fewer anomalous drops than anomalous increases, only the latter will be considered. The horizontal bars in Fig. 12.3 denote anomalous increases. Although Ozawa has two sites in Numazu, one in the north and one in the south, in this figure they are incorporated into one. In Abiko his observations were made in two wells—one of which had a greater diameter than the others. These observations have also been incorporated into one in the figure. As for anomalous increases in groundwater level, it was decided to look at the relationship of a group of such incidents to earthquake events, rather than attempting to directly connect an individual incident to a particular earthquake. Thus an anomalous rise occurring in more than 5 locations in a three-day period was regarded as meaningful. These cases are marked in Fig. 12.2. Open circles ○ denote the earthquakes adopted by Ozawa. How he chose these particular earthquakes is not clear. They were probably taken from newspaper articles. The mag-

THE DEVELOPMENT OF THE EARTHQUAKE PREDICTION PROJECT 297

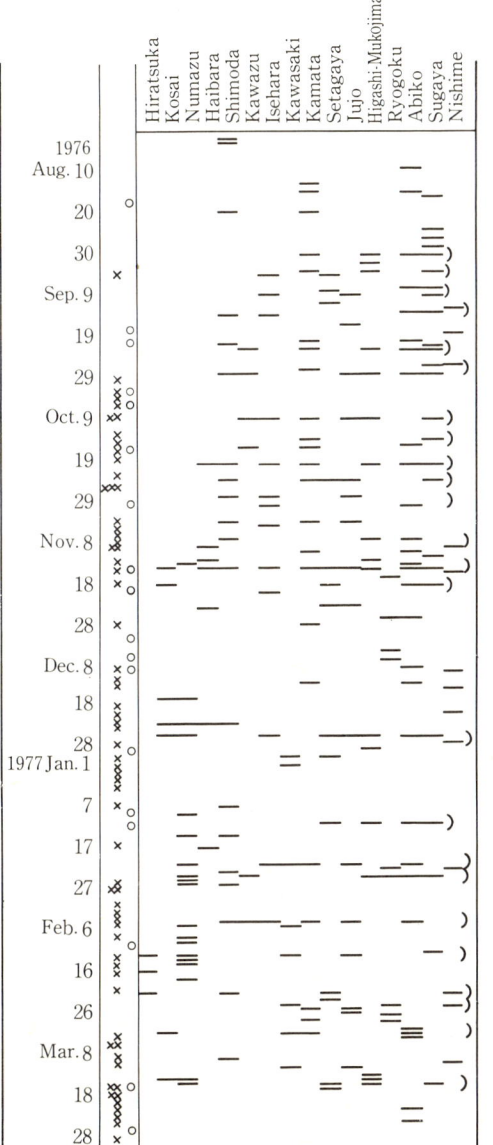

Fig. 12.3 Anomalous groundwater level and earthquakes.

nitude of most of the earthquakes marked with open circles are recorded as $M \geq 3.5$. Comparing the earthquakes with the precursors, there are 14 incidents in which they correspond, 9 incidents of precursors that were not followed by earthquakes, and 3 earthquakes without precursors. If these are, indeed, the facts, they are surprisingly accurate predictions of natural phenomena. Supposing that we hypothesize that an earthquake occurs 9 to 11 days after the precursory phenomena appear: in this case there would be 6 incidents in which the earthquake corresponds with the precursor, 17 precursory incidents without earthquakes, and 10 earthquakes without precursors. Thus Ozawa's method using 23 ± 3 days works better than using 10 ± 1 day. Since no other tests have been done on this subject, however, it is not really clear that 23 ± 3 days is best.

The problem lies in the choice of earthquakes. If we choose from earthquakes in *The Seismological Bulletin* of the Japan Meteorological Agency, only felt ones with their foci in the Kanto, Chubu, and Tohoku districts whose positions may affect the groundwater level, and add them to Fig. 12.2, we get the results marked ×. Earthquakes in Wakayama and off the east coast of Aomori Prefecture are excluded. The magnitude of some of the earthquakes marked × is undetermined while some exceeded $M \doteqdot 3.5$. Taking these earthquakes into consideration, the relationship between earthquakes ×, those marked with open circles, and precursors becomes extremely confusing. In short, in order for Ozawa's method to win the recognition of the experts, he must clarify which earthquakes he proposes to predict. Generally speaking, the greater or closer the earthquakes are, the more readily the precursors appear. Therefore, when precursors appear it is necessary to make it clear that they relate only to earthquakes that are within a certain distance from the observation points and that are greater than a certain magnitude. Otherwise, in view of the innumerable felt earthquakes in the Kanto district alone, it is unclear as to which earthquakes have been predicted from observed precursors. These are some of the issues that must be clarified if this method is to graduate from the ranks of the amateurs.

The importance of statistical analysis was mentioned earlier. It should not be too difficult to improve. The first step in statistics is to treat all data equally, whether or not they fit one's theory. Recent bulletins published by amateur groups have begun to include figures concerning earthquake predictions made and actual earthquakes occurring. This is a desirable trend indeed.

Another way to improve statistical accuracy is to systematically establish as many observation sites as possible. The Chinese discovered, using this strategy, that precursory phenomena tend to concentrate more and more around the earthquake foci as the earthquake approaches.

The relationship between earthquakes and anomalies in animal behavior, plants, and groundwater has not yet been defined either physically or chemically. It is true, however, that various anomalies have been reported before earthquakes since ancient times. At present, however, these anomalies are only meaningful as supplements to geophysical observations, such as these of the earth's geomagnetism, current, crustal movement, etc. The Chinese success in earthquake prediction was the result of diligent geophysical observation. Reports of anomalous phenomena by the public did more than just supplement these observations, particularly just before the event. If such observations can be systematically planned and carried out in Japan, and plenty of data accumulated, they may provide a clue that will help clarify the earthquake mechanism.

In any case, a comprehensive analysis of the accumulated data from systematic observations should be made, since the observation of any one phenomenon tends to lack accuracy. An office that can solicit and receive reports of anomalous phenomena is needed, as is an organization to process and analyze the data in response to the public effort. Fortunately, the JMA began in 1978 to receive and record reports of anomalous phenomena. Even though the analysis of data takes time, it is desirable to accumulate and preserve this information.

12.5 Reflections and Suggestions

The earthquake prediction program began as a research program. The word "research" was deleted from the title at the time the Second Five-Year Project began, and since the Tokai Coast Earthquake Scare of 1976 it has begun to take on the aspect of a routine operation. And yet the inclusion of operational elements did not begin as a result of progress in research. Rather, they became necessary because great earthquakes invariably involve the public. Therefore the Earthquake Assessment Committee for the Tokai Area cannot be expected to make prediction completely successful.

People have become increasingly aware of the need to include social elements in the earthquake prediction process. Admittedly, at the beginning such research was almost nonexistent. The earthquake prediction project began as a scientific investigation. In short, the intention of the project was to advance the earth sciences. The element of disaster prevention—which anyone would think of—was not included. Needless to say, the relationship between an earthquake forecast and its socioeconomic consequences are extremely important for disaster prevention. In the U.S.A., this aspect of earthquake prediction was taken up immediately. A new organization is needed in Japan to respond to this need. Research in this field needs to begin immediately, but existing earthquake prediction groups have no expertise in this field. Since some groups are starting to examine the social consequences of earthquake prediction, organized research within the context of the Earthquake Prediction Project, and open communication among these groups, should begin as soon as possible.

Earthquake prediction research did not have a glamorous beginning. It became established with the help of the Matsushiro swarm earthquakes and the Tokachi offshore earthquake which occurred early in its history. The foundation of geophysical research rests entirely on long-term observations of excellent quality. Moreover, such observations have little or no meaning unless they continue for more than several decades. In order to maintain these activities, it is vital to have a fund that can meet the cost increases that are bound to occur during such a lengthy period. Adequate observation activities cannot be maintained within the present government budget system. For this reason, appropriate guiding principles in regard to funding were developed and applied as needed. The Earthquake Prediction Project has progressed smoothly so far, thanks to these principles, but at this point it is important to pause and take stock. Has something critical been overlooked in the attempt to obtain a budget for the Project? At the 10-year mark since the publication of the Blueprint, has the achievement of the past been examined carefully with a view to setting new goals? Earthquake prediction is primarily for the sake of the public. Has this fact always been taken into consideration when making plans?

It has become increasingly apparent that if one's project has to do directly with earthquake prediction or if it is officially included in the Earthquake Prediction Project, money is easily attainable. Does not such a practice lead everyone to apply in the same area, thus promoting projects that lack originality? It may be academically desirable for many to study the same subject, thus competing with one another, but earthquake prediction cannot be achieved solely through the observation of tilt or any other single area of research. Various and sundry observations are necessary, and the results of these observations must be comprehensively analyzed if they are to be used effectively for earthquake prediction. It may be wise to make similar observations throughout Japan from the viewpoint of a routine operation; and yet, in terms of research, other types of observation should also be encouraged. Thorough discussions are needed between researchers and relevant organizations if research responsibilities are to be delegated so that the various aspects of earthquake prediction can be examined in greater depth. Unfortunately, the present trend is in the opposite direction. I hope that this is not due to scholars whose immediate concern is the funding of their own projects, nor to a lack of new ideas among the experts.

Long-term observation involves routine work which is often thought incompatible with research. Since geophysics cannot exist without such observations, however, cooperation between these two aspects is essential.

A researcher may find a clue that will aid him in his work while making routine observations. Truly original research can be done based on routine observations. In fact, one of the abilities that is required of a researcher is the ability to find research topics that are both applicable to his current environment and related to earthquake prediction.

Academic work and choices of research subjects should be free of restrictions under normal circumstances. Given the present deep involvement of earthquake prediction with disaster prevention, however, it may be necessary to require that certain investigative research—especially that which would utilize existing data—be undertaken. This duty may fall to those who are already involved in earthquake prediction. The comprehensive examination of seismic gaps mentioned in Chapter 2, and a nationwide investigation of active faults and research on their activity, would meet this criterion.

Earthquake prediction should exist solely for the benefit of the people. Unfortunately, the people have either been left in ignorance or frightened by sensational media reports in the past. One of the significant tasks of earthquake prediction research will be to increase people's understanding of the earthquake prediction process (especially as it relates to the idea of probability) and to inform them of the present state of the art. Ways must also be found for the public to participate in prediction and observation activities. So far journalism has been the primary source of earthquake information. Until about a year ago, this information was mostly of a sensational nature. Recently, however, journalists have improved their understanding both of earthquakes and of the profound influence of the media on the public. As a result, sensational reporting has virtually disappeared and has been replaced by correct information. Unfortunately, erroneous reporting in Shizuoka caused an aftershock scare in the wake of the Izu Oshima offshore earthquake of January 14, 1978, in spite of these favorable developments. Experts, too, should devise a way to express their opinions that can be understood correctly by the public, rather than single-mindedly proclaiming their theories to reporters.

Earthquake prediction cannot survive as a commercial enterprise. Even if new instruments are invented, the demand for them will be minimal. Furthermore, the knowledge of industrialists at present is no match for the accumulated knowledge of scientists. There are a few cases in which instruments developed with government grants were inoperable, or were unable to provide reliable data. Earthquake prediction cannot be achieved just by infusions of money, either. People are the assets in this research, not money. There are only several hundred experts who are more or less involved in earthquake prediction. This is quite a different situation from the JMA's weather forecast staff of 6,500. Moreover, experts cannot be trained quickly. A 10- to 20-year project is needed that will create positions and train the people to fill them.

Earthquake prediction is now regarded as a social problem and thus attracts a lot of attention. Thus everybody wants to get on the bandwagon. Earthquake prediction, however, is a serious, low-key research field requiring the long-term accumulation and analysis of facts. It is dangerous to forget that some very down-to-earth research is involved. If one person predicts an earthquake and another one says it won't happen, even if the earthquake occurs the first person is not necessarily the better scientist. At present neither person has enough theoretical knowledge nor enough data to convince the other with his prediction. I hope that reliable earthquake prediction will materialize in the near future and that I will have made at least a small contribution to it. It seems, however, that most scientists are vying as to who can contribute the most by "outguessing" the others as to when the earthquake event will occur.

We are beginning to formulate a way to incorporate amateur observations into earthquake prediction research. A few accurate observations and a greater number of less accurate observations tend to balance one another in geophysical observation and, in fact, work rather effectively together. Three conditions must be met, however, if amateur observations are to make an effective contribution: first, sufficiently accurate and inexpensive instruments must be developed; second, observation tasks must be entrusted to outside observers; and third, an organization must be established to gather reports from these observers. Efforts should be made to achieve this goal. Such efforts will help to develop people's understanding of earthquake and their interest in disaster prevention, which in turn will prove to be an effective tool in minimizing earthquake destruction in the future.

References

Honda, H., 1933: Notes on the mechanism of deep earthquakes, *Geophys. Mag.*, **7**, 257–267.
Matuzawa, T., 1928a: Observation of some of recent earthquakes and their time-distance curves (part 1), *Bull. Earthq. Res. Inst.*, **5**, 1–28.
———, et al., 1928b: On the forerunners of earthquake-motions of certain earthquakes, *Bull. Earthq. Res. Inst.*, **4**, 85–106.
———, 1929a: Observation of some of recent earthquakes and their time-distance curves, *Bull. Earthq. Res. Inst.*, **6**, 177–229.
——— et al., 1929b: On the forerunners of earthquake-motions (II), *Bull. Earthq. Res. Inst.*, **7**, 241–.
Rikitake, T., 1975: Earthquake precursors, *Bull. Seismol. Soc. Amer.*, **65**, 1133–1162.
Scholz, Ch H., L. R. Sykes and Y. P. Aggarwal, 1973: Earthquake prediction—A physical basis, *Science*, **181**, 803–809.

Suyehiro, S., 1966: Difference between aftershocks and foreshocks in the relationship of magnitude to frequency of occurrence for the great Chilean earthquake of 1960, *Bull. Seismol. Soc. Amer.*, **56**, 185–200.

Tsuboi, C., 1933: Notes on the mechanical strength of the earth's crust, *Bull. Earthq. Res. Inst.*, **11**, 275–277.

Tsubokawa, I., 1973: On relation between duration of precursory geophysical phenomena and duration of crustal movement before earthquakes, *J. Geodetic Soc. Japan*, **19**, 116–119.

Wadati, K., 1928: Shallow and deep earthquakes, *Geophys. Mag.*, **1**, 162–202.

INDEX OF EARTHQUAKES

(1) The earthquakes with established names are listed according to the names by which they are best known. The rest are listed according to their epicentral areas. (The locations given in parentheses can be found in the reference map in the end papers.)

(2) There may be some discrepancies in M (magnitude) depending on the method of calculation. Estimates differ for ancient earthquakes, for example, and different countries may use measurement scales that differ slightly.

A. Earthquakes in Japan and Environs (Before 1872)

? 679	Tsukushi (Kyushu district): M 6.7
Nov. 29, 684	Saikaido, Nankaido, Tokaido (Off Kyushu, Shikoku, Tokai districts): M 8.4
May 12, 701	Tango (Northern Kyoto Prefecture): M 7.0
June 5, 745	Mino (Gifu Prefecture): M 7.9
? 818	Kanto district: M 7.9
? 841	Izu (Peninsula): M 7.0
July 10, 863	Etchu, Echigo (Toyama, Niigata Prefectures): M 7.0
July 13, 869	Sanriku coast (Off Iwate and Miyagi Prefectures): M 8.6
Aug. 26, 887	Goki-Shichido (Off Central and Southwest Honshu): M 8.6
Aug. 26, 887	Northern Shinano (Nagano Prefecture): M 7.4
Dec. 17, 1096	Kinai, Tokaido (Central Honshu): M 8.4
Feb. 22, 1099	Nankaido (Off Shikoku district): M 8.0
Oct. 9, 1257	Southern Kanto district : M 7.0
Aug. 3, 1361	Kinai, Tosa, Awa (Off Kinki district and Shikoku): M 8.4
Jan. 21, 1408	Kii, Ise (Wakayama, Mie Prefectures): M 7.0
Sept. 20, 1498	Throughout Tokaido (Off Tokai, Kanto districts): M 8.6
Apr. 4, 1520	Kii (Peninsula): M 7.0
Sept. 4, 1596	Bungo (Oita Prefecture): M 6.9
Sept. 5, 1596	Kinai (Kyoto and its vicinity): M 7.0
Feb. 3, 1605	Saikaido, Nankaido, Tokaido (Off Kyushu, Shikoku, Tokai districts): M 7.9
Dec. 2, 1611	Sanriku coast (Off Iwate and Miyagi Prefectures): M 8.1
Nov. 26, 1614	Echigo Takada (Niigata Prefecture): M 7.7
Oct. 18, 1644	Ugo Honjo (Akita Prefecture): M 6.9
Oct. 31, 1662	Hyuga, Osumi (Miyazaki, Kagoshima Prefectures): M 7.6
Apr. 13, 1677	Rikuchu (Tohoku district): M 8.1
Dec. 29, 1685	Iyo (Ehime Prefecture): M 5.9
Jan. 4, 1686	Aki, Iyo (Hiroshima, Ehime Prefectures): M 7.0
June 19, 1694	Noshiro (Akita Prefecture): M 7.0
Dec. 12, 1694	Tango (Northern Kyoto Prefecture): M 6.1
Dec. 31, 1703	Genroku earthquake (Off Kanto district): M 8.2

Oct. 28, 1707   Hoei earthquake (Off Kanto, Tokai, Shikoku districts): M 8.4
Oct. 29, 1707   Aftershock of Hoei earthquake (Yamanashi Prefecture)
May 13, 1717   Hanamaki (Iwate Prefecture): M 7.6
Dec. 18, 1723   Chikugo (Western Fukuoka Prefecture): M 6.2
Oct. 7, 1731   Iwashiro (Fukushima Prefecture): M 6.6
May 20, 1751   Echigo, Etchu (Niigata, Toyama Prefectures): M 6.6
Oct. 31, 1762   Sado Island (Niigata Prefecture): M 6.6
Mar. 8, 1766   Tsugaru (Western Aomori Prefecture): M 6.9
Aug. 29, 1769   Hyuga, Bungo (Miyazaki, Oita Prefectures): $M.$ 7.4
May 21, 1792   Mt. Unzen (Nagasaki Prefecture): M 6.4
Feb. 8, 1793   Ajigasawa earthquake (Western Aomori Prefecture): M 6.9
June 29, 1799   Kaga (Ishikawa Prefecture): M 6.4
Dec. 9, 1802   Sado Island: M 6.6
July 10, 1804   Kisakata earthquake (Yamagata and Akita Prefectures): M 7.1
Sept. 25, 1810   Ugo (Akita Prefecture): M 6.6
Dec. 18, 1828   Echigo (Niigata Prefecture): M 6.9
Aug. 19, 1830   Kinai (Kyoto and its vicinity): M 6.4
Feb. 9, 1834   Ishikari (Hokkaido): M 6.4
Apr. 22, 1841   Suruga (Shizuoka Prefecture): M 6.4
Apr. 25, 1843   Kushiro, Nemuro (Hokkaido): M 8.4
May 8, 1847   Zenkoji earthquake (Northern Nagano Prefecture): M 7.4
May 13, 1847   Echigo-Kubiki (Niigata Prefecture): M 6.5
Dec. 23, 1854   Ansei Tokai earthquake (Off Tokai district): M 8.4
Dec. 24, 1854   Ansei Nankai earthquake (Off Shikoku district): M 8.4
Dec. 26, 1854   Western Iyo (Ehime Prefecture): M 7.0
Nov. 11, 1855   Ansei Edo earthquake (Tokyo and its vicinity): M 6.9
Apr. 9, 1858   Hida, Etchu, Kaga, Echizen (Western Hokuriku district): M 6.9
Apr. 9, 1858   Kaga, Echizen (Ishikawa, Fukui Prefectures): M 6.9

B. Earthquake in Japan and Environs   (After 1872)

Mar. 14, 1872   Hamada earthquake (Shimane Prefecture): M 7.1
July 23, 1886   Shinetsu border (Border of Nagano and Niigata Prefectures): M 6.1
July 28, 1889   Kumamoto (Prefecture): M 6.3
Jan. 7, 1890   Saikawa Basin (Nagano Prefecture): M 6.3
Oct. 28, 1891   Nobi earthquake (Aichi and Gifu Prefectures): M 8.4
Sept. 7, 1893   Chiran (Kagoshima Prefecture): M 6.4
Mar. 22, 1894   Southwestern coast of Nemuro Peninsula (Hokkaido): M 7.9
June 20, 1894   Northern part of Tokyo Bay: M 7.5
Oct. 22, 1894   Shonai earthquake (Yamagata Prefecture): M 7.3
Jan. 18, 1895   Kasumigaura (Ibaragi Prefecture): M 7.2
June 15, 1896   Meiji Sanriku earthquake (Off Iwate and Miyagi Prefectures): M 7.6
Aug. 31, 1896   Rikuu earthquake (Border of Akita and Iwate Prefectures): M 7.5
Jan. 17, 1897   Northern Nagano Prefecture: M 6.3
Feb. 20, 1897   Sendai offshore (Miyagi Prefecture): M 7.8
Apr. 3, 1898   Mishima (Yamaguchi Prefecture): M 6.8
Apr. 23, 1898   Iwate (Prefecture) offshore: M 7.8
May 26, 1898   Near Muikamachi (Niigata Prefecture): M 6.7
Aug. 10, 1898   Near Fukuoka City (Fukuoka Prefecture): M 6.5
Nov. 13, 1898   Central Kiso River Basin (Aichi Prefecture): M 6.5
Mar. 7, 1899   Southeastern Kii Peninsula: M 7.6

INDEX OF EARTHQUAKES 307

May 12, 1900   Northern Miyagi Prefecture: M 7.3
Jan. 30, 1902   Sannohe district (Aomori Prefecture): M 7.4
May 28, 1902   Kushiro offshore (Hokkaido): M 7.4
Aug. 10, 1903   Western part of Mt. Norikura (Gifu Prefecture): M 5.7
May 8, 1904   Near Muikamachi (Niigata Prefecture): M 6.9
June 7, 1905   Sea near Oshima: M 7.0
Apr. 21, 1906   Near Hagiwara (Gifu Prefecture): M 7.1
Aug. 14, 1909   Gono (Anegawa) earthquake (Shiga and Aichi Prefectures): M 6.9
July 24, 1910   Mt. Usu (Hokkaido): M 6.5
Sept. 8, 1910   Onishika (Hokkaido): M 5.6
Jan. 12, 1914   Sakurajima (Kagoshima Prefecture): M 6.1
Mar. 15, 1914   Akita Senpoku earthquake (Akita Prefecture): M 6.4
Mar. 28, 1914   Hiraga-gun (Akita Prefecture): M 5.8
May 23, 1914   Izumo (Shimane Prefecture): M 6.3
July 14, 1915   Kurino, Yoshimatsu (West of Mt. Kirishima, Kagoshima Prefecture)
Nov. 26, 1916   Kobe (Hyogo Prefecture): M 6.3
Nov. 11, 1918   Omachi earthquake (Nagano Prefecture): M 6.1
Nov. 1, 1919   Near Miyoshi (Hiroshima Prefecture): M 5.9
1920~   Near Wakayama (swarm earthquakes)
Dec. 8, 1921   Ryugasaki (Ibaragi Prefecture): M 7.1
Dec. 8, 1922   Chijiwa Bay (Nagasaki Prefecture): M 6.5 and 5.9
Sept. 1, 1923   Great Kanto earthquake: M 7.9
Jan. 15, 1924   Tanzawa Mountain Range (Kanagawa Prefecture): M 7.2
May 23, 1925   Northern Tajima earthquake (Northern Hyogo Prefecture): M 7.0
July 4, 1925   Miho Bay (Shimane Prefecture): M 6.3
Aug. 10, 1925   Hita (Northern Oita Prefecture)
Mar. 7, 1927   Northern Tango earthquake (Northwestern Kyoto Prefecture): M 7.5
Aug. 6, 1927   Off Miyagi Prefecture: M 6.9
Oct. 27, 1927   Sekihara earthquake (Central Niigata Prefecture): M 5.3
Feb. 13~May, 1930   Ito swarm earthquakes (Shizuoka Prefecture)
Nov. 26, 1930   Northern Izu (Peninsula) earthquake: M 7.0
Dec. 20, 1930   Near Miyoshi (Hiroshima Prefecture): M 6.0
Feb. 17, 1931   Near Urakawa (Hokkaido): M 6.8
Mar. 9, 1931   Off southeastern coast of Aomori Prefecture: M 7.6
Sept. 21, 1931   Western Saitama (Prefecture) earthquake: M 7.0
Mar. 3, 1933   Sanriku offshore earthquake (Off Iwate and Aomori Prefectures): M 8.3
Sept. 21, 1933   Noto Peninsula (Ishikawa Prefecture): M 6.0
Mar. 21, 1934   Izu Amagi Mountain (Shizuoka Prefecture): M 5.5
July 11, 1935   Shizuoka earthquake (Shizuoka Prefecture): M 6.3
Feb. 21, 1936   Kawachi-Yamato earthquake (Nara, Osaka Prefectures): M 6.4
Nov. 11, 1936   Near Wakamatsu City (Fukuoka Prefecture, swarm earthquakes)
Nov., 1936   Osarizawa, Hanawa (Akita Prefecture) (swarm earthquakes)
Jan. 12, 1938   Tanabe Bay offshore (Wakayama Prefecture): M 6.7
May 23, 1938   Shioyazaki offshore (Fukushima Prefecture): M 7.1
May 29, 1938   Near Lake Kussharo (Hokkaido): M 6.0
May 1, 1939   Oga Peninsula earthquake (Akita Prefecture): M 7.0
Apr. 6, 1941   Near Susa (Yamaguchi Prefecture): M 6.2
July 15, 1941   Near Nagano City (Nagano Prefecture): M 6.2
Nov. 19, 1941   Hyuga Sea (Miyazaki Prefecture): M 7.4
Mar. 4 and 5, 1943   Tottori (Prefecture) offshore: M 6.1 and M 6.1

308 INDEX OF EARTHQUAKES

Sept. 10, 1943　Tottori (Prefecture) earthquake: M 7.4
Dec. 7, 1944　Tonankai (Off Tokai and Shikoku districts) earthquake: M 8.0
Jan. 13, 1945　Mikawa earthquake (Southern Aichi Prefecture): M 7.1
Dec. 21, 1946　Nankai earthquake (Shikoku and Southern Kii Peninsula offshore): M 8.1
June 15, 1948　Upper Hidaka River (Hokkaido): M 7.0
June 28, 1948　Fukui (Prefecture) earthquake: M 7.3
July 19, 1948　Aftershock of Fukui earthquake
Dec. 26, 1949　Imaichi earthquake (Tochigi Prefecture): M 6.4 and 6.7
Aug. 22, 1950　Near Mt. Sambe (Shimane Prefecture): M 5.3
Mar. 4, 1952　Tokachi offshore earthquake (Hokkaido): M 8.1
Mar. 7, 1952　Daishoji offshore earthquake (Fukui Prefecture): M 6.8
July 18, 1952　Yoshino earthquake (Nara Prefecture and its vicinity): M 7.0
Oct. 19, 1955　Futatsui earthquake (Akita Prefecture): M 5.7
Sept. 30, 1956　Southern Miyagi Prefecture: M 6.1
Jan. 31, 1959　Teshikaga (Hokkaido): M 6.2 and 6.1
Dec. 26, 1960　Odaigahara earthquake (Nara and Mie Prefectures): M 6.0
Feb. 2, 1961　Near Nagaoka (Niigata Prefecture): M 5.2
Feb. 27, 1961　Hyuga Sea (Miyazaki Prefecture): M 7.0
Aug. 19, 1961　Northern Mino earthquake (Gifu Prefecture): M 7.0
Jan. 4, 1962　Kii Peninsula earthquake (West coast of Wakayama Prefecture): M 6.4
Apr. 30, 1962　Northern Miyagi (Prefecture) earthquake: M 6.5
May 7, 1964　Oga Peninsula offshore (Akita Prefecture): M 6.9
June 16, 1964　Niigata earthquake (Near Awashima Island, Niigata Prefecture offshore): M 7.5
Aug. 3, 1965～　Matsushiro swarm earthquakes (Nagano Prefecture)
Nov. 23, 1965　One of the above earthquakes: M 5.0
Sept. 14, 1967　One of the above earthquakes: M 5.1
Feb. 21, 1968　Ebino earthquake (Miyazaki Prefecture): M 6.1
May 16, 1968　Tokachi offshore earthquake (Hokkaido): M 7.9
Sept. 21, 1968　Northern Nagano Prefecture: M 5.3
Sept. 9, 1969　Central Gifu Prefecture: M 6.6
Jan. 21, 1970　Southern Hokkaido: M 6.7
Oct. 16, 1970　Southeastern Akita Prefecture: M 6.2
Jan. 5, 1971　Atsumi Peninsula offshore (Aichi Prefecture): M 6.1
June 17, 1973　Nemuro Peninsula offshore earthquake (Hokkaido): M 7.4
Nov. 25, 1973　Central Wakayama Prefecture: M 5.9
May 9, 1974　Izu Peninsula offshore earthquake (Shizuoka Prefecture): M 6.9
June 29, 1974　Deep-focus earthquake in western part of Japan Sea: M 7.7
Apr. 21, 1975　Western Oita Prefecture: M 6.4
Nov. 21, 1975　Southern coast of Aichi Prefecture: M 4.3.
Dec. 1975　Microearthquake at Yamasaki Fault (Hyogo Prefecture)
June 16, 1976　Eastern Yamanashi Prefecture: M 5.5
July 26, 1976　Ise Bay (Mie Prefecture): M 4.1
Aug. 18, 1976　Kawazu (Izu Peninsula): M 5.4
Dec. 27, 1976　Near Hamamatsu (Shizuoka Prefecture): M 3.5
Aug. 7, 1977　Near Wakayama City (Wakayama Prefecture): M 4.7
Sept. 30, 1977　Yamasaki Fault (Hyogo Prefecture): M 4
Nov. 6, 1977　Near Nebukawa (Kanagawa Prefecture): M 3.9
Jan. 14, 1978　Izu Oshima offshore earthquake: M 7.0

June 12, 1978   Miyagi (Prefecture) offshore earthquake: M 7.4

C. Earthquakes in Other Parts of the World

Apr. 18, 1906   San Francisco earthquake, U.S.A.: M 8.3
May 22, 1960   Chile earthquake: M 8.3
Mar. 28, 1964   Alaska earthquake, U.S.A.: M 8.4
Mar. 8, 1966   Yingtai, Hebei Province, China: M 6.8
Mar. 22, 1966   Yingtai, Hebei Province, China: M 7.2
Apr. 26, 1966   Tashkent, Uzbek, U.S.S.R.: M 5.5
June 28, 1966   Parkfield, California, U.S.A.: M 6.5
Aug. 30, 1967   Luhuo, Sichuan Province, China: M 6.8
Apr. 4, 1968   Borrego Mountain, California, U.S.A.: M 6.5
July 18, 1969   Bohai Bay, China: M 7.4
Jan. 5, 1970   Donghai, Yunnan Province, China: M 7.7
Jan. 10, 1971   Western New Guinea: M 8.1
Feb. 9, 1971   San Fernando earthquake, California, U.S.A.: M 6.4
Feb. 1971   Tashkent, Uzbek, U.S.S.R.: M 4.8
July 20, 1972   Sitka Island offshore, Alaska, U.S.A.: M 7.1
Feb. 6, 1973   Luhuo, Sichuan Province, China: M 7.9
Feb. 21, 1973   Point Mugu, California, U.S.A.: M 5.2
June 22, 1973   Near Hollister, California U.S.A.: M 3.9
May 11, 1974   Yongshan–Daguan, Yunnan Province, China: M 7.1
Nov. 28, 1974   Near Hollister, California, U.S.A.: M 5.2
Dec. 22, 1974   Liaoning Province, China: M 4.8
Feb. 4, 1975   Haicheng earthquake, Liaoning Province, China: M 7.3
Dec. 9, 1975   San Andreas Fault: M 2.4
May 29, 1976   Longling, Yunnan Province, China: M 7.5 and 7.6
July 28, 1976   Tangshan, Hebei Province, China: M 7.8
Aug. 16, 1976   Songpan–Pingwu, Sichuan Province, China: M 7.2
Aug. 23, 1976   Songpan–Pingwu, Sichuan Province, China: M 7.2
Nov. 7, 1976   Yunnan–Sichuan Border, China: M 6.9
Dec. 13, 1976   Yunnan–Sichuan Border, China: M 6.8
Jan. 10, 1977   New Guinea: M 8.0
Mar. 4, 1977   Vrancea, Rumania: M 7.2
Aug. 19, 1977   Sumbawa Island: M 7.9

GENERAL INDEX

Note: Unless otherwise indicated, research groups, committees, and other organizations are located in Japan.

A_2-type precursors, 254
active faults, 8, 50–58; and geochemical research, 214–15; and small earthquakes, 73; and surface deformation, 35; designation of in California, 51; location of in Japan, 52–54; observation of, 70, 72–73, 293–94, 301; systems of, 40–42
aftershock activity, 20, 67, 73, 82; off Nankai and Tokai coasts, 74
Aggarwal, Y. P., et al., seismic wave velocity research, 90, 93–94
Aichi Southern Coast earthquake (1975), anomalous crustal movement before, 138, 172
Aki, Keiichi, 5
Akita Senpoku earthquake (1914), aftershock of, 20
Alaska earthquake (1964): groundwater changes connected with, 176; terrace formation during, 50: uplift during, 43
amateurs, seismological research by, 294–95
amplitude (A): defined, 63; relation of to magnitude, 64–65
Anatolia Fault System (Turkey), 55
ancient earthquakes. See historical earthquakes
animal behavior, as precursor, 217, 243, 295, 298
Ansei Edo earthquake (1855), 13, 26
Ansei Tokai and Nankai earthquakes (1854), 25, 121: fault model for, 267;
focal zone of, 272
Asada, Toshi, 65; evaluation standard for seismic activity, 82
Atsumi Peninsula offshore earthquake (1971), anomalous crustal movement before, 138, 171, 292

bench marks. See leveling survey
"Blueprint" of the Earthquake Prediction Project Research Group (1962), 104, 135, 283–86, 289, 292
Boso Peninsula: seismic gap south of, 67; study of by Imamura, 22; terrace formation on, 36, 49; uplift and subsidence of, 41, 43–44, 115, 117–18, 121

California: active faults in, 51; shallow earthquakes in, 83. See also San Andreas Fault
Central Gifu earthquake (1969), anomalous crustal movement before, 138–39, 170, 292
Central Wakayama earthquake (1973), anomalous crustal movement before, 138, 171–72
Chile earthquake (1960), 43
China: earthquake prediction in, 6, 210–13, 255–59, 299; electromagnetic research in, 217, 233, 243, 280; explosion seismology in, 207; geomagnetic research in, 227–28. See also Haicheng earthquake; Tangshan earthquake
chronological gradation, 37

continental plate, movement of, 79–81
converters, detection of vertical movement using, 144–45, 147
Coordinating Committee for Earthquake Prediction (CCEP), 250, 267–68, 272
cracks, as earthquake scars, 32
creep phenomenon, 151
crust. *See* crustal movement; crustal strain; crustal stress; crustal structure
crustal movement: and new landforms, 32–33; anomalies in, as earthquake precursors, 70, 123–26, 161–74, 250–55, 260; measurement of, 38–39, 139–49, 291–92; migratory, 139; normal, 111–15, 125–26; observation of, 73, 133–35, 276–79; speed of, 8, 43, 140–41. *See also* subsidence; tilting; uplift
crustal strain, 1, 6, 131, 133; and earthquake prediction, 119–22; critical level of, 120; early study of, 4; in Sagami and Suruga Bays, 121
crustal stress, distribution of, 81, 87
crustal structure, and microearthquake activity, 78–81

Daishoji offshore earthquake (1952), anomalous crustal movement before, 138, 166, 279
damage, earthquake-related, 26, 28–29; reliability of reports on, 14
deep-focus earthquakes, 6–7; and the continental plate, 81; distribution of, 66
dilatancy, 131, 239, 241–42
dilatancy diffusion model, 89–92, 99, 123, 134, 228, 250, 253, 261, 290
Disaster Prevention Research Institute (Kyoto University), 78, 84, 89

earthquake. *See* deep-focus earthquakes; great earthquakes; historical earthquakes; inland earthquakes; intermediate-focus earthquakes; interplate earthquakes; microearthquakes; Pacific coast earthquakes; shallow earthquakes; small earthquakes; swarm earthquakes; *and see* names of specific earthquakes
Earthquake Assessment Committee for the Tokai Area, 249, 268, 272, 285, 294, 299
Earthquake Prediction Project, 70, 104, 283–284, 286–87, 291, 294, 300; facilities of, 288. *See also* "Blueprint"
Earthquake Prevention Plan, 222
Earthquake Research Institute (University of Tokyo), 4, 73, 222, 231
Eastern Yamanashi earthquake (1976), anomalous crustal movement before, 138, 157, 173
elastic wave theory, 4, 7, 103
electric potential, earth's, 220, 228–231
electric resistance, variations accompanying earthquakes, 231–35
electrokinetic effect, 239–41
electromagnetic phenomena, 6, 217–46, 280; and earthquake prediction, 242–43; historical observations of, 218–22
Enshu Sea, seismic gap in, 67
epicenters, distribution of, 67–69, 72
Ewing, James, 3
explosion seismology, 8, 65–66, 78, 89, 97–99; and seismic wave velocity, 101; in China, 207
extensometer: horizontal vault type, 142, 153; quartz tube type, 148–49, 155

fault: formation of, 1, 131; length of in relation to magnitude, 33–34, 47. *See also* active faults; fault displacement; fault models; fault plane; seismic faulting; *and see* names of specific faults
fault displacement, 32–34, 108
fault models, 4–7
fault plane, movement along, 259
focus. *See* hypocenter
foreshocks, as precursors, 57–58, 85–86, 119, 260, 273–75
frequency distribution, 64, 67–68
Fukui earthquake (1948): aftershocks following, 65–66; electromagnetic changes connected with, 221; seismic wave velocity variations before, 89
Fushimi earthquake (1596), groundwater changes reported in connection with, 177
Futatsui earthquake (1955), anomalous

GENERAL INDEX   *313*

crustal movement before, 167

Garm district (USSR), seismic wave research in, 89, 93
Genroku earthquake (1703), 22, 55; terrace formation during, 49
geochemical precursors, 6, 9, 211–16. *See also* groundwater level; helium isotope ratio; radon concentration
geodetic survey, 104, 140, 265
geodimeter, 6, 291
Geographic Research Institute, 10; traverse surveys conducted by, 76–77
geomagnetism, study of, 217, 265, 294
gravity surveys, 104
great earthquakes, recurrence of, 3–4, 15
groundwater level, anomalous changes in, 6, 9, 175–216, 220, 279–80, 295–97
Gutenburg-Richter formula, 64–65, 87

Hagiwara, Takahiro, 4
Haicheng earthquake (1975): anomalous uplift before, 123; geomagnetic anomalies before, 228; prediction of, 255–56
Hamada earthquake (1872), anomalous crustal movement before, 135, 140–41, 163
Hamamatsu vicinity earthquake (1976), anomalous crustal movement before, 138, 173
Hayakawa, Masami, 89
helium isotope ratio, study of, 215–16
historical earthquakes, 11–29, 33
historical records: and earthquake recurrence intervals, 48–49; credibility of, 13, 15; types of, 12; uses of in earthquake prediction, 8, 11–16, 21
historical seismology, 11–29
Hoei earthquake (1707), 24–25
Hollister (California) earthquake (1974), precursory geomagnetic anomalies, 225–26, 229
Holocene marine terraces, 49–50
Honda, Hirokichi, 4–5
horizontal vault observation, 136–37, 139–41

hot springs, earthquake-associated changes in, 177–78
hypocenter: and seismic wave velocity measurement, 92, 95–97, 100; determination of, 67, 69, 96–97
Hyuga Sea earthquake (1961), anomalous crustal movement before, 138, 167

Imaichi earthquake (1949), aftershocks, 65
Imamura, Akitsune, 3–4, 6, 8, 14; observation stations built by, 22; prediction of 1946 Nankai earthquake, 21–24; prediction of 1927 Northern Tango earthquake, 128; use of survey data for prediction, 103
inland earthquakes: 7–8, 61, 78; classification of, 24; prediction of, 9, 79
instruments. *See* converters; extensometer; geodimeter; light wave range finder; proton magnetometer; strainmeter; tiltmeter
intermediate-focus earthquakes, 79
International Seismological Centre, 93
interplate earthquakes, Pacific coast, 81, 253; prediction of, 263
Ise Bay earthquake (1976), strain changes before, 159
Ishimoto-Iida formula for frequency, 64–65, 87
Izu earthquake (841), 48
Izu Oshima offshore earthquake (1978), 301; anomalous uplift before, 125, 127; geochemical anomalies before, 279–280; seismic wave velocity and, 102; strain changes before, 159–60, 276; suddenness of, 249
Izu Peninsula offshore earthquake (1974), 56, 118, 125, 176; anomalous crustal movement before, 138, 172; geomagnetic variations related to, 223; triangulation survey before, 120

Japan Meteorological Agency, observation and prediction activities, 4, 6, 10, 19, 64, 66, 90, 93, 141
Japan Sea coast, microearthquake activity along, 74, 77, 79

Kanto earthquake (1923), 55; aftershocks following, 20, 82; anomalous crustal movement before, 138, 163–64 291; as research impetus, 4; long-term precursory crustal movement and, 252; recurrence interval of, 253; revision surveys after, 106; uplift by 42–44, 108, 115, 117

Kawasumi Map of earthquake recurrence cycles, 14, 56

Kii Peninsula: crustal movement in, 23, 115, 117; revision surveys of, 107

Kinai earthquake (1830), luminescence before, 221

landforms: creation of by earthquakes, 32; measuring displacement of, 36

landslides, as earthquake scars, 31

Land Survey Department (Army), 105–106; prewar surveys by, 103

Large-Scale Earthquake Countermeasures Act (1978), 249

leveling surveys, 104, 111, 123–25; establishment of network for, 105–6

Liaoning Province (China), earthquake prediction research in, 256–58

light wave range finder, use of in triangulation surveys, 108, 115, 126

Longling earthquake (1976), geomagnetic anomalies connected with, 228

luminescence, 221, 243

magnetic surveys, 222

magnitude (M): and crustal movement, 33–34; and fault length, 47; classification of earthquakes by, 25, 66; defined, 63; estimation of, 18

magnitude-dependent precursors, 250–55, 264, 290

Matsushiro swarm earthquakes (1965–), 178, 300; anomalous tilting before, 138, 169, 276, 292; fault formation during, 33; geochemical changes and, 176, 215; geomagnetic phenomena related to, 222, 228

Median Tectonic Line, 42, 47; displacement along, 39–40, 43

Meiji era: development of seismology during, 3, 14; geodetic surveys during, 140

Meiji Sanriku earthquake (1896), 27

Microearthquake Research Group (Kyoto University), 67, 70, 78

microearthquakes: as short-term precursor, 9; distribution of, 72–77, 79, 81; explosion seismology research on, 65–66; observation of, 5, 293

Mikawa earthquake (1945): aftershocks following, 67; anomalous crustal movement before, 165; foreshocks connected with, 58

Milne, John, 3, 220

Mitaka rhomboid base lines, movement of, 115–17, 291

Miura Peninsula: subsidence of, 115, 117–18; terrace formation in, 36

Miyagi offshore earthquake (1978), 249; short-term prediction of, 264–65

Moho discontinuity, 78

Muroto Cape, 23–24; and Nankai earthquakes, 117–18, 121; vertical movement of, 122

Musya, Kinkichi, historical materials collected by, 14, 21

Nagano earthquake (1968), seismic wave velocity change at the time of, 293

Nagaoka earthquake (1961), anomalous crustal movement before, 167

Nakano, Hiroshi, 3

Nankai earthquake (1946), 4, 19, 56; aftershock activity following, 67, 74; anomalous crustal movement related to, 43–44, 108–9, 117–19, 121, 165–66; geomagnetic changes associated with, 177, 219–20; long-term precursory crustal movement and, 252; prediction of, 21, 23; revision survey after, 107–8; seismic wave velocity variations before, 89

Nankai earthquakes, 3; prediction of, 8; recurrence of, 16–17, 48; relation to Tokai earthquakes, 17–18, 122

Nankai Trough, seismic activity along, 19, 56, 74

National Research Center for Disaster Prevention, 286

Nebukawa earthquake (1977), strain changes before, 159

Nemuro Peninsula offshore earthquake (1973): aftershock activity following,

67; anomalous crustal movement before, 138, 171, 278; geomagnetic variations and, 223
Neodani Fault, microearthquakes around, 67
New Zealand, 50–51
Niigata earthquake (1964), 15, 55; aftershock activity following, 67; anomalous crustal movement before, 123, 127, 138, 168, 251, 279; damage caused by, 286; geomagnetic variations and, 223; long-term precursory crustal movement and, 252
Nobi earthquake (1891), 48, 106; aftershocks following, 58
Northern Gifu earthquake (1858), aftershock following, 20
Northern Izu earthquake (1930), 48, 56; anomalous crustal movement before, 164; critical strain before, 120; displacement caused by, 111; foreshocks preceding, 58; revision surveys following, 106, 120
Northern Mino earthquake (1961), anomalous crustal movement before, 168, 251–52
Northern Miyagi earthquake (1962), aftershock activity following, 67
Northern Tajima earthquake (1925), 128
Northern Tango earthquake (1927), 120, 128; aftershock activity following, 73, 82, 118; anomalous crustal movement before, 164; displacement caused by, 111; revision surveys following, 106
Noshiro earthquake (1694), anomalous crustal movement before, 162

observation network, 4, 6, 9, 70–71; for seismic wave velocity research, 94; strainmeter, 149–51
observation wells, horizontal vault-type, 154
ocean floor, seismic activity of, 23, 32
ocean plate, movement of, 79–80; seismic wave velocity of, 81
oceanic trench, and seismic activity, 21, 74, 77, 81, 83
Odaigahara earthquake (1960), anomalous crustal movement before, 138, 167
Omi earthquake (1967), anomalous crustal movement before, 169
Omori, Fusakichi, 3, 14
Oshima Island, explosion seismology experiments on, 97–99. See also Izu Oshima earthquakes

P wave velocity, 79–81, 90. See also V_P/V_S ratio
Pacific coast earthquakes, 3, 8–9, 17, 61
Philippine Sea plate, 74
piezo-magnetic effect, 220, 236–39
plate tectonics, 5–7, 18, 80, 260, 289. See also continental plate; ocean plate; Philippine Sea plate
precursors: categorization of, 250; long-range, 61–129; short-term, 131–248. See also animal behavior; crustal movement; foreshocks; geochemical precursors; seismic gaps; seismic wave velocity
precursory period, 61
prediction, two-stage model of, 261–66
proton magnetometer, 218, 222

radiocarbon dating, 37, 40, 49
radon concentration, 6, 9, 85, 178–79, 206–9, 280; in Chinese earthquake prediction, 212–13
recurrence, 43–46, 48, 54–55; calculating probability of, 56–57
Research Group for Explosion Seismology, 65–66, 78
Richter, Charles F., 63–64
Rikitake, Tsuneji, prediction strategy of, 254–55
Rikuu earthquake (1896), 58, 76; fault formation after, 20
rock-crushing experiments, 7, 90–91
rupture, mechanism of, 133–35

S waves, 7, 100–1. See also V_P/V_S ratio
Sado earthquake (1802), anomalous crustal movement before, 163
Sagami Bay, horizontal strain in, 122
Sagami Trough, movement along, 115–16
San Andreas Fault: and seismic waves, 103; magnetic anomalies along,

225, 228–29, 243; small earthquakes along, 73
San Fernando earthquake (1971), anomalous uplift before, 123
San Francisco earthquake (1906), 73, 103
Sanriku earthquke (869), electromagnetic phenomena and, 221
Sassa, Kenzo, seismic wave research of, 89, 164, 167
scars, 31–58
Scholz, C. H., et al., and dilatancy diffusion model, 89–91, 290
seismic energy, 47
seismic faulting, theory of, 259–60
seismic gaps, 8–9, 67, 82, 301; as earthquake precursors, 83–85; in Tohoku district, 76; in Tokai area, 275
seismic intensity, scale of, 19
seismic wave velocity: measurement of, 96, 99–100; research on in U.S.A., 93–94; research on in USSR, 89–90; variation in as precursor, 101–2, 251. See also V_P/V_S ratio
seismic waves, 3, 6; generation of, 103. See also P waves; S waves
seismograph, 63–64, 96, 100; first uses of, 11
Seismological Society of Japan, 289; 1977 meeting of, 89–90, 264
seismology: development of in Meiji era, 3–4; in prewar Japan, 4–5; postwar, 6–8
seismotectonics, 262–63, 265
Sekihara earthquake (1927), anomalous crustal movement before, 123, 164
Sekiya, Kiyokage, 3
shallow earthquakes, 6–7; distribution of, 66
Shida, Toshi, 4
Shinanogawa Fault System, earthquakes along, 55
Sitka Island (Alaska) earthquake (1972), geomagnetic anomalies associated with, 226
slip, relation to magnitude, 33–34
small earthquakes, 63–66; and active faults, 73; and great earthquakes, 82–83
Songpan–Pingwu earthquake (1976), electromagnetic phenomena associated with, 233, 235
Southeastern Akita earthquake (1970), 76; aftershock activity following, 67; anomalous crustal movement before, 138, 170–71; geomagnetic variations and, 223
specific resistivity, presursory variation in, 280, 294
State Seismological Bureau (China), 256, 258
strain: accumulation of in crust, 61, 139, 141, 157; changes in and earthquakes, 159–60; observation network for, 141–42; velocity of, 47
strainmeter, 104, 145–51, 161; borehole type, 8, 131; embedded type, 142–44, 157; Sachs–Evertson volume type, 292
subcrustal earthquakes, distribution of, 73–74
subduction zone, 74
subsidence: as earthquake trace, 31–32; as long-term precursor, 117; of Pacific coast peninsulas, 115
surveys, 6, 8, 104, 136, 264–65; history of, 104. See also gravity surveys; leveling surveys; magnetic surveys; telemeter surveys; traverse surveys; triangulation surveys
swarm earthquakes, 33, 74, 85–87. See also Matsushiro swarm earthquakes

Tangshan earthquake (1976): geomagnetic anomalies before, 228; prediction of, 210–212
Tashkent earthquake (1966): geomagnetic anomalies associated with, 227; radon concentration changes in conjunction with, 209–10
Tayama, Minoru, 14
telemeter surveys, 5, 80
terraces: and fault activity, 35–36; estimating the age of, 38; formation of, 36–37, 49–50
tidal observation stations, 106
tidal waves, and earthquake prediction, 15
tilting: anomalous, before earthquakes, 43, 104, 135–37, 276; as earthquake trace, 31–32

GENERAL INDEX *317*

tiltmeter, 104, 142; bore-hole type, 291; water tube type, 153, 292
Tohoku district, microearthquake activity in, 79–80
Tokachi offshore earthquake (1952), anomalous crustal movement before, 166
Tokachi offshore earthquake (1968), 67, 286
Tokai district: cyclical occurrence of earthquakes in, 18–19; seismic gap zone in, 275; strain observation network in, 141, 155–56
Tokai earthquake (1707), damage records for, 13, 19
Tokai earthquake, imminence of, 18–19, 56, 103–4, 121, 247, 249, 267–69, 272–75
Tokai earthquakes, relation to Nankai earthquakes, 17–19, 122
Tonankai earthquake (1944), 4, 19, 56, 122; aftershock activity following, 74; and prediction of 1946 Nankai earthquake, 21, 24; fault model for, 267; precursory crustal movement associated with, 117, 135, 138, 164–65, 252, 276, 291; surveys before, 103; prediction of, 3, 8
Tottori earthquake (1943); aftershock activity following, 73, 82; crustal movement connected with, 110–11, 120–21, 125, 138, 164, 276; seismic wave velocity variations before, 89
traverse surveys, 76–77, 281
triangulation surveys, 104–105, 107, 113, 115; after 1943 Tottori earthquake, 110–11, 121; in Izu Peninsula, 120; postwar, 125
Tsuboi, Chuji, 4, 64
Tsukushi earthquake (679), 22
tsunamis, and earthquakes, 23–24, 27, 32

uplift, earthquake-associated, 31–32, 40–41, 43, 45, 124–25
U.S.A.: observation of earth's current, 228–29; research on electromagnetic phenomena, 217, 232–34, 243; research on geomagnetic anomalies, 225–26, research on seismic wave velocity, 93–94
U. S. Geological Survey, 225, 243
USSR: observation of earth's electric potential, 229–30; research on electromagnetic phenomena, 217, 233, 235, 243; research on geomagnetic anomalies, 226–27; research on radon concentration, 209–10; research on seismic wave velocity, 89–90

V_P/V_S ratio, changes in, 89–90, 92–95, 290, 293
volcanoes, and earthquake occurrence, 24–25, 51, 74, 78

Wadati, Kiyoo, 4
Wadati diagram, 92–95
Western Saitama earthquake (1931), and absence of precursory movement, 125
Western Tsugaru (Ajigasawa) earthquake (1793), anomalous crustal movement before, 162

Yamasaki Fault: anomalous crustal movement along, 138, 172; 1977 earthquake along, 84–85
Yanyuan earthquakes (1976), geomagnetic anomalies connected with, 228
Yingtai earthquake (1966), 210
Yoshino earthquake (1952), anomalous crustal movement before, 138, 166–67

Zenkoji earthquake (887), 48
Zenkoji earthquake (1847), 13, 26, 48
Zisin (Earthquake), journal, 21

Map of Japan

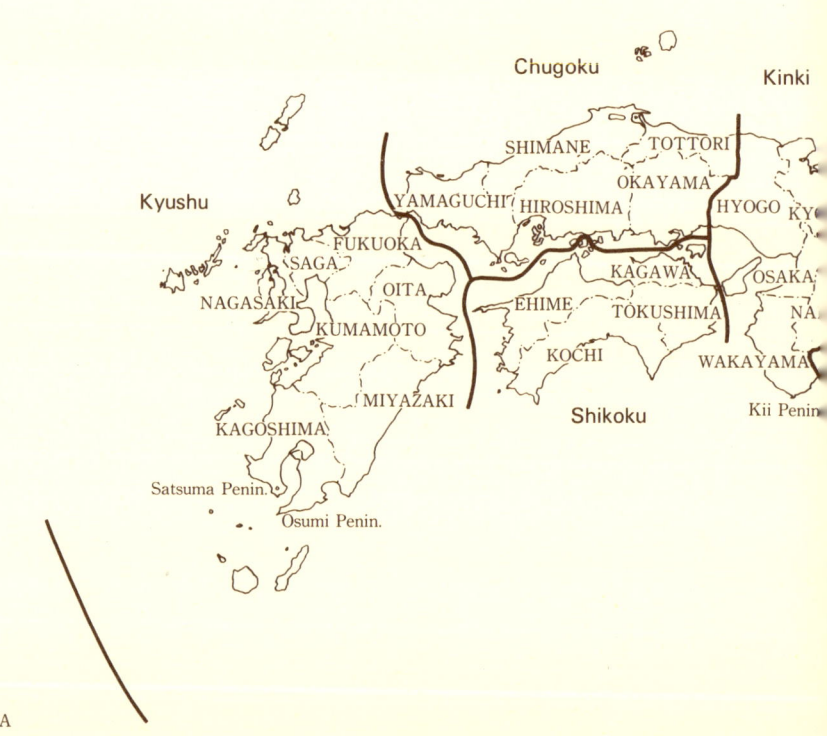